UNIVERSITY OF CAMBRIDGE
ORIENTAL PUBLICATIONS
NO. 4

PUBLISHED ON BEHALF OF THE
FACULTY OF ORIENTAL STUDIES

THE AVESTAN
HYMN TO MITHRA

UNIVERSITY OF CAMBRIDGE
ORIENTAL PUBLICATIONS PUBLISHED FOR THE
FACULTY OF ORIENTAL STUDIES

1 *Averroes' Commentary on Plato's Republic*, edited and translated by E. I. J. ROSENTHAL
2 *FitzGerald's 'Salaman and Absal'*, edited by A. J. ARBERRY
3 *Ihara Saikaku: The Japanese Family Storehouse*, translated and edited by G. W. SARGENT
4 *The Avestan Hymn to Mithra*, edited and translated by ILYA GERSHEVITCH
5 *The Fusūl al-Madanī of al-Fārābī*, edited and translated by D. M. DUNLOP
6 *Dun Karm, Poet of Malta*, texts chosen and translated by A. J. ARBERRY; introduction, notes and glossary by P. GRECH
7 *The Political Writings of Ogyū Sorai*, by J. R. McEWAN
8 *Financial Administration under the T'ang Dynasty*, by D. C. TWITCHETT
9 *Neolithic Cattle-Keepers of South India: a Study of the Deccan Ashmounds*, by F. R. ALLCHIN
10 *The Japanese Enlightenment: a Study of the writings of Fukuzawa Yukichi*, by CARMEN BLACKER
11 *Records of Han Administration. I Historical Assessment*
12 *Records of Han Administration. II Documents*, by M. LOEWE
13 *The Language of Indrajit of Orcha*, by R. S. McGREGOR

ALSO PUBLISHED FOR THE FACULTY
Archaeological Studies in Szechwan, by T.-K. CHENG

THE AVESTAN
HYMN TO MITHRA

WITH AN INTRODUCTION
TRANSLATION AND COMMENTARY

BY

ILYA GERSHEVITCH

CAMBRIDGE
AT THE UNIVERSITY PRESS
1967

CAMBRIDGE UNIVERSITY PRESS
Cambridge, New York, Melbourne, Madrid, Cape Town, Singapore, São Paulo

Cambridge University Press
The Edinburgh Building, Cambridge CB2 8RU, UK

Published in the United States of America by Cambridge University Press, New York

www.cambridge.org
Information on this title: www.cambridge.org/9780521050715

© Cambridge University Press 1959

This publication is in copyright. Subject to statutory exception
and to the provisions of relevant collective licensing agreements,
no reproduction of any part may take place without the written
permission of Cambridge University Press.

First published 1959
Reprinted 1967
This digitally printed version 2008

A catalogue record for this publication is available from the British Library

ISBN 978-0-521-05071-5 hardback
ISBN 978-0-521-05226-9 paperback

CONTENTS

PREFACE	*page* vii
ABBREVIATIONS	xi
INTRODUCTION	1
I The earliest period	3
II Zarathuštrianism	8
III Zoroastrianism	13
IV Christensen's theory	22
V Mithra's functions	26
VI Ahura	44
VII Mithra's companions	58
VIII The Western Mithras	61
TEXT AND TRANSLATION	73
COMMENTARY	149
CRITICAL APPARATUS	301
ADDENDA	319
INDEXES	333

DEDICATED TO

H. W. BAILEY

PREFACE

The interest in Mithras is so great that one would expect the ancient Zoroastrian hymn in praise of his Iranian forbear Mithra to range, in translation, among the more widely known literary products of antiquity. Mithra, apart from being the ancestor of Mithras, has even for himself a claim to public attention in a world torn by dissension, as he is the god of the treaty, more especially of the international treaty (see below, pp. 26 *sq.*). Yet the hymn which reveals the character of the god is all but unknown outside specialist circles. The reason is not a shortage of useful and complete translations, of which there have been no fewer than eight: one each into English and French, and six into German. The English version, by James Darmesteter, appeared in 1883;[1] it was superseded nine years later by the same translator's rendering of the hymn into French.[2] The German versions date from Windischmann's in 1857 (an admirable pioneering achievement) to Lommel's and Hertel's, both of which appeared in 1927; the standard translation is the one by Wolff (1910), which is entirely dependent on Bartholomae's *Altiranisches Wörterbuch* (1904).

If despite so many versions the importance of the hymn for Mithraic studies is not yet generally appreciated, it is because the few experts who have scrutinized it have made only selective use of the information it provides. The poem has been treated as if it were a secondary source, out of which each student of Mithra need consider only such data as seemed relevant to his own notion of the god. No attempt has been made to approach every part of the hymn as a meaningful record of Mithra's character, and the whole poem as a consistent description of a single, well-defined

---

[1] *Sacred Books of the East*, vol. XXIII: *The Zend Avesta*, part II, pp. 119–58.
[2] In two instances, to which I failed to draw attention in the Commentary below, Darmesteter abandoned in the French version the correct interpretation he had printed in the English edition. They are his rendering of the end of st. 142 by 'self-shining like the moon, when he makes his own body shine', and, taken over from Windischmann, of *huyāyna* in st. 116 by 'wife and husband'.

PREFACE

god. To encourage such attempts is the prime purpose of the present book.

The existing translations differ in the rendering of a number of crucial passages, the interpretation of which greatly affects the definition of Mithra's character. Such lack of agreement has not helped to inspire confidence in the hymn. Merely, therefore, to provide one more translation, and argue a case on the strength of interpretations different from everybody else's, would have added to the sense of insecurity, and to the delay in recognition of the significance and authority of the hymn. Since each previous translator had offered only his own opinion, without reference to differing views, what seemed called for at this stage was a survey of the interpretations available, and a discussion of their merits on the basis of the text. Accordingly I have quoted in the Commentary for each moot point all the alternative interpretations known to me—regardless of whether I agree with any of them— provided they do not rest on wild emendations or on disregard of Avestan grammar.[1] Thus the layman will not simply have thrust on him yet another personal interpretation, but will be in a position to assess the present state of understanding of the hymn and evaluate its contents by himself. Should he wish to penetrate to the source of any divergence of opinion, even an elementary acquaintance with the Avestan language[2] will enable him to compare the alternative translations with the text and variants, which are here for the first time transliterated *in toto* from Geldner's edition.

The text printed below reproduces Geldner's readings even when the translation facing it is based on variants he had relegated to the Apparatus. Whenever such a divergence arises attention has been drawn to it in the Commentary. In the treatment of the text I have been extremely conservative, not daring to assume that it requires to be corrected except in the cases mentioned below, p. 25 with footnote. Several new features in the present translation

---

[1] On the other hand I have not usually justified my own rendering when it agrees with the standard translation of Bartholomae-Wolff, unless their view has been challenged.

[2] To be derived with the least effort from A. V. Williams Jackson's *Avesta Grammar*, Stuttgart, 1892 (anastatic reprint 1933).

represent attempts on my part to interpret the text as it stands, instead of following previous translators in resorting to emendations. The translation has been kept strictly literal, to ensure that my understanding of the text is expressed without ambiguity. A few of the notes may strike specialists as being of an elementary character. These were written with a view to meeting the needs of students for whom the book may serve as an introduction to Avestic studies—in the absence of an up-to-date reader. In the Commentary it has proved impracticable to separate the remarks on religious history from the textual and linguistic considerations which prompted them. It is hoped that the subject index will remedy this inconvenience.

The Introduction is intended for orientation, not as an exhaustive discussion of all the opinions that have been voiced on Mithra's place in Zoroastrianism. For a recent survey of the study of early Zoroastrianism the reader may now consult M. Duchesne-Guillemin's informative Ratanbai Katrak lectures, *The Western Response to Zoroaster*, which were published before October, 1958, when a set of first proofs of the present book was sent to the author. M. Duchesne-Guillemin devotes much space to the theory of M. Dumézil on the Aməša Spəntas and Mithra, a theory which he anticipates I would 'carefully avoid' (*op. cit.* p. 104). It is in fact mentioned only in passing, below, p. 48, n., the reason being that I was not concerned with interpretations of the Avestan religion in support of which the Avestan evidence has to be explained away. I shall have more to say on M. Dumézil's theory in a review of M. Duchesne-Guillemin's book which will be published in *BSOAS*, XXII (1959), 154 *sqq.* Meanwhile the reader may be referred to the outright rejection of the Vedic part of M. Dumézil's theory by J. Brough, *BSOAS*, XXI (1958), 395 *sqq.*, XXII (1959), 69 *sqq.*, and P. Thieme, *Mitra and Aryaman* (see below, p. XIV) on grounds which also apply to M. Dumézil's approach to the Zoroastrian religion.

Thieme's monograph, on the other hand, is a work of great importance for Mithraic studies. With his intimate knowledge of the Rig Veda Thieme has placed on a firmer basis than I was able to provide (see below, pp. 29 *sq.*) Meillet's definition of the Vedic

PREFACE

Mitra as the hypostasis of the treaty. In his analysis of Mitra Thieme has drawn freely on the Avestan hymn to Mithra, with full justification; whenever he has offered a new interpretation of a Mithra Yašt passage it will be discussed below, in the Addenda.

The manuscript of the present book was completed in 1955, and accepted during my absence in Iran by the Publications Committee of the Faculty Board of Oriental Studies. After my return to Cambridge in October, 1956, the text had to be left unchanged except for minor alterations and insertions in square brackets. The Addenda were written in the summer of 1958.

I.G.

CAMBRIDGE
January 1959

ABBREVIATIONS

A	*Afrīnakān*, see *Wb.* p. viii (1).
AAWL	*Abhandlungen der Akademie der Wissenschaften und der Literatur*, Wiesbaden.
Akkad.	Akkadian.
AKM	*Abhandlungen für die Kunde des Morgenlandes*.
APAW	*Abhandlungen der Preussischen Akademie der Wissenschaften*.
Ap. I.	*Altpersische Inschriften*, von Ernst Herzfeld (1938).
Aram.	Aramaic.
Arm.	Armenian.
Arm. Gr.	*Armenische Grammatik*, von H. Hübschmann (1897).
Av.	Avestan.
Aw. Eb.	*Awestisches Elementarbuch*, von Hans Reichelt (1909).
B.	Buddhist.
Bal.	Baluči.
BBB	*Ein manichäisches Bet- und Beichtbuch*, von W. Henning, *APAW*, 1937.
Beh.	Behistun Inscription, see Kent.
Benv.	E. Benveniste.
BSL	*Bulletin de la Société de Linguistique de Paris*.
BSO(A)S	*Bulletin of the School of Oriental (and African) Studies*.
Bth.	Christian Bartholomae.
Bth.–Wo.	See Wo.
Chr.	Christian.
Comp.	*Les composés de l'Avesta*, par Jacques Duchesne-Guillemin (1936).
Da.	(*a*) James Darmesteter; (*b*) his translation of *Yt* 10 in *ZA*, II (1892), 444 *sqq*.
Duch.	Jacques Duchesne-Guillemin.
El.	Elamite.
Essai	*Essai sur la langue parthe*, par A. Ghilain (1939).
Ét. Ir.	*Études iraniennes*, par James Darmesteter, 2 vols. (1883).
EVP	*An Etymological Vocabulary of Pashto*, by Georg Morgenstierne (1927).
F	*Frahang i oīm*, see *Wb.* p. ix (6).
Future Life	*The Zoroastrian Doctrine of a Future Life*, by Jal Dastur Cursetji Pavry (1929).
G	*Gāsānbār*, see *Wb.* p. viii (1).

ABBREVIATIONS

Gdn.	(a) Karl F. Geldner; (b) his translation of Yt 10 in KZ 25 (1881), 484 sqq.; (c) Gdn.'s edition: Avesta, The Sacred Books of the Parsis (1896).
GGA	Göttingische Gelehrte Anzeigen.
GIP	Grundriss der iranischen Philologie.
GMS	A Grammar of Manichean Sogdian, by Ilya Gershevitch (1954).
Gt. Bd.	The Bûndahishn, edited by T. D. Anklesaria (1908).
H	Haδōxt Nask, see Wb. p. ix (5).
Handbuch	Handbuch der Zendsprache, von Ferdinand Justi (1864).
Hoc.	Hochgottglaube im alten Iran, eine religions-phänomenologische Untersuchung, von Geo Widengren (1938).
Htl.	(a) Johannes Hertel; (b) his translation of Yt 10 in Die Sonne und Mithra im Awesta (Indo-Iranische Quellen und Forschungen, Heft IX), 1927, pp. 132 sqq.
Hzf.	(a) Ernst Herzfeld; (b) his translation of parts of Yt 10 in Zoroaster and his World (1947), II, 422–94 passim.
IE	Indo-European.
IEW	Indogermanisches etymologisches Wörterbuch, von Julius Pokorny, parts 1–9, pp. 1–864 (1949–55, as yet incomplete).
IF	Indogermanische Forschungen.
IIFL	Indo-Iranian Frontier Languages, by Georg Morgenstierne, 2 vols. 1929–38.
IIJ	Indo-Iranian Journal.
IIr.	Indo-Iranian.
Inf.	Les infinitifs avestiques, by E. Benveniste (1935).
JAs.	Journal Asiatique.
JRAS	Journal of the Royal Asiatic Society.
Kent	Roland G. Kent, Old Persian (1950); consult for Beh. and NR.
Khot.	Khotanese.
KZ	Zeitschrift für vergleichende Sprachforschung auf dem Gebiete der indogermanischen Sprachen.
Lesebuch	Die zoroastrische Religion (Das Avestā), von Karl F. Geldner, 1926 (Religionsgeschichtliches Lesebuch, herausgegeben von A. Bertholet, No. 1); contains translation of parts of Yt 10 on pp. 18 sqq.
LEW	Lateinisches etymologisches Wörterbuch, von A. Walde; 3., neubearbeitete Auflage von J. B. Hofmann (1938–55).
Lo.	(a) Herman Lommel; (b) his translation of Yt 10 in (Die) Yäšt's (1927), pp. 67 sqq.
Mages	Les Mages dans l'Ancien Iran, par E. Benveniste (1938).

ABBREVIATIONS

Man.	Manichean.
MHCP	*The Manichaean Hymn Cycles in Parthian*, by Mary Boyce (1954).
MIr.	Middle Iranian.
Mir. Man.	*Mitteliranische Manichaica aus Chinesisch-Turkestan*, von F. C. Andreas, aus dem Nachlass herausgegeben von W. Henning, *SPAW*, parts I (1932, pp. 175 *sqq.*), II (1933, pp. 294 *sqq.*), and III (1934, pp. 849 *sqq.*).
Morg.	Georg Morgenstierne.
MPers.	Middle Persian.
MSS	*Münchener Studien zur Sprachwissenschaft*.
N	*Nīrangastān*, see *Wb*. p. viii (2).
NGGW, NGWG	*Nachrichten von der Gesellschaft der Wissenschaften zu Göttingen*.
Noms d'agent	*Noms d'agent et noms d'action en indo-européen*, by E. Benveniste (1948).
NPers.	New Persian.
Npers. Et.	*Grundriss der neupersischen Etymologie*, von Paul Horn (1893).
NR	OPers. inscriptions at Naqš-i Rustam, see Kent.
NTS	*Norsk Tidsskrift for Sprogvidenskap*.
Ny	*Nyāyišn*, see *Wb*. p. viii (1).
OE	Old English.
OHG	Old High German.
OInd.	Old Indian.
OIr.	Old Iranian.
OIr. Lit.	*Old Iranian Literature*, by Ilya Gershevitch (*Handbuch der Orientalistik*, herausgegeben von B. Spuler, 4. Band: *Iranistik*, 2. Abschnitt: *Literatur*).
OK	*Ostiranische Kultur im Altertum*, von Wilhelm Geiger (1882).
OLZ	*Orientalistische Literaturzeitung*.
OPers.	Old Persian.
Origines	*Origines de la formation des noms en indo-européen*, by E. Benveniste (1935).
Orm.	*Ormazd et Ahriman*, par J. Duchesne-Guillemin (1953).
Orm.	Ormuri.
Oss.	Ossetic.
P	See *TSP*.
Pahl.	Pahlavi.
Parth.	Parthian.
Pers. St.	*Persische Studien*, von H. Hübschmann (1895).
Phl. Riv.	Pahlavi Rivāyats.

ABBREVIATIONS

Ps.   *Bruchstücke einer Pehlevi-Übersetzung aer Psalmen*, von F. C. Andreas, aus dem Nachlass herausgegeben von Kaj Barr (*SPAW*, 1933, pp. 91 *sqq.*).
Pš.   Pashto.
Rel.   *The Persian Religion according to the chief Greek texts*, by E. Benveniste (1929).
Rel.   *Die Religionen des alten Iran*, von H. S. Nyberg (1938).
Rel. (*Zar.*)   *Die Religion Zarathustras*, von Herman Lommel (1930).
RV   Rig Veda
Smith, Maria W.   *Studies in the Syntax of the Gathas of Zarathushtra together with Text, Translation, and Notes*, by Maria Wilkins Smith (*Language Dissertations* published by the Linguistic Society of America, No. IV, 1929).
Sogd.   Sogdian (cf. B., Chr., and Man.).
SPAW   *Sitzungsberichte der Preussischen Akademie der Wissenschaften.*
Spi.   (*a*) Friedrich Spiegel; (*b*) his translation of *Yt* 10 in *Avesta*, vol. III, *Khorda-Avesta* (1863), pp. 79 *sqq.*, and commentary in *Commentar über das Avesta*, vol. II (1868), pp. 546 *sqq.*
St., sts.   Stanza, stanzas.
Studien   *Studien zum Avesta*, von Karl Geldner (1882).
T.Br.   *Taittirīya Brāhmaṇa.*
Thieme   *Mitra and Aryaman*, by Paul Thieme. *Transactions of the Connecticut Academy of Arts and Sciences*, 41 (1957), pp. 1-96.
TMMM   *Textes et monuments figurés relatifs aux mystères de Mithra*, 2 vols. (1896-9), by F. Cumont.
TPS   *Transactions of the Philological Society.*
TSP   *Textes Sogdiens*, édités, traduits et commentés par E. Benveniste (1940). Individual texts of this collection are quoted by their numbers preceded by *P*.
UGE   *Untersuchungen zur Geschichte von Eran*, von J. Marquart, 2 parts (1896-1905).
Ved.   Vedic.
Vend.   *Vendidad*, see *Wb.* p. viii (1).
Vič.   *Vičarkart i dēnīk*, see *Wb.* p. x (p).
Visp.   *Visprat*, see *Wb.* p. viii (1).
VJ   *Vessantara Jātaka*, Texte Sogdien édité, traduit et commenté par E. Benveniste (1946).
Vrtra   *Vrtra et Vrθragna*, par E. Benveniste et L. Renou (1934).
Vyt.   *Vištāsp Yašt*, see *Wb.* p. ix (8).

ABBREVIATIONS

Wb.	*Altiranisches Wörterbuch*, von Christian Bartholomae (1904).
Wi.	(*a*) Friedrich Windischmann; (*b*) his translation of *Yt* 10 (1857), in *AKM*, i (1859), pp. 1 *sqq*.
Wn.	(*a*) Jakob Wackernagel; (*b*) his *Altindische Grammatik*, vols. I (1896) and II, 1 (1905).
Wn.–Debr.	*Altindische Grammatik*, vols. III (1929–30) and II, 2 (1954), by Jakob Wackernagel and Albert Debrunner.
Wo.	*Avesta*, übersetzt auf der Grundlage von Chr. Bartholomae's Altiranischem Wörterbuch, von Fritz Wolff (1910).
Wx.	Waxi.
Y	*Yasna*, see *Wb.* p. viii (1).
Yaγn.	Yaγnōbi.
Yäšt's	*Die Yäšt's des Awesta*, übersetzt und eingeleitet von Herman Lommel (1927).
YAv.	Younger Avesta(n).
Yd.	Yidγa.
Yt	*Yašt*, see *Wb.* p. viii (1).
ZA	*Le Zend-Avesta*, 3 vols. (1892–3), par James Darmesteter. *Yt* 10 is translated in vol. II, pp. 444 *sqq*.
Zar.	*The Hymns of Zarathustra*, translation of the Gâthâs by Jacques Duchesne-Guillemin, translated from the French by Mrs M. Henning (1952).
ZDMG	*Zeitschrift der deutschen morgenländischen Gesellschaft*.
ZII	*Zeitschrift für Indologie und Iranistik*.
Zor.	*Zoroastre*, par Jacques Duchesne-Guillemin (1948).
Zor. Prob.	*Zoroastrian Problems in the Ninth-Century Books*, by H. W. Bailey (1943).
ZWb.	*Zum altiranischen Wörterbuch*, von Christian Bartholomae (1906).

INTRODUCTION

INTRODUCTION†

I. THE EARLIEST PERIOD

THE Avestan hymn (*Yašt*) to Mithra, to be dated approximately in the second half of the fifth century B.C., is the one extensive ancient literary record we have of the attributes, habits, equipment, companions, and cult of the Iranian god whose worship was destined to spread into Europe as far as Britain some five to six hundred years after the hymn was composed. In the Avesta itself, beside the hymn, there are a few more references to Mithra which help to complete the picture. Outside the Avesta ancient Iranian evidence on the god, contemporaneous with, or earlier than the hymn, is confined to the appearance of his name in Old Persian inscriptions of Artaxerxes II (405–359) and Artaxerxes III (359–338), and as a component of Iranian personal names attested in Old Persian, Greek, Aramaic, Akkadian and Elamite sources. The earliest Greek reference to Mithra is found in Herodotus (I, 131), who apparently confused him with the goddess Anāhitā (see Benveniste, *Rel.* 27 *sq*.). The Iranian Manichean texts composed from the third century A.D. onwards allow us a glimpse of the evolution which Old Iranian Mithra had undergone in the mythology of various Iranian nations (see below, pp. 40 *sq*.), but here, too, information is scanty. The Zoroastrian books written in Pahlavi in the ninth century A.D. take comparatively little interest in Mithra, and add little to what is known of him from older sources.

The documentation of the prehistory of the Iranian Mithra is scarce, but it does make it clear that the god was worshipped by Indo-Iranian tribes not less than ten centuries before the Avestan hymn was composed. In Western Asia his bare name appears in a list of five gods by whom a treaty with the Hittite king Šuppiluliumaš was sworn in the early fourteenth century B.C. by Mattiwaza, an Indo-Iranian ruler of the Hurrian kingdom of Mitanni in

† This Introduction is a revised version of a lecture delivered in the University of Manchester in March 1955.

INTRODUCTION

Western Mesopotamia.† The five divinities are Mitra, Varuṇa, Indra, and the two Nāsatyas, all of whom figure prominently in the the Rig Veda, the earliest literary product of the Indo-Aryan branch of the Indo-Iranian group of tribes. Even assuming that the first Rigvedic hymns are not earlier than the Mitanni treaty,‡ we may say that the cult of Mitra was by then so well established among Indo-Iranian tribes that it had very likely been in existence for a long period before.

In the Rigvedic hymns, whose composition extended over many centuries, Mitra is mentioned more than two hundred times, yet the information the texts offer on the god is exasperatingly meagre. This appears to be due mainly to the predilection of Rigvedic poets for invoking Mitra together with Varuṇa in a compound *mitrā-varuṇā* (meaning 'Mitra and Varuṇa') of the type grammarians call dvandva. What the poets say of *mitrāvaruṇā* does not substantially differ from the view they take of Varuṇa. Consequently it is not easy to distinguish Mitra's share in the association of the two gods. A. A. Macdonell (*A Vedic Reader*, pp. 118 *sq.*, 134 *sq.*) has so conveniently arrayed the main Vedic facts concerning *mitrāvaruṇā* on the one hand, Varuṇa on the other, that we cannot do better than quote him *in extenso*, printing in italics certain details to which we shall return:

Mitrāvaruṇā. This is the pair most frequently mentioned next to Heaven and Earth. The hymns in which they are conjointly invoked are much more numerous than those in which they are separately addressed. As Mitra (III, 59) is distinguished by hardly any individual traits, the two together have practically the same attributes and functions as Varuṇa alone. They are conceived as young. Their eye is the sun. Reaching out they drive with the rays of the sun as with arms. They wear glistening garments. They mount their car in the highest heaven. Their abode is golden and is located in heaven; it is great, very lofty, firm, with a thousand columns and a thousand doors. *They have spies* that are wise and cannot be deceived. They are kings and universal monarchs. *They are*

---

† See Götze and Christensen in Iwan v. Müller's *Handbuch der Altertumswissenschaft, Kulturgeschichte des alten Orients*, vol. III, 1, pp. 57 *sq.*, 79, 91 *sq.*, 209 *sq.*

‡ See H. W. Bailey *ap.* E. B. Ceadel, *Literatures of the East*, p. 100, who is prepared, at a guess, to assign the beginnings of Rigvedic versification to the eighteenth century B.C.

also called Asuras, who wield dominion by means of *māyā* 'occult power', a term mainly connected with them. By that power they send the dawns, make the sun traverse the sky, and obscure it with cloud and rain. They are rulers and *guardians of the whole world*. They support heaven, and earth, and air.

They are lords of rivers, and they are the gods most frequently thought of and prayed to as *bestowers of rain*. They have kine yielding refreshment, and streams flowing with honey. They control the rainy skies and the streaming waters. They bedew the pastures with ghee (= rain) and the spaces with honey. They send rain and refreshment from the sky. Rain abounding in heavenly water comes from them. One entire hymn dwells on their powers of bestowing rain.

Their ordinances are fixed and cannot be obstructed even by the immortal gods. They are *upholders and cherishers of order* (read 'Truth'). They are *barriers against falsehood, which they dispel, hate, and punish. They afflict with disease* those who neglect their worship.

Varuṇa. Beside Indra (II, 12) Varuṇa is the greatest of the gods of the Rig Veda, though the number of the hymns in which he is celebrated alone (apart from Mitra) is small, numbering hardly a dozen.

His face, eye, arms, hands, and feet are mentioned. He moves his arms, walks, drives, sits, eats, and drinks. His eye with which he observes mankind is the sun. He is far-sighted and *thousand-eyed*. He treads down wiles with shining foot. He sits on the strewn grass at the sacrifice. He wears a golden mantle and puts on a shining robe. *His car*, which is often mentioned, shines like the sun, and *is drawn by well-yoked steeds*. Varuṇa sits in his mansions looking on all deeds. The Fathers behold him in the highest heaven. *The spies of Varuṇa are sometimes referred to*: they sit down around him; they observe the two worlds; they stimulate prayer. By the golden-winged messenger of Varuṇa the sun is meant. Varuṇa is often called a king, but especially a universal monarch (*samrāj*). The attribute of sovereignty (*kṣatrá*) and *the term Asura* are *predominantly applicable to him*. His divine dominion is often alluded to by the word *māyā* 'occult power'; the epithet *māyín* 'crafty' is accordingly used chiefly of him.

Varuṇa is mainly lauded as upholder of physical and moral order. He is a great lord of the laws of nature. He established heaven and earth, and by his law heaven and earth are held apart. He made the golden swing (the sun) to shine in heaven; he has made a wide path for the sun; he placed fire in the waters, the sun in the sky, Soma on the rock. The wind which resounds through the air is Varuṇa's breath. By his ordinances the moon shining brightly moves at night, and the stars placed up on high are seen at night, but disappear by day. Thus Varuṇa is lord of light both by day and by night. He is also *a regulator of the waters*. He caused the rivers to flow; by his occult power they pour swiftly into the ocean

INTRODUCTION

without filling it. It is, however, with the aerial waters that he is usually connected. Thus he makes the inverted cask (the cloud) to pour its waters on heaven, earth, and air, and to moisten the ground.

Varuṇa's ordinances being constantly said to be fixed, he is preeminently called *dhṛtávrata* 'whose laws are established'. The gods themselves follow his ordinances. His power is so great that neither the birds as they fly nor the rivers as they flow can reach the limits of his dominion. He embraces the universe, and the abodes of all beings. *He is all-knowing*, and his omniscience is typical. He knows the flight of the birds in the sky, the path of the ships in the ocean, the course of the far-travelling wind, beholding all the secret things that have been or shall be done, *he witnesses men's truth and falsehood*. No creature can even wink without his knowledge.

As a moral governor Varuṇa stands far above any other deity. *His wrath is aroused by sin, the infringement of his ordinances, which he severely punishes.* The fetters (*páśās*) with which he binds sinners are often mentioned, and are characteristic of him. On the other hand, Varuṇa is gracious to the penitent. He removes sin as if untying a rope. He releases even from the sin committed by men's fathers. He spares him who daily transgresses his laws when a suppliant, and is gracious to those who have broken his laws by thoughtlessness. There is in fact no hymn to Varuṇa in which the prayer for forgiveness of guilt does not occur. Varuṇa is on a footing of friendship with his worshipper, who communes with him in his celestial abode, and sometimes sees him with the mental eye. The righteous hope to behold in the next world Varuṇa and Yama, the two kings who reign in bliss.

On Varuṇa we now have the first volume of an important work by Lüders, which is being published posthumously. Although so far only part of Lüders' detailed argumentation is available, his main conclusion is anticipated in the Introduction to the first volume. It is that Varuṇa is essentially the god in charge of *ṛtá* 'Truth'. One of Lüders' merits consists in the conclusive proof he has offered that the meaning of *ṛtá* is indeed 'Truth'. For *aša-*, the Avestan etymological equivalent of Ved. *ṛtá*, this meaning was long ago established by Andreas and maintained by Lommel,† but most scholars chose to perpetuate the old translation of *aša-* by '(cosmic) order', assuming that this was the meaning of Ved. *ṛtá*. It is thus not only to Vedic but also to Avestic studies that Lüders has rendered a great service. For obviously no true understanding

† See Lüders, *Varuṇa*, 27, n., and below, p. 153, n. ‡.

of Indo-Iranian religious thought is possible unless it is realized that its key conception, expressed by *ṛtá/aša*, means 'Truth'.

Lüders' clear-sighted insistence on what had previously been understood only perfunctorily, the dominant role of Truth in Indo-Iranian religion, has enabled him to present Varuṇa in a new and immediately convincing perspective. Because transcendental Truth is situated in a primordial spring inside the highest heaven, water everywhere is the 'womb of Truth' (pp. 25 *sq.*). This is why also Varuṇa, the guardian of Truth, is to be found in the waters. Accordingly the ancient Indo-Aryans swore their oaths by water, invoking Varuṇa, who was present in the water to guard Truth and witness the validity of the oath (pp. 28, 30 *sq.*). Thus Lüders was able to penetrate to the essential definition of Varuṇa as the god of oath (*Eidgott*), which provides the clue to his character of an avenger of falsehood, and to the spies and thousand eyes by which he detects the infringers of Truth. In addition, as Varuṇa had his seat in the waters, where Truth is situated, he was bound to take charge of them; in the naturalistic interpretation of the Vedic pantheon he thus became a water-god.

The Vedic Mitra is so closely connected with Varuṇa that, as Lüders remarks (p. 37), he must be homogeneous (*eines Wesens*) with him. The characteristics of *mitrāvaruṇā* quoted above from Macdonell show that Mitra is closely associated with Varuṇa in the task of punishing falsehood after detecting it by means of spies, as well as in controlling the waters. However, as Lüders remarked (pp. 12 *sq.*, 28, 38), Mitra's seat is not in water, but apparently in fire, since pacts were presumably concluded in front of blazing fire. This is the only trait which makes a palpable difference between Mitra and Varuṇa. In addition, the unusual qualification *yātayájjana* is applied not only to Mitra and Varuṇa as a pair, but also to Mitra alone. The verb *yātáyati*, which is also found with Mitra as subject, means, according to Lüders (p. 38), 'to call to account (*zum Vergleich bringen*)', cf. *ṛṇám yātáyati* 'to settle a debt'; hence *yātayájjana* is defined by Lüders as either 'he who calls people to account', or 'he who causes them to settle their due'. Both the compound and the finite form *yātáyati* occur in the one and only Rigvedic hymn which is dedicated to Mitra (III, 59).

INTRODUCTION

Although its nine stanzas are not very informative, they may be quoted here in K. F. Geldner's translation (*Der Rig-Veda*, vol. I, pp. 406 *sq.*)† [see Addenda]:

1. Der sich Mitra [Freund] nennt, eint die Menschen. Mitra erhält Erde und Himmel. Mitra gibt auf die Völker Acht, ohne die Augen zu schliessen; dem Mitra opfert die schmalzreiche Spende!

2. Mitra! Der Sterbliche, der (für dich) eine Labung hat, soll den Vorrang haben, der dir, o Aditisohn, mit einem Gelübde dient. Von dir beschützt, wird er nicht erschlagen noch ausgeraubt; nicht trifft ihn Ungemach von nah oder fern.

3. Frei von Krankheit, an der Opferspeise uns gütlich tuend, mit aufgestemmten Knieen, soweit die Erde reicht, dem Gelübde des Aditisohnes nachzukommen suchend, möchten wir in der Gunst des Mitra stehen.

4. Dieser verehrungswürdige, freundliche Mitra ist als gutregierender König geboren, als Meister. An dieses opferwürdigen Gunst, an seinem glückbringenden Wohlwollen möchten wir teilnehmen.

5. Dem großen Aditisohn soll man mit Verneigung nahen. Er eint die Menschen, ist dem Sänger freundlich. Diesem hochgeschätzten Mitra opfert diese erwünschte Spende ins Feuer!

6. Gewinnbringend ist die Gunst des völkererhaltenden Gottes Mitra, ruhmglänzend seine Herrlichkeit.

7. Der weitbekannte Mitra, der an Größe über den Himmel, an Ruhm über die Erde reicht;

8. Dem Mitra sind die fünf Völker ergeben, dessen Macht überlegen ist. Er trägt alle Götter.

9. Mitra [Freund] bei Göttern und Āyu's, hat er für den Barhis legenden Mann die Speisen bestimmt, deren Gelübde (als Fastenspeisen) erwünscht ist.

II. ZARATHUŠTRIANISM

Before turning to the Avestan Mithra we must dwell on the character and origin of the Avesta, and in particular the *Yašts*

† In his introductory note to the hymn, which he describes as 'ganz farblos', Geldner states that 'im ganzen Lied wird mit dem Appellativbegriff von *mitrá* "Freund" gespielt'. On this subjective view see p. 30 below. The reader may note that 'freundlich' in sts. 4 and 5 translates *suśéva*; for 'Der sich Mitra [Freund] nennt' (st. 1) and 'Mitra [Freund] bei Göttern und Āyu's, hat er... die Speisen bestimmt' (st. 9), Macdonell, *Vedic Reader*, has respectively 'Mitra speaking' and 'Mitra, among gods and mortals, has provided food'. The expressions *jánān yātayati* (st. 1) and *yātayájjana* (st. 5), on which we have seen Lüders' view, are rendered by Geldner as 'eint die Menschen'.

ZARATHUŠTRIANISM

(hymns), of which the hymn to Mithra (Yašt 10) is one.† The Avesta is a collection of sacred writings belonging to two religions, which are conveniently referred to as Zarathuštrianism and Zoroastrianism. The former is the doctrine preached by Zarathuštra (630–553, or 628–551, or 618–541 B.C.), the latter—as we shall see in the next chapter—an Iranian religious κοινή which includes Zarathuštrianism and began to be formulated in Avestan language in the second half of the fifth century B.C. Formally Zarathuštrian and Zoroastrian writings are easily distinguished, in that the former are composed in the archaic *Gāthic* dialect of Avestan, the latter in what by contrast is called the *Younger Avestan* idiom.

The only religious tenets which can be reliably ascribed to Zarathuštrianism are those explained or implied in Zarathuštra's own words as handed down in the Gāthās, the *yeṅhē hātąm* prayer (see below, Commentary, note 6⁴), and perhaps certain parts of the *Yasna Haptaṅhāiti* (cf. *OIr. Lit.* §24). Essentially these tenets pertain to an ethical dualism tempered by a monotheism which is centred in Ahura Mazdāh. Zarathuštra's dualism moves on two planes: on one the opposing terms are *Aša* 'Truth' and *Drug* 'Falsehood', on the other *Spənta Mainyu* 'the Incremental Spirit', and *Angra Mainyu* 'the Fiendish Spirit'. In Younger Avestan texts Spənta Mainyu sometimes appears as 'creator' (cf. *Yt* 10.143, see Lommel, *Rel.* 21, and below, pp. 56 *sq.*); this notion goes back to Zarathuštra, in whose system Ahura Mazdāh, the father of Spənta Mainyu (*Y* 47.3), is *through* Spənta Mainyu (*Y* 44.7) the creator of all things. We may safely say that Spənta Mainyu, though a distinct Entity, is an aspect or organ, namely the creative organ, of Ahura Mazdāh.

Spənta Mainyu's relation to Ahura Mazdāh in the Gāthās is closely paralleled by that of six Entities who also behave like aspects or organs of Ahura Mazdāh, and are created by him.‡ Their

---

† In some of the remarks in this and the next chapter I have partly summarized, partly expanded, the relevant exposition in *OIr. Lit.*, to which the reader is referred for other views and bibliography.

‡ The term 'Entities' for these aspects, or organs, is one Zarathuštra himself used in referring to them (see below, Commentary, note 6⁴). That the first four Entities were created by Ahura Mazdāh is stated in the Gāthās (see Lommel, *Rel.* 30 *sq.*, and cf. the passages with *dāmi-* quoted below, p. 168). No such Gāthic statement is found in respect of Haurvatāt or Amərətāt; however, on the

INTRODUCTION

names are (1) *Vohu Manah* 'Good Mind', (2) *Aša* 'Truth', (3) *Xšaθra* 'Power', (4) *Ārmaiti* 'Devotion', (5) *Haurvatāt* 'Wholeness', and (6) *Amərətāt* 'Life'.† In the Younger Avesta these Entities are collectively referred to as *Aməšā Spənta* 'the Incremental Immortals', and it is made clear that each of them is in charge of an 'element', viz. respectively cattle, fire, metal, earth, water, and vegetation. Outside Avestan, at a later period, etymological equivalents of Av. *aməšā spənta* (older form *$amṛtā^h$ $spantā^h$*) are found with the simple meaning 'element': in Iranian Manicheism MPers. (')*mhr'spnd* and Sogd. *mrδ'spnd* are common nouns designating the five light elements (ether, wind, light, water, fire), and Chr. Sogd. *mrd'spnty* corresponds to τὰ στοιχεῖα of Galatians iv. 3.

It is often said that the term *Aməšā Spənta*, which is not found in the Gāthās, is a Younger Avestan invention, and the meaning 'element' of this term represents a YAv. 'materialization' of the spiritual view Zarathuštra had taken of the Entities 'Good Mind', 'Truth', 'Power', etc., as organs of Mazdāh. Duchesne-Guillemin goes even so far as to consider the YAv. 'elemental' treatment of the Aməša Spəntas a return to polytheism: see below, p. 165. This view is untenable. Lommel, *Rel.* 123 *sqq.*, has discerningly brought together a sufficient number of Gāthic passages in which one or the other 'Entity' connotes its corresponding 'element', to exclude the possibility of chance. Zarathuštra's use of 'Haurvatāt and Amərətāt' in the sense of 'food' is especially telling, and proves conclusively that the prophet was well aware of the 'elemental' function of the Entities; to the examples quoted by Lommel, *op. cit.* 126 *sq.*, one may add *Y* 44.18 in the interpretation proposed below, note 46[1]. It is, however, not only the YAv. *function* of the Aməša Spəntas which can be traced to the time of Zarathuštra, but also their collective name. For although this name does not happen to

one hand, in addressing Ahura Mazdāh Zarathuštra appears to say that these two Entities are 'thine' (see Commentary, note 46[1]), on the other hand the Younger Avestan information that Aši, daughter of Ahura Mazdāh, is the sister of the Aməša Spəntas (*Yt* 17.2) implies that all Aməša Spəntas are Mazdāh's children. They are described as 'the shapes (*kəhrp-*) he assumes' in *Yt* 13.81.

† On the meaning of *Amərətāt* see P. Thieme, *Studien zur indogermanischen Wortkunde und Religionsgeschichte*, 29 *sqq.*

be attested in the surviving verses of Zarathuštra, the prophet does refer to the Entities collectively, and in such a manner as to leave scarcely any doubt that he thought of them both as Entities (*hātąm*) and Aməša Spəntas. The arguments which lead to this conclusion are set forth in note 6^4 below, where the reader will find that the conclusion entails an important corollary: the number of Aməša Spəntas, which in the Younger Avesta is given as seven, was seven also for Zarathuštra, namely the six Gāthic Entities quoted above, with Spənta Mainyu at their head.

The corollary at first sight looks surprising. If Spənta Mainyu is the first and chief Entity, where, one asks, is the 'element' over which he presides in his capacity of Aməša Spənta? The answer is not far to seek. *A priori* a list of 'elements' which begins with 'cattle' looks suspect; why is not 'man' at the head? Now, the third chapter of the *Bundahišn*, an encyclopedia of Zoroastrian lore composed in Pahlavi in the ninth century A.D., discusses one by one the Aməša Spəntas in the following terms:† 'The second of the supernatural beings (*mēnōkān*) is Vahuman; out of the empirical creation (*hač gētīkān dahišn*‡) he took as his own (*ō x^{v}ēš paδīrift*) the species of cattle;.... The third of the supernatural beings is Artvahišt; out of the empirical creation he took fire as his own;....' Thus the text continues down to Amurδat, 'the seventh of the supernatural beings', who 'took vegetation as his own'. Preceding this enumeration we read: 'The first of the supernatural beings is Ohrmazd and those three creators;§ out of himself he took (*or* made) the foundation of man (*bun-i martom hač x^{v}at grift* (or *kart*)), among the material beings; ...and he took (*or* made) man as his own creation (*u-š martom ō x^{v}ēš grift* (or *kart*) *dahišn*).'

Here Ohrmazd is treated as heading a list of seven supernatural beings, of which the other six, the Aməša Spəntas, are each correctly related to their corresponding empirical 'element'. Since

† The *Bundahišn* passage, the latest translation of which is Nyberg's in *JAs.* 1929, 233, was duly noted in this connection by Lommel, *Rel.* 106 *sq.*, who, however, failed to see the relevance to the problem of the Aməša Spəntas of Ohrmazd's and man's position in this list.

‡ *dahišn*, Henning's reading. On *mēnōkān* and *gētīkān* cf. below, note 13^1.

§ That is, Ohrmazd plus three times Ohrmazd, inasmuch as he presides over the first, eighth, fifteenth, and twenty-third days of the month, as explained at the beginning of the third chapter of the Bundahišn.

INTRODUCTION

Ohrmazd, too, is related by a similar wording to something which belongs to the empirical world, the conclusion is hard to escape that this something, man, is the 'element' peculiar to Ohrmazd, who therefore in this list fulfils the function of an 'elemental' Aməša Spənta. This situation cannot straightway be projected into the Avestan system, in which Ahura Mazdāh never appears as an Aməša Spənta.† What has happened becomes clear directly we recall that the original Zarathuštrian antithesis between the two Spirits, the Incremental (Spənta Mainyu) and the Fiendish (Angra Mainyu), was in late Zoroastrianism simplified to one in which the opponent of the Fiendish Spirit (Ahriman) is Ahura Mazdāh (Ohrmazd) himself, the reason being, as Lommel has made clear (*Rel.* 18 *sqq.*), that Ahura Mazdāh gradually absorbed the functions and even the name of his creative organ Spənta Mainyu. Since first, as explained at the end of note 6^4, Spənta Mainyu is Zarathuštra's chief Aməša Spənta, secondly in the *Bundahišn* man is treated as first 'element' under the care of Ohrmazd, and thirdly Ohrmazd has absorbed at least one other important function of Spənta Mainyu's, that of being Ahriman's opposite number, we are safe in concluding (*a*) that in the *Bundahišn* Ohrmazd as 'supernatural being' in charge of man is fulfilling a function originally decreed to the Aməša Spənta Spənta Mainyu, and (*b*) that the 'element' under the care of the Gāthic Spənta Mainyu is man.

We may now recapitulate that part of Zarathuštra's doctrine which we need to grasp firmly for purposes that will become clear as we go on: a dualism on two planes, involving opposition respectively between 'Truth' and 'Falsehood' on the one hand, and Spənta and Angra Mainyu on the other; a monotheism centred in Ahura Mazdāh, who has created, or emanated, seven supernatural aspects of himself, the Entities; the empirical counterparts of the Entities are the 'elements' man, cattle, fire, metal, earth, water, plants, as representatives of which the Entities are called Aməša Spəntas.

† I share Bartholomae's view that the expression 'Ahura Mazdāh and the other Aməša Spəntas' in st. 139 of the Mithra Yašt should not be taken to include Mazdāh among the Aməša Spəntas, cf. note 139^1. The opposite view is taken by Lommel, *Rel.* 35.

III. ZOROASTRIANISM

Zoroastrianism, to a fair sample of which the greater part of this book is devoted, is a mixed religion whose ingredients are:

(1) Zarathuštrianism;

(2) the cult of certain non-Zarathuštrian divinities who are either (*a*) an Indo-Iranian inheritance, since they have equivalents in the Vedas (e.g. Mithra, Haoma, etc.), or (*b*) have no counterpart in the Vedas, and may therefore be considered peculiarly Iranian (e.g. Anāhitā, Drvāspā, the hypostasis of fortune (x^v*arənah-*), etc.);

(3) certain Zarathuštrian notions (e.g. *aši-*, *sraoša-*) recast as divinities on the pattern of the divinities in (2).

The authors of the Younger Avesta insist that all the ingredients of this mixed religion should be regarded as having been recommended by Zarathuštra himself. A device they commonly use to create this impression is to introduce matter which more often than not the prophet would have rejected wholeheartedly, by the prefatory remark: 'Thus Ahura Mazdāh told Zarathuštra.' Such a device might be held to point to the mixed religion itself being an artificial product, concocted without serious attempts to avoid internal contradictions, by a Zarathuštrian clergy that was eager to please and annex at all costs communities worshipping Mithra, Anāhitā, Tištrya, etc. Against such an interpretation one may argue that, however willing the priests may have been to make compromises with what from Zarathuštra's point of view can only be described as the enemy, they would have had to take care not to lose face with their own communities, who would not lightly take for 'Zarathuštrian' today what yesterday they were told had been condemned by the prophet. In a delicate compromise of this kind one would have expected reformers and authors of sacred texts to proceed with tact, and aim at a syncretistic religion rather than a mere juxtaposition of incompatibles.

There seems to be only one plausible reason why the priestly authors, who lacked neither literary skill nor loyalty to their prophet, should not have attempted to unify the motley scripture they were composing beyond placing Zarathuštra's signature all over it: the authors of the fifth- to fourth-century Zoroastrian scripture are not the authors of the religious mixture it reflects. Their task was merely to compose texts for an existing mixed religion, whose character it was beyond their power, or wish, to alter. This task of 'codification' was undertaken by Zarathuštrian priests because they alone had the skill to do so, having been brought up in the highly developed literary tradition which we first meet in Zarathuštra's poems. They had enough literary sources at their disposal. Mazdāhism was abundantly represented in the works of Zarathuštra and his immediate successors, of which they were the jealous custodians. Sacred texts dedicated to Mithra, Tištrya, and other 'pagan' divinities were no doubt also in circulation, worded most likely in an archaic idiom. But the mixed religion as such had no scripture to represent it, and probably did not yet belong to any particular denomination. Here lay the incentive for the Zarathuštrian authors: by supplying the mixed religion with a scripture, and presenting it as having been revealed by Ahura Mazdāh to Zarathuštra, they could establish the claim that they alone were its legitimate priestly representatives.

In attempting to trace the origin of the religious mixture, we must bear in mind that the Avestan scripture is reliable evidence of the religious experience of only one Iranian people, namely the one whose language Avestan was. This people occupied a country called Aryana Vaējah, which partly or wholly coincided with the Greater Chorasmian state abolished by Cyrus (559–530) (see Henning, *Zoroaster*, 42 *sq.*). Of the political organization of this Eastern Iranian state, which according to Henning centred around Marv and Herat, we have no direct evidence, but there is indication in the Avesta, discussed in note 145[1] below, that at some time of its pre-Achaemenian history, which may well be more ancient than that of the Median empire, the state in question was a federation under the control not of a single ruler but a council of leading representatives of its member nations.

With the absorption of Greater Chorasmia in the Achaemenian empire, religious developments in that country were bound to be influenced by the religious climate of Persis and especially the Achaemenian court. In *OIr. Lit.* §14 the opinion was expressed that the religious beliefs which Darius (522–486) professes in his inscriptions derive ultimately from Zarathuštra's teachings. Common to both are the exclusive worship of Ahura Mazdāh; the direct and spiritual approach to the god, which Darius sums up in the remarkable sentence 'Ahura Mazdāh is mine, I am Ahura Mazdāh's'; the insistence on the nefarious role played by Falsehood (OPers. *Drauga*, corresponding to Zarathuštra's *Drug*); the view that Truth (OPers. *Arta*, to be restored by conjecture, see note 103[1]) is the supreme possession attainable both in life and hereafter (cf. note 2[3]). What one misses in Darius are references to the Fiendish Spirit, to the Incremental Spirit and the other Aməša Spəntas, and to Zarathuštra. In themselves such omissions on inscriptions of a political character would be of little account, but they do seem significant in this case, especially the last one, because not even Herodotus mentions Zarathuštra's name, and the religion he describes as that of the Persians is not the one which Zarathuštra had preached (see Benveniste, *Rel.* 29 *sqq.*).

Let us try to imagine how Darius might have acquired what notions he seems to have had of Zarathuštra's religion. Brought up far from Aryana Vaējah,† very likely unfamiliar with the Gāthic language, surrounded by Magian priests who would oppose any religious reform, Darius, one would think, obtained his Zarathuštrian initiation from a second-hand source. As this must have been one he particularly trusted, we may avail ourselves of information given by Herodotus (I, 209 *sqq.*), which has not hitherto been used in this connection: when in his fatal campaign against the Massagetae Cyrus had crossed the Araxes, the suspicion arose in him that Darius, who being scarcely twenty years old had been left behind in Persis, was plotting against his crown; Cyrus revealed his suspicion to Hystaspes, who at once recrossed the Araxes and returned to Persis to keep watch on his son; shortly afterwards Cyrus fell in battle.

It is obvious that to reach the Massagetae Cyrus and with him Hystaspes had to cross Chorasmia. Here is a historically attested link between Zarathuštra's country and Darius. What Vištāspa brought back and imparted to young Darius need not have been

† Darius presumably grew up in Persis, of which his father Hystaspes (OPer. Vištāspa) was governor (Herodotus, III, 70).

INTRODUCTION

more than the bare outlines of Zarathuštra's religion. The prophet's name would scarcely interest Darius as much as the fact that this was the religion of Aryana Vaējah, 'the expanse of the Iranians' (see Benveniste, *BSOS*, VII, 267 *sq.*), the region where Iranians first established a political and cultural centre. In Ahura Mazdāh, the sole god of the official religion of Aryana Vaējah, Darius would see the true 'god of the Iranians', as the Elamite version of the Behistun Inscription (column III, 77) calls him. In placing himself under his protection Darius may well have felt that he had secured the best possible support for his plan to impose Iranian rule on the world.† With this interpretation one understands why Zarathuštra's name appears nowhere in Achaemenian records: the ancient Persians who held his religion thought of themselves as Mazdāh-worshippers, not as Zarathuštrians.‡

The introduction of Mazdāhism into Persis as a factor to be reckoned with coincides, as far as we can tell, with Darius' accession to the throne and a general slaughter of Magian priests, the μαγοφονία, which for a long time was commemorated in yearly celebrations.§ Darius' treatment of the Magi alone suggests that their religion was different from his. This impression is increased to certainty by Darius' rebuilding of 'places of worship' which the Magus Gaumāta had destroyed (*Beh.* §14), and Herodotus' account of the Persian religion as interpreted by Benveniste, *Rel.* 32 *sqq.* If the Magi nevertheless made a remarkable recovery and remained the only priests known to have officiated at the Achaemenian court,

---

† When I submitted this theory to Professor Henning he remarked that one might go beyond relying on one or more visits of Hystaspes to Chorasmia, and assume that the branch of the Achaemenian family to which he belonged was of Eastern Iranian origin. Such an assumption goes well with the fact that the name Hystaspes = Vištāspa was borne both by Darius' father and by the Eastern Iranian king of Chorasmia whom Zarathuštra converted to his creed. Loyalty towards Chorasmia would incline both Darius and his father to view with sympathy the religion which had become prevalent in that country.

‡ The OPers. loanword in Aramaic, *mzdyzn* 'Mazdāh-worshipper', in which *yzn* = *yazna* (against Av. *yasna*, NPers. *jašn*) is best explained as the Median form of the OIr. word for 'worship', is, of course, no proof that the cult of Mazdāh was taken over by the Persians from the Medes. Once the form *yazna- had become current in OPers., it would be used also in compounds that did not yet exist in the days of Median hegemony. Against the theory that the worship of Mazdāh is attested in Assyrian records see A. Ungnad, *OLZ*, 1943, 193 *sqq.*

§ See Henning, *JRAS*, 1944, 133 *sqq.*

it is clear that despite their own views they had agreed to serve Darius' religion. However, as Mazdāhism is not the creed which Herodotus describes them as holding, it cannot have been the only creed they served. Thus we reach the conclusion that the Magi were a clergy of all denominations, a class of professional priests who officiated in the service of several if not all forms of Iranian worship that were practised in Western Iran. Herodotus reports that it was not lawful to offer sacrifice except in the presence of a Magus, whose function, once he had presented himself, was to chant a θεογονία. If, as assumed by Benveniste, *Rel.* 31, the θεογονία was the Median equivalent of an Avestan Yašt, the Magus would have in his repertoire different hymns pertaining to the worship of different Iranian divinities; he was thus equipped to chant a hymn to Mithra, or Ahura Mazdāh, or Vrθraγna, or Anāhitā, etc., according as each worshipper would instruct and pay him to do.

With this interpretation of the meagre data at our disposal, the Magus of the time of Darius and Xerxes appears to be the precursor of the Zoroastrian priest we meet in the Younger Avesta: in st. 137 of the Mithra Yašt Ahura Mazdāh reveals to Zarathuštra that a man will secure for himself Mithra's protection if he instructs and pays a priest to 'utter Mithra', that is, if we imitate Herodotus' terminology, to chant a *Μιθραγονία. What is true of Mithra no doubt also applied to Anāhitā and the other divinities of the Iranian pantheon. The gods did not each have their own priests, but were all served by one professional priesthood, which in Western Iran was recruited from the tribe, or caste, of the Magi; in Aryana Vaējah, up to a certain time, an eclectic non-Zarathuštrian priesthood may have performed a part similar to, but less prominent than, that of the Magi in the West, while the Zarathuštrian priests served Ahura Mazdāh exclusively; later the Zarathuštrian priests monopolized the priestly profession by becoming 'Zoroastrian', that is, by including in their ministry also the non-Zarathuštrian gods. When Darius introduced Ahura Mazdāh into Persis the god was presumably treated by the Magi as a mere addition to their motley pantheon, and Darius, in the absence of a Persian Zarathuštrian clergy, had no objection to their

INTRODUCTION

taking over the service of his god. In deference to Darius and Xerxes the Magi would perfect their acquaintance with Mazdāhian theology, possibly drawing information from emissaries of the Zarathuštrian community of Aryana Vaējah. This theology, which would scarcely appeal very much to Magian taste, did not alter their pantheistic outlook, as Herodotus' account of the Persian religion clearly indicates (cf. above, p. 3): Mithra and Anāhitā were still to them as deserving of praise as Ahura Mazdāh. Obviously, then, although by the middle of the fifth century they had certainly heard of Zarathuštra, the Magi had no reason to consider themselves 'Zarathuštrians', the more so as even worshippers who in Aryana Vaējah would be called 'Zarathuštrians', in Persis were apparently merely known as 'Mazdayaznians'.

By now the reader will have detected the trend of our argument: not an unprincipled Eastern Iranian Zarathuštrian priesthood is responsible for the contamination of the prophet's doctrine with incompatible elements, but the special conditions in which Mazdāh as sole god was exported to Persis in the last quarter of the sixth century, to be returned to Aryana Vaējah as *primus inter pares* some fifty to eighty years later. In about 441 a reform of the Old Persian calendar on the Egyptian pattern took effect (see S. H. Taqizadeh, *Old Iranian Calendars*, pp. 13, 33, and *passim*). It is likely that on the same occasion the so-called 'Zoroastrian' names of the months and days were introduced.† They are called 'Zoroastrian' because they coincide with the calendar names of the Younger Avesta, whose mixed religion they appear to represent. The introduction under Artaxerxes I (465–425) of 'Zoroastrian' calendar names is generally thought to mean that this king had adopted Zoroastrianism as official state religion. Thus the Zoroastrian, mixed religion of the Younger Avesta is supposed to have conquered the Achaemenian court as early as the middle of the fifth century, and experienced the triumph of having its calendar imposed on the Persian empire,

† It must be admitted that the 'Zoroastrian' calendar *names* may have been introduced at any time within a century after 441, the date when the calendar itself was reformed. There is, however, nothing to disprove the simpler assumption, which we may therefore provisionally take for granted, that the 'Zoroastrian' names were officially inaugurated simultaneously with the reformed calendar.

and this without even the latest Avestan texts displaying the slightest awareness of such remarkable success, or of the very existence of the Achaemenians and their empire. A serious objection to this view is that Avestan Zoroastrianism, to be able to conquer the Achaemenian court by 441, must have been in full swing in its home country Aryana Vaējah several decades earlier, say by 480. Thus the original Zarathuštrian community is allowed little more than 60 years from the death of its revered prophet to develop, apparently out of sheer perversity, a theology which is an insult to his intentions. There is no need to face such an impasse. It is only to be expected that the reformed calendar and its names would travel from Persepolis to Aryana Vaējah, and not in the opposite direction. Most probably the Younger Avestan calendar is what was imposed on Aryana Vaējah, as on other provinces of the empire, by court decree, and the calendar names are an agreed compromise hammered out at court between Artaxerxes I and his advisers, among whom Magian opinion was no doubt strongly represented. Mazdāhism, by then the traditional religion of the royal family, obtained the lion's share of the names, but other Iranian cults cherished by the Magi also received their due.

If this view is adopted, the religious situation reflected in the calendar names must be taken primarily as the one obtaining in Persis at the time when the names were introduced: a coexistence of various cults among which Mazdāhism was the most important. There is, however, no reason to think that in the second half of the fifth century the religious situation of Aryana Vaējah was substantially different, except for the presence in that country of an enterprising Zarathuštrian clergy in charge of local Mazdāhism. The Zarathuštrian priests may be presumed to have striven for a long time to discourage the worship of gods other than Mazdāh, and they appear to have met with some measure of success.† But

---

† This can be inferred from the words spoken by Mithra in sts. 53 *sq.*, on which Windischmann, *AKM*, 1, 38 (quoted with justified approval by Cumont, *TMMM*, 1, 226, n. 13) remarked: 'Die Klage Mithra's, daß er nicht mit namengenanntem Opfer verehrt werde wie die anderen Yazata's, deutet wohl auf eine Zeit hin wo der Mithracultus noch nicht allgemein und den übrigen Culten ebenbürtig war.' The only amendment this statement invites is to read 'nicht mehr' instead of 'noch nicht'.

INTRODUCTION

one century of Persian domination, during which the Magian attitude to religion had a good chance of spreading to Aryana Vaējah, will have produced a revival of the pantheistic outlook which Zarathuštra had endeavoured to eradicate. If we rely on the date 441 B.C., we may say that the Zarathuštrian clergy of Aryana Vaējah will have noticed shortly afterwards from the new calendar names promulgated in their country that the court no longer insisted on the exclusive worship of Ahura Mazdāh. It would then be an anachronism to say that the calendar names 'represent' Zoroastrianism; they would merely reflect a coexistence of various cults, a *de facto* religious situation, not a religious system. Only after the Zarathuštrian priests of Aryana Vaējah had 'translated' this coexistence of religions into terms of a single religious system, claiming that it was the one intended by Zarathuštra, would the Achaemenian calendar names happen to represent the Zoroastrian religion. If, on the other hand, we distrust the linking of the calendar names with the date 441, we may allow the Zarathuštrian priests of Aryana Vaējah another forty years to realize that the times had changed and the rights of other gods beside Ahura Mazdāh could no longer be ignored: for Artaxerxes II (405–359) had fallen in with the Magian outlook to such an extent as to mention in his inscriptions Mithra and Anāhitā on a par with Ahura Mazdāh.

Thus it was either, and preferably, soon after 441 B.C., or at the turn of the fifth century, that the Zarathuštrian priests of Aryana Vaējah had the inspiration of turning the religious mixture they saw had received official sanction into the mixed religion we call Zoroastrianism, by supplying it with a scripture composed in the language of Zarathuštra, as spoken in their days. Such a thing, we may take it, would never have occurred to the conservative Magi, whose θεογονίαι were probably recited in archaic Median language, understood by only a few worshippers. In Aryana Vaējah, too, hymns to Mithra and other non-Zarathuštrian divinities will have been recited by non-Zarathuštrian priests in archaic Avestan language, lost, or almost lost, on the worshippers. The latter, wishing to place themselves under the protection of these gods, had no choice but to employ such priests. But when the Zarathuštrian priests produced a scripture relating to Mithra, Vərəθraɣna, etc., a scripture which every Aryana Vaējahian could understand, and which moreover purported to emanate from the great sage who had written the celebrated Gāthās, the fate of the non-Zarathuštrian priests was sealed, and the success of Zoroastrianism assured.

The history of early Zoroastrianism can now be seen as a give-and-take game played by the Zarathuštrian priesthood of Aryana Vaējah on the one hand and the Magi on the other. The original Zarathuštrian doctrine was introduced by Darius into Western Iran as Mazdāhism. The Magi, perhaps grudgingly, accorded it an important place beside their traditional cults. Their eclectic standpoint gradually prevailed, and weakened the loyalty to Mazdāh of the royal house. Eventually Artaxerxes I, or at the latest Artaxerxes II, officially came down on the side of the Magi. The reaction of the Zarathuštrian priests was resourceful in the extreme. They adopted the eclectic standpoint, and by intensive literary activity converted it into a system purporting to be Zarathuštra's. The logical weakness of the system was no obstacle to its success; its compilation was not a case of thinking out what would persuade, but merely of reproducing what everybody believed or was ready to believe, and giving it a semblance of unity. Thus the Magi were defeated at their own game. What they had exported as a plurality of cults and beliefs came back to them as Zoroastrianism: the same religious situation, in the one case a juxtaposition of loose ends and bits, with at most the description 'Iranian' to hold it together, in the other, a 'system' exploiting to the full the propagandistic value of Zarathuštra's authority, whose fame was growing the more he receded into legendary antiquity.

There was only one remedy the Magi could adopt, short of committing suicide; by resorting to it they in the long run despoiled the Zarathuštrian priests of the fruits of their toils, and turned defeat into victory: they declared Zarathuštra to have been a Magus, and claimed to be the true heirs and custodians of his doctrine. This was most likely the occasion when the legend was invented, reported in Pahlavi and some classical sources, that Zoroaster's native place was in Media, the homeland of the Magi (cf. Jackson, *Zoroaster*, 189 *sqq.*). Once adopted in self-defence by the Magi as their chief credential, Zoroaster's name became

INTRODUCTION

famous throughout Western Asia, and soon penetrated to the Greeks. If we array our facts and theory around the year 441 as the date when the religious mixture was so prominently advertised as to induce the Zarathuštrian priests to provide it with a scripture, we may expect that by the end of the fifth century the Magi will have been driven to acknowledging Zoroaster as their prophet. This dating squares well with the familiarity of fourth-century Greeks with Zoroaster's name, and Herodotus' failure to mention it.† To Persian Mazdayasnians, of course, the prophet's name was well known long before the Magi usurped it, as the use of the name Spitamas in the time of Artaxerxes I shows (see Henning, *Zor.* 28, n. 1). But foreign observers could not be expected to penetrate to the remote origins of what was merely one of the numerous Persian cults. To them the Persian religion quite rightly consisted in what the Magi did and said, and they became aware of Zoroaster's importance only after the Magi cared to assure them of it.

IV. CHRISTENSEN'S THEORY

The composition of the Avestan Yašts, which constitute the earliest part of the Younger, Zoroastrian, Avesta, is accordingly best assigned to about 430–420 B.C.‡ If our reconstruction of the circumstances in which the Zoroastrian scripture arose is correct, it follows that Christensen's generally adopted theory on the composition of the Yašts is, to say the least, unnecessary. The case against it has been argued in *OIr. Lit.* §§ 35 sq. Here we may take a glance at Christensen's analysis of the Mithra Yašt (*Les Kayanides*, p. 15):

(a) *Restes du Yašt original*: §§ 2–48, 60–72, 75–87, 95–8, 104–14, 123–5, 140–5.

(b) *Anciennes additions zoroastriennes*: §§ 1, 49–59, 73–4, 88–94, 99–103, 115–19.

(c) *Additions récentes*: §§ 120–2.

† We shall do well to share Marquart's opinion, *Philologus*, Suppl. VI, 531, that 'auf Xanthos den Lyder darf man sich für Zoroaster nicht berufen'; cf. also *ibid.* p. 608, n. 353, end.

‡ The later parts of the Younger Avesta, some of which, especially the *Vendidad*, show distinct Magian influence, do not fall within the purview of this Introduction; cf. *OIr. Lit.* §§ 40–2.

Presumably the author would have wished to add 126–36 to (*a*), and 137–8, perhaps also 139, to (*b*).

It is true that the passages under (*a*) partly speak of Mithra in terms which were quite likely to have been used before Zarathuštra was born, while (*b*) and (*c*) consist partly (by no means entirely) of statements which only a Zoroastrian would make. From this we have no right to conclude that the passages under (*a*) in their extant form are of earlier composition than the passages under (*b*) or (*c*). Would it be sound method to conclude from the episodes of Roman history which Vulcanus wrought on Aeneas' shield (*Aeneid*, VIII, 626 *sqq.*) that the Aeneid was a poem written by a Trojan refugee 'avec anciennes additions romaines'? If Virgil could get his facts about the Trojan *débâcle* more or less correctly (who is to check the details?), and present them in poetic garb with anachronistic additions inspired by his personal bias, why could not an Iranian priestly poet of about 430 B.C. do likewise in respect of Mithra? He, too, no doubt had sources at his disposal—chiefly, one imagines, ancient verses praising Mithra, which the non-Zarathuštrian priests were still reciting in archaic Avestan language (cf. above, p. 20); he may even have extracted from a guileless visiting Magus the chief characteristics of Mithra as presented in the Magian *Μιθραγονία. When he had collected his material he composed a hymn out of it, weaving in an occasional allusion to Zarathuštra as he had been ordered to do, but remaining conscious that his verses would be understood by worshippers of Mithra who knew what to expect and would brook no lies.

The poet's very conscientiousness, which it would be wrong to call callousness or stupidity, disproves Christensen's whole idea of what 'Zoroastrianization' implies. If this process meant that an existing 'pagan' hymn to Mithra was rearranged, offensive bits expunged, and Zarathuštrian bits interpolated, so as to make it palatable to genuine Zarathuštrians, it is inconceivable that a Zarathuštrian priest would have let the statement stand, that Ahura Mazdāh had worshipped Mithra (cf. below, pp. 52 *sq.*). Even apart from this extreme, what is clear to us must have been even clearer to Zarathuštrian priests: the only way of making Mithra, Anāhitā, etc. fit Zarathuštra's theology was to ignore them altogether. Not being prepared to go thus far, they wisely refrained from attempting to alter their 'pagan' character. One may guess that if they conceived at all of such a term as 'Zoroastrianization', its meaning to them was 'continuation of Zarathuštra's revelation'. Such a definition does not have to imply deliberate deceit, not at any rate on the part of the men who composed the Younger Avesta. To them Mithra, Anāhitā, etc., being acknowledged by respectable persons from the Great King downwards, had become respectable divinities. All respectable beliefs in their opinion originated from Zarathuštra. The Zarathuštrian church, having regrettably lost Zarathuštra's supposed teachings about Mithra, Anāhitā, etc.,† had a duty to collect them wherever they could be found and formulate them in Zarathuštra's own language,‡ so as to ensure their being correctly handed down to future generations. With such an attitude it did not matter if existing beliefs concerning Mithra contradicted what Zarathuštra had said. These beliefs, too, were now held to have originated from Zarathuštra, and the prophet's authority was sufficient to set at rest any misgivings aroused by contradictions.

Christensen's interpretation of the structure of Yašt 10, unlikely on theoretical grounds, could be maintained only if a linguistic difference were discernible between his group (*a*) and the two later ones. This Christensen believed to be the case, relying on a linguistic analysis of the 'additions zoroastriennes', which seemed to display the faulty grammar and lax construction characteristic of parts of the Avesta that are obviously of later composition than the bulk of the so-called great Yašts. However, a renewed effort to understand the 'Zoroastrian' stanzas of the Mithra Yašt reveals that in grammar and syntax they are as correct by Younger Avestan standards as the 'pagan' stanzas, with which they have every appearance of being contemporaneous. Even the metre is the same in both, except that stanzas 1 and 122 have longer lines. There is thus little enough reason to rely on a chronological distinction between 'pagan' and 'Zoroastrian' parts of this Yašt, the more so

† The loss was conceivably imputed to the upheaval resulting in Aryana Vaējah from Cyrus' conquest, as in Sasanian times the loss of part of the Avesta was said to be a result of Alexander's invasion of Iran (see Bailey, *Zor. Prob.* 151 *sqq.*). ‡ Cf. above, p. 20.

as we shall see (pp. 51 *sqq.*) that even the definition of what is 'pagan' and what is 'Zoroastrian' is by no means as simple and clear-cut as Christensen assumed. If on the other hand it is admitted that the 'pagan' stanzas are contemporaneous with the 'Zoroastrian' ones, who should have composed the former if not the Zoroastrian poet responsible for the latter? Perhaps one of the non-Zarathuštrian priests whose existence we have surmised on p. 20? By what coincidence should such priests, who must have clung conservatively to their archaic hymns for centuries, have wished to translate them into everyday speech at the very time when their Zarathuštrian rivals needed the hymns in this idiom, not because they approved of them, but, according to Christensen, because they wished to purge them of offensive matter and 'Zoroastrianize' them?

The present study of the Mithra Yašt has convinced me that there is no valid reason to think that this remarkable long poem is the result of successive 'Zoroastrianizations' over a period of centuries of one or several 'pagan' hymns to Mithra. Apart from two ancient glosses in sts. 128 and 129, one formula common to all Yašts (second half of st. 6), one apparently misplaced line (see note 92[1]), and a few grammatical lapses due to the hazards of tradition,† the Mithra Yašt can very well be understood as the work of one author, or small group of authors, who composed it in the second half of the fifth century. If any parts of the hymn are later interpolations they are indistinguishable from the remainder. It is therefore unprofitable, and liable to lead to absurd results, to carve up the hymn into layers whose respective dates range from the eighth century B.C. to the Sasanian era, as Herzfeld did. The hymn will be found to make no less good sense when taken as a single piece of literature, than when interpreted as a collection of disparate bits of verse and prose composed at different periods and hap-

---

† Such as singular instead of plural or *vice versa* (*aṇhayeiti:rāšayente* 21; *azaite:hištante* 38; *nō:azəmnąm* 86; *išavascit̰:bavaiti* 39; *haməraθāδa:haməraθē* 71; *daiṅhāvō:yātayeiti* 78; *vīspə:arəδəm* 100), plural instead of dual (*fravazǎnte* 119), nominative instead of accusative or *vice versa* (*vaṇhuš* 140, *asānasča* 136, *razištąm čistąm* 126), or otherwise wrong endings (*vərəθrajanō* and *upamanō* 9, *akatarəm* 26, *tā* 48, *uγra* 107, *dašinəm* 126). A misspelling is found in 34 (*harəθā*).

25

INTRODUCTION

hazardly strung together and rearranged by successive compilators who lived at a distance of centuries from each other. If it is thought that I have erred on the side of credulity, I would plead that it is preferable to rely on what the text appears to say, than on what modern opinion believes it to be concealing, on the strength of unprovable and unlikely theories.

V. MITHRA'S FUNCTIONS

The Avestan common noun *miθra-*, which is formally indistinguishable from the name of the god, means 'contract'. It is repeatedly found in the Mithra Yašt,† in which Mithra is primarily the god of contract.‡ The contract which Mithra supervises with his thousand perceptions (sts. 35, 107), thousand ears (sts. 7, 91, 141), ten thousand eyes (*ibid.*), and ten thousand spies (sts. 24, 60),§ punishing its infringement and rewarding its observance, must be taken to include all forms that exist in society, not only of agreed but also of involuntary relationship, such as that which obtains between brothers or father and son (st. 116). Even engagements undertaken with 'owners of Falsehood', who are sure to break their part of the bargain, must be honoured (st. 2). The partners of contractual relation are graded in a curious scale (sts. 116 *sq.*). In it the contract marked 'thousand', the 'thousandfold contract between two countries', reflects the most interesting aspect of Mithra's function

---

† Sts. 2 *sq.*, 45, 109, 111, 116 *sq.*; cf. also *Vend.* 4.2 *sqq.*, where conditions of contract (*miθra-*) are discussed which Lüders, *SPAW*, 1917, 347 *sqq.*, has shown to go back to Indo-Iranian times.

‡ Cf. sts. 3, 17–23, 26 *sq.*, 38, 45, 48, 62, 72, 80, 82, 104 *sq.*, 107–9, 111.

§ Lommel has suggested (*Oriens*, VI (1953), 331) that Mithra's ten thousand eyes and spies are the stars. Although he admits that the Mithra Yašt does not confirm this theory, he quotes in its support the epithet *stəhrpaēsah-* 'star-decked', which Mithra's chariot bears in st. 143. It would be strange if Mithra's spies, who are expressly stated to be 'on every height' (st. 45), were stuck to his chariot. In any case these 'stars' are likely to be mere precious stones, see below, note 136[3], sect. (4). The correct interpretation of Mithra's (and the Vedic *mitrāvaruṇā*'s) 'spies', which Lommel in my opinion has not succeeded in shaking, is the one proposed by Lüders, *Varuṇa*, 35: 'Wie ein König hat Varuṇa seine Späher.... Das sind nicht die Sterne...; diese Späher haben keine physische Grundlage. Es sind die Späher, mit deren Hilfe der indische und schon der arische König über das Tun und Treiben seiner Untertanen wacht.'

MITHRA'S FUNCTIONS

as guardian of the contract. He is the guarantor of orderly international relations, the god of the international treaty. A ruler, says the text, according as he honours or fails to honour the treaty will appease through Mithra the mind of a ruler he has antagonized or incense even a ruler he has not antagonized (sts. 109, 111). Mithra protects or destroys the countries according as they respect Mithra—representative of the covenant—or are 'defiant' (st. 78). 'Defiance' on the part of a country earns it Mithra's implacable revenge in st. 27. No doubt, too, the 'anti-mithrian' countries which in st. 101 are made to feel Mithra's club had been guilty of violating the treaty. Mithra's epithet *karšō.rāzah-* 'director of (boundary) lines' (st. 61), if this is the correct interpretation, also pertains to his tutelage of international relations.

The first condition for a country to be able to honour its treaties is that its internal affairs should be well regulated, the authorities obeyed, revolutions averted. In the fulfilment of this condition Mithra understandably takes an active part. On the one hand he 'smashes' the authorities that are false to him as representative of the treaty (st. 18), on the other hand we read in *Yt* 13.94 *sq.*:

> *ušta.nō zātō āθrava*
> *yō spitāmō zaraθuštrō;*
> . . . .
> *iδa apąm vījasāiti*
> *vaŋuhi daēna māzdayasniš*
> *vīspāiš avi karšvąn yāiš hapta;*
> *iδa apąm miθrō yō vouru.gaoyaoitiš*
> *fraδāṯ vīspā̊ fratəmatātō daḣyunąm,*
> *yaozaintīšča rāmayeiti;*
> *iδa apąm ⟨apąm⟩ napā̊ suro*
> *fraδāṯ vīspā̊ fratəmatātō daḣyunąm,*
> *yaozaintīšča nyāsāite.*

In fulfilment of (our) wish a priest is born to us, Zarathuštra the Spitamid...; henceforward the good Mazdayasnian Religion will pervade all seven climes; henceforward grass-land magnate Mithra will further all supreme authorities (lit. 'councils of premiers', see note 145[1]) of the countries, and pacify (the countries) that are in turmoil; henceforward the strong Grandson of the Waters will further all supreme authorities of the countries, and hold down (the countries) that are in turmoil.

INTRODUCTION

The appearance in this connection of the Grandson of the Waters will occupy us again below, pp. 59 *sq*. Here we may note that Mithra's 'pacification' of countries in turmoil is connected with the world-wide spread of the Religion, which requires conditions of peace. We shall find below, note 64[7], that it is very likely Mithra's furtherance of peace and political and social order which constitutes the subject of a pledge he appears to have given to the Religion. It looks as if the god of contract is himself bound by the 'contract of the Mazdayasnian Religion', which in st. 117 receives the highest mark ('ten thousand') of all contracts.

The question now to consider is whether the Avestan Mithra's association with the contract is a secondary development due to a fortuitous identity of his name with a word for 'contract', or represents a, or the, primary function of the god. Theoretically the first alternative is quite possible. There are several bases **mi-* in Indo-European, to any of which the suffix *-tro-* could be added, and a further homonym could result from the addition of the suffix *-ro-* to a base **mit-*.† But obviously, unless the study of all ancient sources on Mithra/Mitra reveals that the god's primary function cannot have been the guardianship of contract, or/and that another function of Mithra's must be the primary one, it would be unreasonable to reject as not genuine the equation Mithra = contract, which the Avesta repeatedly states in terms which could not be clearer. Let us survey the four ancient sources, namely (1) the Mitanni reference to Mitra, (2) the Rigvedic evidence, (3) the Roman Mithras, and (4) the evidence relating to the Iranian Mithra, with a view to finding an answer to this question.

(1) shows us Mitra as guarantor of the treaty, but in the company of other gods; there is no way of telling whether the Mitanni did not see in him primarily a god in charge of something else. (3) is inconclusive in the opposite sense. Figurative art is not a reliable means of expressing abstract notions, and one would not care to deduce from Mithras' handshake with the Sun (cf. Cumont, *TMMM*, I, p. 173), that he was connected with the contract. On

---

† For various suggestions on the etymology of the god's name cf. Wn.-Debr. II, 2, 701, F. B. J. Kuiper, *Acta Orientalia*, XII, 238, T. Burrow, *TPS*, 1949, 53, n. 2.

the whole, one would think, custodianship of contract was at most a secondary function of the Roman Mithras. His chief function apparently was one which he shared with the Avestan Mithra, that of a giver of light and life. Let us call this function *B*, and the guardianship of contract *A*. There is no logical development which might lead from *B* to *A*, but we shall presently see (pp. 30 *sq.*) that by a simple argument Iranian worshippers of Mithra could have extracted *B* from *A*. As the Roman Mithras is derived from the Iranian Mithra (see pp. 61 *sqq.*), it would seem that he greatly developed the latter's function *B*, which in the Avesta is at the incipient stage, and reduced, or altogether shed, function *A*. If it is denied that function *B* was developed from *A*, the only alternative is to hold both as pertaining to Mithra from the beginning. Such an assumption, however, is worth maintaining only if the Rigvedic evidence supports it. This we shall see is not the case. The conclusion is that though the Roman Mithras cannot be shown to have held function *A*, he cannot be used to disprove that this was the original function of his Iranian forbear.

With regard to (2) the situation is that, if we did not have the Avestan evidence, no more could be said of the Rigvedic Mitra than that he 'calls people to account' and is somehow homogeneous with Varuṇa, the god of oath (cf. above, p. 7). But as soon as the Avestan identification of Mithra with the contract is taken into consideration, it becomes clear that the contract is also the Rigvedic Mitra's domain. One understands at once not only why he 'calls people to account', an occupation in which the Iranian Mithra is constantly engaged in the Avestan Mithra Yašt, but also why his personality is all but merged in Varuṇa's. Both Varuṇa, the greater of the two, and Mitra watch over the observance of Truth, one in respect of the oath, the other in respect of the contract. Accordingly it is their joint task to detect and combat Truth-infringers; this is why Mitrāvaruṇā, like the Avestan Mithra, 'have spies,...are guardians of the whole world,...are barriers against falsehood, which they...punish,...afflict with disease† those who neglect their worship' (cf. the statements printed in italics above, pp. 4 *sqq.*). Once Mitra was associated with Varuṇa in the per-

† Cf. what is said of Mithra in st. 110.

petual watch over Truth against those who fail to honour it, he naturally began to share some of Varuṇa's other attributes. Hence both the Avestan Mithra and the Rigvedic Mitrāvaruṇā are 'bestowers of water' (st. 61, cf. also st. 100), and a link is seen between the Avestan Mithra's omniscience (sts. 24, 60, etc.) and possession of ten thousand eyes, and Varuṇa's being 'all-knowing' and 'thousand-eyed', as well as between Mithra's chariot drawn by four coursers (sts. 68, 125) and Varuṇa's car 'drawn by well-yoked steeds'.

It is true that rather confusingly the Vedic language has a common noun *mitrá* 'friend', which is formally indistinguishable from the name of the god.† But the circumstances suggest that this is a case of accidental homonymity. For although in a few Rigvedic and Avestan passages Mitra/Mithra is friendly enough to men, the same can be said of most gods. The one defining Rigvedic epithet of Mitra, *yātayájjana*, has scarcely anything to do with 'friendship', and to say that the Avestan Mithra, who is 'both wicked and very good to men' (st. 29), represents 'the Friend' and not, as the text persistently suggests, the contract, means to replace, on mere etymological grounds, a vivid and unmistakably characteristic identification of a god by a colourless description that would fit almost any Avestan divinity except the Fiendish Spirit and the daēvas. In any case, to counterbalance the bias to which the noun *mitrá* 'friend' might incline us, there is evidence of the existence of a Rigvedic noun *mitrá* 'contract' (cf. Bailey, *TPS*, 1953, 40), etymologically identical with the Avestan noun *miθra-* quoted above, p. 26. It is then tempting to accept Lüders' opinion (cf. above, p. 7), that while Varuṇa as god of oath watched over Truth in water, Mitra as god of contract did the same in fire, Indo-Aryan contracts being for this reason concluded in front of blazing fire, as oaths were sworn in the presence of water. [See Addenda.]

Having satisfied ourselves that the Rig Veda, despite its reticence on Mitra, does support the theory that the god's original function was to watch over the contract, we may return to the Iranian evidence, our source (4) (cf. above, p. 28). In discussing (3) we said that a simple argument could have led from Mithra's function

† On Ved. *mitrá* 'friend' cf. below, p. 41, n. §.

MITHRA'S FUNCTIONS

A to *B*. Let us now pursue its links. As guardian of the covenant and watcher of contract-breakers Mithra is said in the initial formula of each *karde* of the Mithra Yašt to be sleepless and ever-waking (st. 7); elsewhere he 'watches in darkness' (st. 141), that is, at night (cf. also st. 95), and is described as 'the caretaker who without falling asleep protects and observes the creatures of Mazdāh' (st. 103). It is clear, then, that Mithra is not only up and about all day, but also all night. Consequently in the morning he is up before anybody else, including the sun, that proverbially early riser. Such early habits may well have suggested to Mithra's worshippers, who would understandably have wanted to *see* Mithra in some natural phenomenon, that he was the light of daybreak, which *precedes* the appearance of the sun. That this is the case would seem to be stated in st. 142, where Mithra '*in the morning brings into evidence the many shapes*, the creatures of the Incremental Spirit, as he lights up his body, being endowed with own light like the moon'. As, however, the words printed in italics are not certainly a correct translation of the Avestan text (cf. note 142[1]), the passage can only be used as corroboration of other evidence. But we may note in passing that Mithra's epithet *hvāraoxšna-* (see note 142[2]) 'endowed with own light' is not necessarily an *ad hoc* invention of the poet to explain how Mithra could appear in the form of light: the epithet may be an ancient relic, a spark, as it were, of the fire in which, according to Lüders (see above, p. 7), Mithra once resided. [See Addenda.]

The kind of light Mithra represents is luckily defined in unambiguous terms in *Vend.* 19.28, a passage discussed below, note 143[6]: 'The flashing, shining dawn of the third night flares up; Mithra, the keeper of good watch, approaches the mountains where Truth breathes freely; the sun rises.' One can hardly complain that this statement lacks precision in the timing of Mithra's appearance. The oft-quoted st. 13, in which Mithra is 'the first... to approach across the Harā, *in front of* the...sun,...the first to seize the gold-painted mountain-tops', represents thus merely a translation of Mithra's time-table into terms of space: because he is the light that appears *earlier* than the sun, he travels *in front* of the sun. It is a plausible guess that in Western Mithraism

INTRODUCTION

the same situation received the opposite interpretation: the light of daybreak is *past* by the time the sun rises; hence on Mithraic reliefs Mithras stands on the sun's chariot *behind* the sun.†

Once the light-giving part of function B is explained, its life-giving part easily follows from what has been said. In the Avesta Mithra is defined as a life-giver by his epithets *puθrō.dā-* 'bestower of sons' (st. 65), *uxšyaṯ.urvara-* 'making plants grow' (st. 61), and *gayō.dā-* 'bestower of life' (st. 65). Now, seen from the practical point of view of the Iranians, the granting of sons is on the same level as the dispensation of fat and herds (st. 65), the provision of wide pastures (st. 112), or the appointment of richly furnished and well-staffed houses (st. 30). In st. 108 riches and fortune, health and 'property that affords much comfort' are thought of in association with 'noble progeny hereafter'. The provision of material comfort‡ and of sons must be viewed as part and parcel of Mithra's care for the nation's welfare and prosperity, which create conditions of internal stability, thus leading to treaty-abiding international relations (cf. above, pp. 27 *sq.*). The epithet *puθrō.dā-* is accordingly a by-product of Mithra's concern for the stability of contractual relations, which in the case of long-term stipulations anyway depends on the availability of sons to carry out their fathers' obligations. The epithet *uxšyaṯ.urvara-* can be traced indirectly to the same source, if the history of Mitra/Mithra is borne in mind. The ability to make plants grow obviously results from Mithra's function of 'replenisher of waters' and 'rain-pourer' (st. 61), which in its turn is due to his Truth-watching association with Varuṇa, who had become a water-god (cf. above, pp. 7, 30). Ultimately, then, even the growing of plants requires of Mithra no more than guardianship of the covenant as original function. Once *puθrō.dā-* and *uxšyaṯ.urvara-* are accounted for, *gayō.dā-* presents no difficulty. A god who bestows progeny, brings down rain and

---

† See the monuments quoted by Cumont, *TMMM*, I, 176, n. 6. The scene is best visible on Nos. 245*c*3 (plate facing p. 346 of vol. II) and 251*d*8 (plate following p. 364 of vol. II).

‡ Below, p. 229, we have argued that the Av. title $x^v\bar{a}\theta r\bar{o}.disya$- 'comfort-assigner' belongs to Miθra by virtue of his concern for the material comfort of his worshippers.

raises vegetation, and is moreover identified with the first light of the morning, which brings back to life the sleeping οἰκουμένη, would almost inevitably be considered a 'giver of life'. [See Addenda.]

Once it is admitted that function B, which may be the only one of the Roman Mithras, but is ancillary to A in the Avesta, can be comfortably derived from A, there is no point in denying what all appearances combine to suggest: that Mithra's original function is to watch over the contract, and nothing else. For what may be considered a third function of the Avestan Mithra, that of a warlord, cannot seriously be held to be an original trait of this god. On the one hand martial demeanour is conspicuously absent in the Rigvedic Mitra; on the other hand the Avestan Mithra, in his capacity as 'punisher of wrong' (st. 35), including the wrong perpetrated by 'defiant' countries (cf. above, p. 27), could hardly help resorting to war as a means of mass punishment.† This need not have turned him into a god of war, but it happened to do so. To infer, as is often done, that the Avestan Mithra, in addition to representing the Vedic Mitra, is also the heir of the Vedic war-god Indra, means to open the door wide to a reckless identification of gods that never had anything to do with each other, thus spelling the confusion, instead of the clarification, of Indo-Iranian religious history. If Mithra is called Mithra, and not Indra, whose name, moreover, duly occurs in Avestan as Indra, then the only possible excuse for grafting the Rigvedic Indra on the Avestan Mithra would be an association in Rigvedic hymns of Mitra with Indra as close as that with Varuṇa. Of such association there is no trace. Inevitably once by secondary process the Iranian Mithra had become a god of war, he would in this respect resemble Indra. Ancient Indo-Iranian martial epithets and items of equipment—such as the thunderbolt/mace (Ved. *vájra*/Av. *vazra*-)—which Indra has in the Vedas, would be attributed to Mithra in his new capacity; the two gods might even share suitable lines borrowed from traditional epic poems. This need not, and, on the available evidence, definitely does not mean that 'the warlike

---

† That the Iranian Mithra's warlike character is derived from his 'force de punir' was clearly seen by Meillet (p. 154) in his decisive article quoted below, p. 43.

INTRODUCTION

character of Mithra...derives from the dominating aspect of Indra'.†

To round off the arguments which point to Mithra's original function as having been the guardianship of Truth in the form of contract, we may briefly consider two pieces of non-Avestan evidence relating to two opposite ends of Iran. One is the oft-quoted Persian habit of swearing μὰ τὸν Μίθρην, ὄμνυμί σοι τὸν Μίθρην (Xenophon), νὴ τὸν Μίθραν (Plutarch) (cf. Cumont, *TMMM*, I, 229, n. 2). The other is found in a Buddhist Sogdian text discovered in Central Asia, the Sogdian version of the *Vessantara Jātaka* (ed. E. Benveniste). It contains the only reference to Mithra so far ascertained in Buddhist Sogdian literature. In Manichean Sogdian texts *myšyy βγyy* (from *Miθra^h baga^h*) is often found as the name of the sun-god and the Manichean Tertius Legatus (see below, p. 40). That Buddhist Sogdians also thought of Mithra as a sun-god is likely, but proof is lacking. All the more interesting is the *VJ* reference to Mithra as god of contract. Prince Sudāšan, determined to obtain Buddhaship, has renounced all his possessions: his royal home, his chariot and equipment, even his children; eventually he is left with only his wife Mandrī in a mountain forest. To test him the Supreme God takes on the appearance of an aged brahmin, and begs to be given Mandrī. The prince agrees. The Supreme God is satisfied with Sudāšan's spirit of sacrifice. Still disguised as brahmin he hands him back Mandrī, but pretends to do so provisionally: Sudāšan is to keep his wife *in trust*, until the brahmin comes back to reclaim her. The intention of the god is to prevent Sudāšan from giving her away again to the next brahmin that comes along. The wording is as follows:

rty nwkr "n'nt' 'yw pr"mn L' (1201) *δwry z'yh šwt rty 'yw pts'r zyw'rt* (1202) *rty ZKw k'z'kh pr'ys rty ZKn swδ'šn* (1203) *KZNH w'β 'zwty*

---

† Thus A. Pagliaro, *Oriental Studies...Pavry*, p. 381, according to whom the liberation of the cows accomplished in the Rigveda by Indra after the slaying of Vṛtra 'is attributed in the Avesta to Mithras (Yašt x 85–6)'. In fact nothing in sts. 85 *sq.* suggests Indra's mythicized, symbolical liberation of rain-cloud cows: solid, terrestrial cows, prosaically captured by 'owners of Falsehood' who, false to the treaty, had abducted them, as in st. 38, implore Mithra to come to their rescue. If he did listen to their call, all he had to do was to knock down the treaty-infringers. There is no dragon to be killed, no mountains are to be pierced, no waters to be expected from the liberated cows.

MITHRA'S FUNCTIONS

'kw 'nyw z'yh s'r ZY (1204) šw'm k'm rty ZKw 'ynch tw' nβ'nt (1205) zynyh pr'yc'm k'm 'wyn δ'm'yγtyh (1206) 'myδry pt'ych ZY 'wyn γrcykt cyt'yty (1207) ZY ZKn wn'kh cytk ZY ZKn y'yh cytk (1208) nβ'nt pr KZNH yw'r 'PZY šw tyw ''δ'k (1209) δβ'r L' δβr'y rty 'γw swδ''šn 'wyh (1210) mntr'yh zynyh n'y's

And now, O Ānanda, the brahmin did not go far; he turned back, reached the hut, and spoke thus to Sudāšan: 'I shall be going to another land, and will leave the woman in your trust (*lit.* with you in trust) *in the presence of Mithra, the Judge of Creation* (on this expression see below, note 92[7]), together with the mountain spirits, the forest spirit, and the spirit of the source, provided (*lit.* for so but) you do not give her away to anybody.' Thereupon Sudāšan took over Mandrī in trust.

Here a proper contract takes place, of which Mithra is witness. Mandrī is given in trust to Sudāšan in the presence of Mithra, on condition that he will not again part with her.

We may now consider Mithra's original function as settled, and turn to what has so long obscured it, namely the Iranian, but not Avestan, notion of Mithra as sun-god, which has often been said to be of pre-Iranian origin, perhaps pertaining to him from the beginning. What encouraged this view was the mistaken idea that already the Avestan Mithra, though in general clearly distinguished from the sun (cf. sts. 13, 51, 90, 118, 145), is on two occasions confused with him. One occasion is claimed† to be the description of Mithra's chariot as being 'the sun-god's', $h\bar{u}...x\check{s}aet\bar{a}i$, in st. 143. It will be seen from our commentary *ad loc.* that actually $h\bar{u}...x\check{s}aet\bar{a}i$ is not in apposition to Mithra, but refers to the sun in his own rights. The other occasion, insisted on by Lommel, *Paideuma*, III, 1949, 210, occurs in st. 136, where Mithra's chariot is described as having one wheel. This could at a stretch serve as confirmation of other proof that Mithra is a sun-god in the Avesta. Even then it would be awkward for the sun to be represented by the wheel of Mithra's chariot. Since, however, there is not a shred of other evidence that the Avestan Mithra is a sun-god, the assertion that he is rests solely on this one wheel. Now, it is a fallacy to think that only the sun can have a one-wheeled chariot. This prejudice is disproved by the Rabelaisian Avestan episode of young Snāviδka

† Most recently by Duchesne-Guillemin, *Anthropologie religieuse*, Suppl. to *Numen*, II, 1955, 99.

INTRODUCTION

(*Yt* 19.43 *sq.*), which it will suffice for our purpose to quote in translation:

...Kərəsāspa..., who killed leaden-jawed Snāviðka whose hands were of stone. Thus was his challenge: 'I am a minor, not yet of age. If I become of age I shall use the earth as a wheel, I shall make the sky the body of my chariot. I shall fetch down Spənta Mainyu from Paradise, and raise Angra Mainyu from disgusting Hell. Both shall pull my chariot, the Incremental Spirit and the Fiendish, provided heroic-minded Kərəsāspa does not kill me.' He was killed by heroic-minded Kərəsāspa towards (=just before reaching) the height of his life, at the rising of his vitality.†

Clearly Snāviðka was not the one to shrink from the thought of making the sun or the moon his second wheel. If he was content with one wheel, the reason must be that Avestan authors considered one wheel sufficient in the case of mythical chariots. Even if originally the one-wheeled chariot was an exclusive characteristic of the sun, by Avestan times such a limitation evidently no longer applied. Accordingly we gather from st. 136 not that Mithra is the sun, but that his chariot is unusual, as may be expected of one that was built by the Incremental Spirit himself (st. 143).

There is yet another prejudice which has prompted some scholars to think that the Avestan Mithra was a sun-god, namely that he is a god of the sky, in Nyberg's opinion not only of the diurnal,‡ but mainly even of the nocturnal sky (*Rel.* 59 *sq.*). In support of this view, apart from Mithra's 'star-decked' chariot (cf. above, p. 26, n. §), which proves no more that its owner is a sky-god than Haoma's 'star-decked' girdle proves such a thing of Haoma, the

† The last two lines read *ava apanəm gayehe* | ...*sānəm uštānahe*, F₁ having a ‡-inch gap before *sānəm*. Bartholomae restored [*fra*]*sānəm*, assumed a strange ellipse of a verb having *ava* as preverb (*Wb.* 163, n. 2), and translated 'he (Kərəsāspa) ⟨prepared for him⟩ expiry of life, destruction of vitality'. It is, however, clear that *apanəm* is governed by *ava*, hence before *sānəm* a governing preposition, say *paiti*, or again *ava*, should be restored. Moreover, Bth.'s interpretation yields a tautology with the preceding *təm janāt...kərəsāspō*, whereas one expects a statement that Snāviðka was killed before coming of age when, at the height of life, with his vitality risen to culmination, he would have been in a position to carry out his threat. Hence we restore [*paiti*] *sānəm*, connect *apanəm* with *apanō.təma-* 'summus' (cf. Bth., *Wb. s.v.*, note), and derive *sānəm* from the base *san-*, on which see note 104⁷.

‡ Note that Zaehner's derivation of Zēbakī *mī* 'day' from *mīr (from Mithra), *Zurvan*, 150, n., is erroneous, since *mī* represents the common Eastern Iranian word for 'day', which in Sogdian appears as *myδδ*, see Morgenstierne, *IIFL*, II, 229, 401.

36

following Avestan passage has been quoted (cf. Nyberg, *Rel.* 301, and Cumont, *TMMM*, I, 183):

asmanəm ('the sky')...
yim mazdå vaste vaŋhanəm ('which M. wears as a dress')
stəhrpaēsaŋhəm mainyu.tāštəm
hačimnō miθrō rašnuča
ārmaitiča spəntaya
yahmāi nōiṭ čahmāi naēmanąm
karana pairi.vaēnōiθe   (*Yt* 13.3)

This Cumont interpreted in the sense that Mithra has 'un vêtement brodé d'étoiles, fait dans le ciel et dont nul ne voit les bouts'. The quotation marks, which are Cumont's, refer to Darmesteter's translation of this stanza (*ZA*, II, 507), which reads: 'vêtement brodé d'étoiles, fait dans le ciel, que revêt Mazda, avec Miθra et Rashnu et avec Ârmaiti, et dont nul ne voit les bouts'. Now, Darmesteter had correctly seen that *miθrō.rašnu* is a compound,† and that the two °*ča* unite Mithra-Rašnu with Ārmaiti in one group governed by *hačimnō*, apposition of *mazdå*. *hačimna-* means 'accompanied by, being with, equipped with', sometimes practically 'having' (cf. sts. 67, 141, and notes 67[2], 84[1]); the noun in the instrumental governed by *hačimna-* never shares in the action of the subject with which *hačimna-* agrees. Darmesteter can therefore only have meant that Mazdāh, being in Mithra-Rašnu's and Ārmaiti's company, wears the sky as a dress. But his wording is unfortunate, and it is understandable that Cumont, who claimed no knowledge of Avestan, should have fallen under a misapprehension whose confusing effect on Mithraic studies is noticeable even in contemporary literature. It is more difficult to excuse the lack of precision in Nyberg's wording, *loc. cit.*: 'Der Himmel...ist Ahura Mazdāhs Kleid, das er neben Mithra, Rašnu und Armaiti trägt.' For since Cumont's days we have had not only Bartholomae's clear entry of *hačimnō* under the heading 'begleiten, gehen mit—; verbunden, vereint sein mit—' (*Wb.* (1905), 1740), but also Lommel's unequivocal and correct translation of the whole passage (*Die Yäšt's* (1927), 112): 'Den der Weise, begleitet von Mithra, Rašnu und der heiligen Frommergebenheit, angelegt hat als einen

† Bartholomae unnecessarily emended the unanimously attested reading *miθrō* to *miθra* (*Wb.* 1185, line 7).

mit Sternen geschmückten, von Geistern geschaffenen Mantel, an dem von niemand die beiden Enden der Hälften gesehen werden.' The notion that the Avestan Mithra was a god of the sky must accordingly be dismissed as a nineteenth-century myth.

We are exceptionally fortunate with Mithra in that he is the subject of a long Avestan hymn that is packed with information. We are thus in a position to sort out the strands of a character who at first sight presents a bewildering variety of sides: contract, war, first morning light, source of life, sun; and there are more (cf. below, pp. 42 *sq.*). Without the Avestan hymn it would be impossible to disentangle the later accretions from what is original. We have so far seen that Mithra's original function is the guardianship of contract, and have considered three accretions, two, war and morning light, fully grown at the Avestan stage, the third, the giving of life, only hinted at in the Avesta, but due, together with the bringing of light, to become the chief discernible characteristic of the Roman Mithras. It remains now to inquire whether the Iranian identification of Mithra with the sun, which is firmly attested in the Christian era, and may have begun earlier, can also be derived from the stage at which Mithra is met with in the Avesta. Luckily in this case, too, the author of the Avestan hymn is sufficiently explicit. Although he in no way identifies Mithra with the sun, his description of the god's movements provides the essential elements out of which, looking back, one may say a sun-worship almost had to develop. For the Avestan Mithra is not only the light which shortly after its appearance is merged in the sun's, but it can be shown that day and night he closely follows, or rather precedes, the sun.

In a most interesting article in *NGGW*, 1923, 1 *sqq.*, supplemented by another in the same periodical, 1928, 195 *sqq.*, Emil Sieg has shown that the Vedic sun, the one wheel of whose chariot is luminous on one side only, turns round on reaching the West, so that the dark side of the wheel faces the earth; he is thus able to rejoin the East unseen by men, though travelling over their heads. It is a credit to Darmesteter's perspicacity that long before Sieg's articles appeared he had contemplated the possibility, in view of the Vedic sun's behaviour, that beginning in st. 95 Mithra 'referait en sens inverse le chemin du jour, revenant du couchant au levant'

(*ZA*, II, 467). Of course there can here be no question of a wheel, luminous on one side only, showing its dark side after sunset. If such a wheel exists at all in Iranian mythology (it has not been noticed), one would require solid proof before attributing it not only to the sun's chariot, but also to Mithra's. In any case Mithra does not require such a device, as he is 'endowed with own light', which he switches on in the morning (st. 142, cf. above, p. 31) and presumably turns off at night.

There is, however, evidence that as in day-time Mithra flies in front of the sun (st. 13) from East to West (st. 67), so once he reaches the West he turns round, as Darmesteter had suspected. No sooner has the poet informed us in st. 95 that after sunset Mithra 'goes along the whole width of the earth', than we are told in st. 100 that Rašnu flies on his left. But when Mithra starts off from Paradise on his reconnoitring expedition (st. 124) Rašnu is on his right (st. 126). It seems only logical to infer, from the reciprocal change of position, that the two gods had reversed the direction of their joint flight. For if Rašnu was travelling on Mithra's right on their westward journey, and each of the two turned round on himself on reaching the West, the return journey would find Rašnu on Mithra's left. Mithra's itinerary and time-table can accordingly be outlined as follows: starting off at, or after midnight(?) from Paradise (which is situated above Mount Harā, see note 50[1]) (sts. 124 *sqq.*), with Rašnu on his right, Mithra arrives in the East (that is, a point east of Mount Harā) at daybreak, whose light is due to him (st. 142; *Vend.* 19.28; cf. above, p. 31). From the East he crosses the Harā (st. 13) and approaches X^{v}aniraθa, the central clime (st. 67); thence, though this is not stated, he continues to the West. There, at sunset (st. 95) he turns round (as shown by Rašnu's now being on his left), and reconnoitres the whole earth (as explained in note 95[1]). Finally, probably towards midnight, he returns to Paradise, taking with him the libations he had been offered (see note 32[1]).

It does not take a wild imagination to see that the god who at the earlier, the Avestan stage, is the first light in the morning, travels west by day, and returns overnight to the East, in time to be visible at daybreak, had a good chance of being identified with the sun at a later stage. The sceptic who may wonder how it is that the

INTRODUCTION

Avestan Mithra travels west by day and east by night if he is *not* the sun, may be reminded that Mithra, as never-sleeping watcher of the covenant and its infringers, has to tour the earth incessantly. In actual fact he is simultaneously present everywhere (see note 95[1], end). This would not, and evidently did not, deter his worshippers from mapping out for him a route and a time-table, which in any case would tend to be linked with the movement of the sun, the most prominent and regular hoverer over the earth. The only daring step to take was the one which translated the relation of Mithra and the sun from time into space (see above, p. 31). Once the *earlier* Mithra had come to be placed *in front* of the sun, he would be assigned the sun's route and further time-table out of sheer mental inertia.

It cannot be our task to pursue the post-Avestan Mithra's fate in detail. The classical and Oriental references to his solar nature have been conveniently collected by Cumont, and requoted countless times since. But we may dwell for a moment on the use to which Mithra was put in the three regional forms of Iranian Manicheism. One of these is attested in the Manichean texts written in the Middle Persian language; the second and third are known from texts written in Parthian and Sogdian respectively. The *terminus post quem* of the Manichean reinterpretation of Iranian gods is approximately given by the date of Mani's death, which, as Henning has argued, took place in A.D. 274 (*Asia Major*, N.S. III, 197 *sqq.*). Now, in Iranian Manicheism Mithra is identified in Sogdian and Parthian with the Tertius Legatus (the leader of the Third Emanation), but in Middle Persian with the Spiritus Vivens (the δημιουργός of the Second Emanation), as shown in the following table:

	Persian	Parthian	Sogdian
Spiritus Vivens	*myhryzd*	*w'd jywndg*†	*w'δδjywndyy*‡ or *'βtkyšpy xwt'w*§
Tertius Legatus	*nryshyzd* ‖	*myhryzd* *nrysfyzd* ‖	*myšyy*¶ *βγyy*

† Lit. 'Spiritus Vivens'.   ‡ Borrowed from Parthian.
§ Lit. 'the Lord of the seven climes', see Henning, *BSOAS*, XII, 314.
‖ The Avestan Nairyō.saṅha, cf. Commentary, note 52[1].
¶ *myšyy* is the Sogd. phonological development of OIr. *miθra-*. Buddhist Sogd. *'myδry*, which we considered above, p. 35, is a historical spelling.

This distribution was convincingly explained by Henning, who discovered it (*OLZ*, 1934, 6 *sqq.*), as reflecting the different developments Mithra, and incidentally Nairyō. saṇha,† had undergone in the mythology of the three Iranian nations. The fact that the Sogdians and the Parthians chose Mithra to represent the Tertius Legatus, who is a sun-god, shows that to them Mithra was a sun-god. As if to confirm this, Sogdian Manicheans actually use the expression *myšyy βyyy* to denote the astronomical sun, and the Parthian word *myhr* (pronounced *mihr*) not only means 'sun',‡ but penetrated with this meaning as a loanword into New Persian.§ Conversely, if the Persians chose Mithra to represent the Manichean Spiritus Vivens, who is a δημιουργός, they may have thought of the ancient god of contract as a δημιουργός. This at any rate was Henning's impression, *art. cit.* p. 7, n. 2, who referred to the oft-quoted τοῦ κόσμου ὃν ὁ Μίθρας ἐδημιούργησε of Porphyrius, *De antro nymph.* 6, at the same time rightly protesting that the Avestan Mithra was in no way a δημιουργός. The bold assertions to the contrary we have tried to refute in note 61[5], but we shall see that as the Avestan Mithra contained the seed of the future sun-god, so his veiled and somewhat heretical-looking association with the 'fashioner' or 'creator' Spənta Mainyu (which will occupy us below, pp. 54 *sqq.*) may have given rise in later times to the belief that in addition to everything else Mithra was a δημιουργός. At any rate, as the Persian Manichees did not assign the role of the sun-god Tertius Legatus to Mithra, there is reason to think that the development of non-solar Mithra as attested in the Avesta into a sun-god originated in Eastern Iran.‖ If in late classical and even later Oriental sources the *Persian* Mithra is described as a sun-god, this will be due, like so much else in Persian culture, to Parthian influence. It will be remembered that from the second century B.C.

---

† The Sogdians assigned to *nr(y)šnx βyyy* the task of the Friend of Light, the leader of the Manichean Second Emanation.

‡ *myhr 'wd m'h* 'sun and moon', *SPAW*, 1933, 549, bottom, as pointed out by Henning, *art. cit.* 6, n. 1.

§ Where it must be distinguished from its homonym *mihr* 'love, friendship, affection, kindness', which belongs to Ved. *mitrá* 'friend' (cf. above, p. 30), Russian милый, etc.

‖ In this connection it is noteworthy that in Yidγa, an Eastern Iranian dialect, the word for 'sun' is *mīra*, from *Miθra*, see Morgenstierne, *IIFL*, II, 69; 228.

INTRODUCTION

to the second century A.D. Parthia was the political and cultural leader of Iran.

We may conclude our survey of the main functions discernible at various periods in Mitra–Mithra–Mithras–Mihr–Miši with a table showing their development:

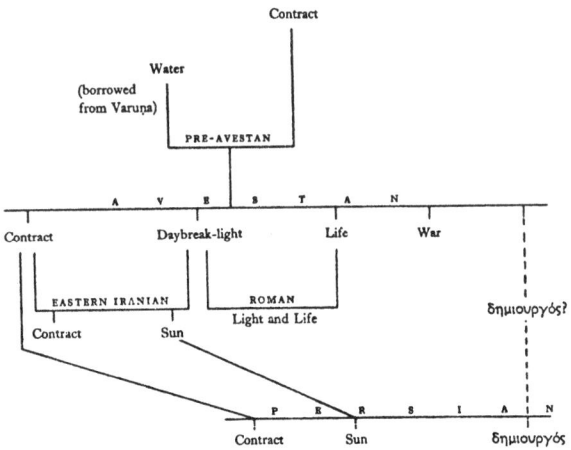

This table does not, of course, exhaust all the functions of Mithra, as a glance at the Mithra Yašt will easily convince the reader. However, what has been omitted seems to be either (*a*) derived, sometimes (*b*) indirectly derived, from one or another of the above functions, or (*c*) deduced from certain ancient epithets of the god, or (*d*) due to his association with Ahura Mazdāh, which will presently occupy us. Under (*a*) we should put Mithra's important function, originally shared with Ahura (see below, pp. 47, 53), of Judge of the world or creation, which it is reasonable to derive from his position as arbiter of the contract. Here, too, belongs his role of Mesites in Zervanism (cf. Zaehner, *Zurvan*, 141 and *passim*) which is not yet apparent in the Avesta. Under (*b*) can be quoted Mithra's political chieftainship (cf. st. 145), which may be a cross between his derivative function of Judge of the world and his supremacy in war; in its turn political chieftainship, combined with supervision of the covenant, may be the source of the authority Mithra has to confer, consolidate, and undo the ruling power of men (see below, p. 60). To (*c*) may belong his concern for the cow (cf. below, p. 54),

MITHRA'S FUNCTIONS

extracted from his epithet *vouru.gaoyaoiti-* 'having, or bestowing wide pastures', in which *gaoyaoiti-* literally means 'cattle-pasture'. The epithet itself may originally have accrued to him as a bestower of water (see above, p. 30, and below, p. 65, n. †), on which pastures thrive, and of wealth (cf. above, p. 32). Similarly Mithra's concern for learning and knowledge (st. 33) may be traced to his epithet *bayanąm aš.xraθwastəma-* 'having the greatest insight among the gods' (st. 141, cf. also st. 107), which he probably had of old as detector of breaches of contract. Finally under (*d*) we should name Mithra's mysterious 'purification of the Religion', a task he shares with Zarathuštra (see note 92[8]), and generally anything that lends Mithra a Zarathuštrian flavour.

Of course the interpretation of epithets and functions of so complex a god as the Avestan Mithra must not degenerate into hair-splitting. Many epithets he probably bears simply because he is a god. Others he may have acquired casually in the course of the ten centuries which preceded the composition of the Avestan hymn, for reasons we can never hope to discover. Theoretically any of Mithra's numerous Avestan epithets may contain an important clue without our being aware of it. Thus one would not pay much attention to the epithet *gayō.dā-* 'giver of life', if it were not for the Roman Mithras, with whom the giving of life seems to be an essential function. But the lesson to be drawn is not that all epithets must be considered as equally essential, in case any of them might one day turn out to be unexpectedly important. For even *gayō.dā-*, though significant in view of the post-Avestan development of the god, sheds only indirect light on the prehistory of the Avestan Mithra (cf. above, pp. 32 *sq.*).

With Mithra, as with every complicated problem, discrimination is the essential condition of progress. This chapter may therefore fittingly end with a belated tribute to the discrimination of Antoine Meillet, who saw through the maze of conflicting data, and with unerring precision established that Mithra's original competence was the contract, and nothing else (*JAs.* 1907, II, 143 *sqq.*). What delayed acceptance of Meillet's unanswerable case must have been his mild insistence on etymological considerations. In matters of religious history linguistic arguments are so distrusted that they

INTRODUCTION

cast a shadow of doubt even on the non-linguistic facts which prove them. Meillet was accordingly disbelieved until recent years, when Thieme† and Lüders vindicated his view. What continued to blur the issue was the mistaken idea that the Avestan Mithra had deep-rooted connections with the sun and the sky, which, it was argued, must have been part of his original divine essence. It is hoped that the present study of the Iranian Mithra will help to dispel the last doubts that Meillet's perspicacity correctly caught the one trait of this versatile god that was his from the beginning.

VI. AHURA

Twice in the Mithra Yašt (sts. 113, 145), and once in the 'condensed hymn' to Mithra (quoted below, note 113[1]), the Avesta presents us with a dvandva *miθra ahura*, which in the later, liturgical texts of the Avesta commonly occurs as *ahura miθra*, with inversion of its components. Of the two sequences the former is the more ancient, for as Duchesne-Guillemin has pointed out (cf. *Comp.* 47 *sq.*), it conforms to the Indo-Iranian rhythmic rule according to which in dvandvas the shorter of two nouns precedes. Once Zoroastrian priests began to consider the implications of the dvandva *miθra ahura*, which had entered their literature as part of the formulary of Mithraic worship, they would feel that *ahura* ought to occupy the first place, and invert the dvandva when using it in the liturgies they subsequently composed. Nevertheless in the Mithra Yašt they handed down the ancient dvandva as they had received it. This is a comforting testimonial to their honesty, which should be heeded by those students of the Avesta who incline to suspect the priests of having falsified or concealed the true nature of the non-Zarathuštrian gods that had been admitted to the Zoroastrian religion. [See Addenda.]

There is general agreement that the dvandva *miθra ahura* cannot be separated from the Vedic dvandva *mitrāvaruṇā* (see above, p. 4). The majority of scholars are even prepared to quote the two dvandvas as proof that the Iranian Ahura Mazdāh is no other than

---

† *Der Fremdling im Rigveda*, 134 *sq.*

the Vedic Varuṇa. This identification is, however, unacceptable. The arguments against it, ably set forth by Lommel, *Rel.* 272 *sqq.*, can be reduced to two. One, an insurmountable obstacle, is the difference in name, the other a streak of craftiness and deceit in the Vedic Varuṇa which is totally absent in Ahura Mazdāh. It has been argued that precisely this repellent streak, or some 'pagan' trait, induced Zarathuštra, or a predecessor of his, to rename Varuṇa, at the same time cutting out the unpleasant sides of his character. But it is doubtful if such surgical operations can be performed except at modern writing-desks. What self-respecting prophet would make his a god he dislikes by lopping off bits which are generally known to belong to him, and changing the god's name to 'Lamb', when everybody knows he is the same old wolf? How could such a prophet go about and praise the meekness of his 'Lamb' without being laughed at? [See Addenda.]

On the other hand, what Ahura Mazdāh and Varuṇa have in common cannot be gainsaid: both are the supreme guardians of Truth, both, and only they, are associated in dvandvas with Mithra/Mitra. In addition, now that Lüders has made it clear that Varuṇa as guardian of Truth is a water-god, we cannot close our eyes to the fact that in the *Yasna haptaṇhāiti*, the oldest part of the Avesta after the Gāthās, the waters are called *ahurānīš*, 'daughters *or* wives of Ahura', just as in the *Taittirīya Brāhmaṇa* and *Saṃhitā* they are said to be Varuṇa's wives (cf. Lüders, *op. cit.* 46). This, then, is the problem: Ahura Mazdāh is certainly a different god from Varuṇa, yet he shares with him several characteristics, including the one which is essential to either: the intimate association with Truth. Truth is such an exclusive and typical object of Indo-Iranian worship that the responsibility for it cannot have been passed on from one god to another as if it were an ordinary function. Only if Ahura Mazdāh, too, had Truth in his charge from the beginning, could he absorb Varuṇa, and render him superfluous.

Insufficient attention has been paid hitherto to the patent fact that in Zarathuštra's system Truth occupies two distinct positions. On the one hand it is the Principle whose opposite is Falsehood. In this capacity it appears also in the Rig Veda. On the other hand —and this is an exclusively Iranian invention—Truth is also the

INTRODUCTION

third† of Ahura Mazdāh's Entities (cf. above, p. 10). In Zarathuštra's Gāthās the Entities are intimately connected with Mazdāh, being, as it were, his organs. The god has charge of the Entity Truth as a man has charge of his right hand. As prime motor of the *Entity* Truth and its companions, Ahura Mazdāh is palpably a different god from Varuṇa, the guardian of the *Principle* Truth. If Zarathuštra's religion were merely a refinement and sublimation of the Varuṇa cult, as is widely held, it would be inconceivable that Truth, the latter's foremost and essential ingredient, should have been debased to occupying the third place in the list of Mazdāh's Entities. On the other hand the original Ahura Mazdāh, who happened to have Truth as one of his organs, could easily displace Varuṇa in the latter's capacity of guardian of Truth. Once deprived of his main *raison d'être*, Varuṇa would inevitably wither away, and whatever else in his nature was deemed worth preserving would be passed on to Mazdāh.

The picture is, however, not complete unless the Fiendish Spirit is brought in. This Spirit is said by Zarathuštra to be the twin brother of the Incremental Spirit (*Y* 30.3), who as we saw above, p. 9, has Ahura Mazdāh as father. The Fiendish Spirit is then, like Lucifer, an angel (more precisely an emanation or organ of Mazdāh's) that had rebelled, and the 'Mazdāhian' part of Zarathuštrianism, as distinct from the 'Varunian' part, resembled the Christian doctrine in that only God, who is good, was primordial, evil being due to insubordination, or 'bad choice', on the part of one of God's agents. This part of Zarathuštra's system is then plain monotheism. The situation latent in the Varunian system is different. There Truth is obviously a primordial principle, since it was not created by Varuṇa, who is merely there to guard it. It may well be that the conclusion that Falsehood, too, is primordial, was not drawn until Zarathuštra made it his own. It is, however, an inevitable corollary of the primordial existence of Truth. For obviously Truth could never have generated Falsehood.

The root of Zarathuštra's dualism should accordingly be seen not in the opposition of the two spirits, since neither of them is primordial, but in the opposition of Truth and Falsehood implicit

† Counting the Incremental Spirit as first, see above, pp. 11 *sq.*

in the Varunian system, which Zarathuštra was probably the first to exploit to the full. Very differently from later Zoroastrianism, which is merely a casual association of partly incompatible elements, Zarathuštra's system thus reveals itself as an admirable syncretism of two religions. One, A, as he found it, consisted in the worship of one god whose organs were seven entities—including 'Truth'—which corresponded to seven 'elements'. The other, V, was essentially concerned with the observance of Truth. In Truth Zarathuštra saw their common denominator. He elaborated the V system by insisting on the primordial character of the opposite of Truth, thus giving his doctrine a dualistic foundation. The A monotheism he adapted to this dualism by introducing the notion of free will, and opposing to god's creative organ a destructive organ or agent, who, being free to do so, had chosen to conform to the opposite of Truth. Naturally not even Zarathuštra could amalgamate a dualism with monotheism without incurring inconsistencies. But the system he achieved displays cohesion and structural balance, and complies with the most exacting rational, ethical, and spiritual aspirations. Clearly lack of understanding is not the reason why Zarathuštra did not aim at rigorous consistency. Henning, *Zor.* 48, has found the right words to explain why a prophet of Zarathuštra's intelligence and sense of mission would be unlikely to come down uncompromisingly on the side of either dualism or monotheism.

In the last quarter of the seventh century B.C., when Zarathuštra was taking stock of the religious position in Aryana Vaējah, we may suppose that beside Ahura Mazdāh and his seven 'elemental organs' Varuṇa also had his worshippers—as well, of course, as Mithra and other gods of the Indo-Iranian pantheon. In those days an archaic Avestan dvandva *miθrā vourunā† will have existed, backed by theological speculations on the joint role of the two gods. One such speculation may have consisted in assigning to **Vouruna the judgeship (*ahū-ratu*-ship) of the supernatural (*mainyava*-) world, to Mithra that of the empirical (*gaēθya*-) world, as we shall see in detail in note 92[6]. Such a division of the

† *Vouruna is the form which the Vedic name *Varuṇa* would assume in the Avestan language, to judge by Av. phonological rules.

world, if correctly attributed by us to seventh-century non-Zarathuštrian circles, would, incidentally, indicate that the physical dualism of the Zarathuštrian system (matter versus spirit), as distinct from its ethical dualism (on which cf. Henning, *Zor.* 45), was *inherited*, not invented, by Zarathuštra.

It is unbelievable that Zarathuštra should have regarded *Vouruna and Mithra with the detestation usually imputed to him by modern scholars. If *he* was responsible for merging *Vouruna in Mazdāh he can scarcely have hated him. Altogether it is unlikely that he would regard his treatment of *Vouruna as a dispossession, as we might be inclined to do. To him, if our theory is correct, *Vouruna would be merely Mazdāh wrongly held by some worshippers to be another god called *Vouruna, Truth being the means which had enabled him to recognize the identity of the two gods. Accordingly he could without compunction pass on to Mazdāh *Vouruna's functions, using Truth, so to speak, as a bridge. Any aspect of *Vouruna's he disliked he could dismiss as having been wrongly attributed to him by his worshippers. With regard to Mithra his position was more delicate. To him he had to close the door, as Mazdāh had no dvandva partner with whom he might have identified *Vouruna's dvandva partner. The prophet was, of course, much too honest to bring in Mithra by the back door, e.g. by substituting him for one of the Aməša Spəntas, or adding him to their number. Such 'substitutions' or 'adjustments', which religious historians have from time to time assumed, might to a limited extent have happened in the popular religious melting-pot of Zoroastrianism; as part of the working method of a prophet of Zarathuštra's stature and integrity they are unthinkable.† On

---

† G. Dumézil's theory (set out in detail in his *Naissance d'Archanges*) that Zarathuštra's Entities represent deliberate replacements of ancient Indo-European gods and their social functions, Vohu Manah having been assigned the place of Mithra, seems to me unrealistic. The few arguments M. Dumézil adduces in support of his theory are extremely weak. Basically the author relies on an interpretation of one of the most obscure Gāthic stanzas, *Y* 29.3, as revealing Aša's disdainful indifference towards the cow. In fact the first two lines of the stanza are just as likely to express solicitous concern for the cow on the part of all the Entities, among whom Aša is specifically named (cf. *JRAS*, 1952, 174 *sq.*), while the last line appears to be uttered either by Ahura Mazdāh, or, as Bartholomae thought, the Fashioner of the cow. [See Addenda.]

the other hand, if out of loyalty to Mazdāh Zarathuštra had to leave out Mithra, it does not follow that he considered him a daēva, as Christensen† and other scholars have suggested (cf. below, pp. 51, 63).‡ As close companion of *Vouruna, whom the prophet had partly identified with Mazdāh, as guardian of the contract, an aspect of Truth on which Zarathuštra laid great store (cf. Y 44.19, 46.5), Mithra had every claim to Zarathuštra's affection. Seeing that the later, Zoroastrian, Sraoša has a certain affinity with Mithra (cf. note 41[1]), it is possible that already in Zarathuštra's lifetime a beginning was made in building up the notion of Discipline (sraoša-) on the pattern of the ever-vigilant Mithra, over whose unavoidable exclusion from Zarathuštrianism even the prophet may have felt a pang of regret. Mithra's austere character must in any case have appealed to the early followers of Zarathuštra; for in an early post-Gāthic text he is clandestinely introduced in Zarathuštrianism as companion of Spənta Mainyu (see below, p. 57).

Having stated our view on the part *Vouruna played in Zarathuštra's syncretism, we now come to the replacement of *Vouruna by Ahura outside Zarathuštrianism. The dvandva *miθra ahura* must have originated in *non*-Zarathuštrian circles, since in the Younger Avesta it is obviously an old fossil, and in the old days no Zarathuštrian would have touched the name of Mithra, let alone dared to make him a companion of Ahura Mazdāh. The improbable alternative that the author of the Mithra Yašt had before him an old fossil *miθra vouruna* in which *he* replaced *Vouruna by Ahura, can safely be discarded. Had he met such a fossil there is no reason why he should have changed it at all, as *Vouruna's name would surely not have been more offensive to Zarathuštrians than Mithra's. In addition, it is doubtful if anybody would still be aware, by the second half of the fifth century, of the subtle ties between Ahura and the defunct *Vouruna. It is thus likely that when Vištāspa in the first half of the sixth century made Zarathuštrianism the official religion of Aryana Vaējah, even those worshippers who continued to cling to the old Indo-Iranian pantheon

† *Essai sur la démonologie iranienne*, 6 sq.
‡ Against the theory that Mithra was a daēva Zaehner, *Zurvan*, 17, rightly adduces the epithet *ahura*-, which he twice (not 'constantly', as Zaehner has it) bears in his hymn (sts. 25, 69).

INTRODUCTION

accepted the official view that *Vouruna and Mazdāh were identical. In this they were helped by the title *ahura*, literally meaning 'Lord', which Mazdāh and *Vouruna had in common.† Feeling perhaps reluctant to go to the extent of replacing *Vouruna by Mazdāh in their hymns, they developed the practice of calling the god simply Ahura 'Lord', thus satisfying both their god and Vištāspa.

Accordingly, one basic difference between Zarathuštrians and the supporters of the ancient pantheon came to consist in this, that the former worshipped as only god Ahura Mazdāh, whereas for the latter Ahura, not entirely Mazdāh, was one god among many they cherished. Thus we understand why the famous dvandva became *miθra ahura*, and not **miθra mazdå*. Thus we also understand the distinction which occupied Benveniste (see below, p. 169, n. ‡), between the Avestan epithets *mazdaδāta-* 'created by Mazdāh', and *ahuraδāta-* 'created by Ahura', of which the latter is only applied to *vərəθraγna-* 'victoriousness' (and its hypostasis, the god Vərəθraγna), and *zam-* 'earth'. The compound *ahuraδāta-*, like the compound *miθra ahura*, contains a fossilized reference to the discarded Iranian *Vouruna, who was evidently regarded as having created both Vərəθraγna and the earth.

As time went by, and Mazdāh gained in venerability through becoming older and ever more widely accepted, even the pantheistic sects probably came to believe that their god Ahura was Mazdāh, and the difference outlined above was reduced to one between monotheistic and pantheistic Mazdāhians. Thus in Aryana Vaējah there developed a religious situation similar to the one we have assumed for Persis (cf. above, p. 19), except with regard to the priests in charge of it. In Persis the same priests officiated for all sects, thereby giving the sects the superficial appearance of belonging to a single religion, whereas in Aryana Vaējah the Zarathuštrian priests remained loyal to monotheistic Mazdāhism until under the impact of developments in Western Iran they translated the existing religious situation into terms of a 'Zoroastrian' church and scripture, which included also pantheistic Mazdāhism.

The above hypothesis, which reduces the basic ingredients of

† The Vedic Varuṇa has the title *ásura* (cf. above, p. 5), which etymologically and semantically corresponds to Av. *ahura-*.

AHURA

Younger Avestan Zoroastrianism to two, monotheistic (viz. Zarathuštrian) and pantheistic Mazdāhism, enables us to redefine the term *daēva*. It always seemed a contradiction in terms that Zoroastrian hymns to Mithra, Anāhitā, etc., expressed strong dislike of daēvas, when one might well suspect that these divinities themselves were counted as daēvas by Zarathuštra. The obvious answer is that they were not. But, in that case, which gods are daēvas? Presumably those who were rejected as evil by pantheistic Mazdāhians from the time when they were still *Vourunians. That it should be a prerogative of worshippers of the *ahura* *Vouruna and the gods with whom he was associated to detest the gods of the daēva class is only natural, seeing that the Vedic Varuṇa is the chief *ásura*. Zarathuštra, who by means of the 'bridge of Truth' had recognized Ahura Mazdāh in *Vouruna (see above, p. 48), could hardly help sharing with *Vouruna-worshippers the detestation of the anti-*Vourunian daēvas. These were therefore equally disliked by Zarathuštrians and the non-Zarathuštrian pantheists whom the Zarathuštrians absorbed in about 430 B.C., following, according to our theory, the Achaemenian religious pattern of the time. Hence, in all likelihood, that pattern also excluded the daēvas from worship; Xerxes' prohibition of their cult† seems to have had a lasting effect. However, on the fringe of the two all-pervading Mazdāh cults, the monotheistic and the pantheistic, the propitiatory worship of daēvas continued to exist, as Zaehner, *Zurvan*, 14 *sqq.*, has aptly pointed out; we shall revert to this subject below, pp. 63 *sq*.

We are now in a position to understand the background not only of the dvandva *miθra ahura*, but of all the references to Ahura Mazdāh in the Mithra Yašt. One weakness of previous exegesis of the hymn is to my mind the exaggerated extent to which Ahura Mazdāh is held to be, historically, the god of the Gāthās, or a 'Zoroastrian' interpretation of that god, although everybody agrees that in the dvandva *miθra ahura* Ahura is Zarathuštra's god only in name. Admitting that Mazdāh's injunction to worship Mithra (sts. 119, 122, 137, 139; cf. also 53, 73) may be 'Zoroastrian', statements to the effect that Mazdāh created Mithra and made him as great as himself (st. 1; cf. also st. 61 with note 61[5]), or built him

† Cf. Roland G. Kent, *Old Persian*, p. 151, §4*b*.

INTRODUCTION

an abode (st. 50), or, worst of all, worshipped him (sts. 123, 140, 143), are so flagrantly and needlessly un-Zarathuštrian, that one does not see how a 'Zoroastrian' poet could have *invented* them. On the other hand, polytheistic Mazdāhians might variously have held Ahura or Mithra to be the more important of the two, according as they were guided by the long-standing supremacy of Ahura (*Vouruna) or the precedence Mithra had in the sacred formula *miθra ahura*. On the assumption that statements such as we have referred to emanate from non-Zarathuštrian sources, we have no reason to be shocked at them, and the Zoroastrian author of our Yašt who accepted them is cleared of all suspicion of disloyalty or stupidity: he could not know that *this* Ahura Mazdāh was *Vouruna, who as an indirect and unintended result of Zarathuštra's activity had come to be called Ahura Mazdāh by his worshippers, and not Zarathuštra's god Ahura Mazdāh.

It is worth dwelling for a moment on the initial statement of the Yašt that Mazdāh created Mithra and made him as great as himself. Mazdāh's creation of Mithra is a belief which, if we substitute Varuṇa for Mazdāh, may well go back to the prehistory of the two gods; for it agrees with the Rigvedic position that although the two gods are, as Lüders put it, 'homogeneous' (see above, p. 7), yet Mithra is inferior to Varuṇa. On the other hand the homogeneity of the two gods, and Mithra's precedence in the dvandva *miθra ahura*, may have been the reasons why in some respects Mithra came to surpass Ahura Mazdāh. It is true that the description of Mithra as *mazištō yazatō* in st. 142 is not an unequivocal indication of the god's supremacy; for although most translators render it by 'the greatest god', Lommel's interpretation of it as meaning 'the very great god' provides a plausible contrast with the formula *mazištō yazatanąm* 'the greatest of the gods', which in *Yt* 17.16 is applied to Ahura Mazdāh. But absolute supremacy is accorded to Mithra in matters of insight, endowment of fortune, and sheer physical prowess. On his insight see p. 43 above. Of fortune he has or gives more than any other supernatural god;† he both grants (sts. 16, 108) and diverts it (st. 27). As regards physical qualities he is the mightiest, strongest, most mobile, fastest, and most

† See *Ny* 1.7, the 'condensed hymn' quoted in note 113[1]; cf. also sts. 67, 141.

victorious of the gods (sts. 98, 135). To non-Zarathuštrians taking such an exalted view of Mithra, the idea that even Ahura Mazdāh had worshipped him, which our author reproduces in sts. 123, 140, and 143, would easily occur. It goes without saying that once a precedent for Ahura Mazdāh's worshipping another divinity was set up inside the new 'Zoroastrian' scripture, the practice could be extended. Hence we also find Ahura Mazdāh worshipping Anāhitā (*Yt* 5.17), Tištrya (*Yt* 8.25), and Vayu (*Yt* 15.2).

On the other hand it should be noted that while Ahura Mazdāh is in *Yt* 17.16 not only *mazištō* 'the greatest', but also *vahištō yazatanąm* 'the best of the gods', Mithra, as we saw (p. 30), is 'both wicked and very good' (*akō vahištasča*) to countries and men (st. 29). Perhaps we may say that compared with Ahura Mazdāh, who is just but preponderantly good, in the sense that he does not personally seek out the wicked for punishment, rather letting them run to their self-appointed doom, Mithra is preponderantly just, and carries out justice himself. Some two or more centuries before our Yašt was composed, we may suppose that Mithra and *Vouruna shared the judgeship (*ahū-ratu*-ship) of the world, as mentioned above, p. 47. In the Zoroastrian system, however, this judgeship is either shared between Ahura Mazdāh and Zarathuštra, or contemplated only in respect of the empirical world, with Mithra as judge (see st. 92 and note 92[6]). If one takes into consideration Mithra's Sogdian title δ*āmext* 'Judge of the world (*or* creation)' (see note 92[7]), as well as his Pahlavi title *dātaβar* 'judge' and possibly the Avestan epithet *rašnu-* (see note 79[1], (2)), it would seem that, except in secondary partnership with Zarathuštra, Ahura's share of the judgeship, viz. that of the supernatural world, was in time allowed to lapse. A possible reason may have been that in supernature there is little scope for a judge, so that Mithra, the fully employed judge of the empirical world, came eventually to be thought of as the divine judge *par excellence*. In this capacity not only is the Avestan Mithra accompanied by the hypostasis of Justice (*Arštāt*, see note 139[2]), but has himself the necessary insight (*xratu-*) to pass equitable judgement, and the physical skill to execute it. Though 'wicked' when necessary, Mithra is not always inexorable; he is merciful (st. 140), and joins his worshippers 'for mercy' (st. 5). His is militant justice with a disposition to forgiveness, and an endearing affection for the unjustly oppressed, the loyal pauper (sts. 84 sq.; cf. note 85[1]), and the unmilked (st. 84) cow dragged away into captivity (sts. 38, 86).†

Before leaving the subject of Ahura Mazdāh, we must consider an Avestan expression which first appears in an early post-Gāthic text, Y 42.2:

    pāyūča ‡ θwōrəštārā yazamaidē
    mazdąmčā zaraθuštrəmčā yazamaidē

and we worship the two, the protector and the fashioner, we worship Mazdāh and Zarathuštra,

and again, in the hymn to Sraoša, Y 57.2:

    sraošəm...yō
    yazata ahurəm mazdąm
    yazata aməšə̄ spəntə̄
    yazata pāyū θwōrəštāra
    yā vīspa θwərəsatō dāmąn

Sraoša...who worshipped Ahura Mazdāh, worshipped the Aməša Spəntas, worshipped the two, the protector (and) the fashioner who fashioned (3rd dual preterite) all creatures.

If we only had Y 42.2 one might have thought that the second line we have quoted contained a chiastic gloss to the dvandva pāyū θwōrəštārā,§ the 'fashioner' being Mazdāh, and the 'protector' Zarathuštra. This interpretation is excluded by Y 57.2, where Mazdāh is clearly distinguished from either member of the dvandva. The Pahlavi commentator of Y 57, having failed to recognize the dual endings, explained the dvandva by pānak brīnkar mihr 'the protector and fashioner, Mithra'. Although his ignorance of Avestan grammar led him astray, it is very likely that the tradition which made him look for Mithra in this expression is correct. As Darmesteter said (ZA, i, 360, n. 5), after referring to

† It need hardly be stressed that Widengren's picture of Mithra as a treacherous ('heimtückischer') god who arbitrarily, without provocation, brings misery and strife (Hoc. 123 sq.), is distorted. It is based on a misconception regarding the speakers of st. 98 (see below, note 68[8]), and a wilful interpretation of sts. 108–11, where Mithra's distribution of blessings and ills is strictly in accordance with merit: see the beginning of st. 108, not quoted and disregarded by Widengren, and cf. sts. 48 and 112.

‡ Bartholomae's emendation. MSS. have °yūšča, °yušča, °yūšča, °yūmča.

§ The °čā of pāyūčā connects the line to the preceding sentence, not pāyū to θwōrəštārā.

54

the dvandva *miθra ahura*: 'il est probable que *pāyū θwōrəštārā* désigne non pas "Mithra", mais "Ahura et Mithra". Le commentaire n'aura pas reconnu Ahura parce qu'il était déjà nommé.' Bartholomae, *Wb.* 889, line 5, agreed with this view, and it is clearly satisfactory to have Mithra described as a 'protector' (cf. sts. 46, 54, 80, and 103), and Ahura as a 'fashioner' = 'creator'. However, while it would be in order for the Pahlavi commentator not to recognize that *θwōrəštārā* refers to Mazdāh, this cannot be said of the Avestan authors of *Y* 42.2 and *Y* 57.2, who must have known what they were talking about; had they thought that *θwōrəštārā* meant 'Ahura Mazdāh', they would not have mentioned Mazdāh as a separate item.

In this connection we shall do well to recall that there are reasons to doubt that the Gāthic word *θwōraštar-* refers to Mazdāh even in *Y* 29.6, where it occurs once more. In the Gāthic stanza Ahura Mazdāh tells the cow: *aṯ zī θwā fšuyantaēčā vāstryāičā θwōraštā tatašā* 'the fashioner has fashioned† thee for the cattle-breeder'. Here, to begin with, the identity of the speaker discourages the assumption that the *θwōraštar-* in question is Ahura Mazdāh. As the same, and other, Gāthic poems mention a divine character called *gə̄uš tašan-* 'Fashioner of the cow', the *θwōraštar-* of *Y* 29.6 who *tatašā*, 'fashioned', the cow is very likely this *gə̄uš tašan-*. His further identity emerges with all desirable clarity from *Y* 47, a poem in which Spənta Mainyu, the Incremental Spirit, is fully named in each stanza except the third and the fourth, where he is simply called 'this Spirit'; it is the third stanza which concerns us here:

ahyā mainyə̄uš tvə̄m ahī tā spəntō
yə̄ ahmāi gąm rānyō.skərəitīm həm.tašaṯ

Thou (Mazdāh) art the incremental father of this Spirit who fashioned for us the bliss-bringing cow.‡

---

† Both *θwarəs-*, of which *θwōraštar-* is a noun of agent, and *taš-*, the verb here used, to which belongs the *gə̄uš tašan-* presently to be mentioned, mean 'to cut, give shape by cutting, fashion, create'.

‡ *yə̄* 'who' inevitably refers to *mainyə̄uš* 'spirit', since *həm.tašaṯ* 'fashioned' is 3rd singular. This view is expressly stated by Bartholomae, *Arische Forschungen*, III, 29, n. (who recognized the allusion to *gə̄uš tašan-* in these two lines), and Tavadia, *ZDMG*, 1953, 322, n. 10, and accepted by Lommel (*Rel.* 20, and *NGWG*, 1935, 127). Lentz (same article and page as Tavadia) wants to refer *yə̄*

INTRODUCTION

Once we obtain the equation *θwōrəštar* = *gə̄uš tašan-* = Spənta Mainyu, we can put to good use Wackernagel's etymological identification, last discussed by Manu Leuman,† of *θwōrəštar-* with the name of the Vedic god Tvaṣṭar (from **tvarṣtar-*). If Leumann is right in defining the Avestan *gə̄uš tašan-* as the Vedic creative god Tvaṣṭar 'newly named and specialized' (p. 84), then the expression *gə̄uš tašan-* is in Avestan a replacement of *θwōrəštar-*. It follows that Av. *θwōrəštar-* is not only the linguistic equivalent of Ved. Tvaṣṭar, but was, at least in pre-Gāthic times, the name of the same creative genius. *Y* 29.6 and *Y* 47.3 combined show that for Zarathuštra this genius was Spənta Mainyu. Of the two names of the genius, **Θwōrəštar* and Spənta Mainyu, the former is the older, on the evidence of Ved. Tvaṣṭar. Thus we reach the conclusion that in the monotheistic cult of Mazdāh as practised before Zarathuštra accomplished his syncretization (cf. above, p. 47), the first Aməša Spənta, the creative one, was the ancient Indo-Iranian **Tvaṣtar*. Historically, then, 'Spənta Mainyu' ('Incremental Spirit') may be taken as an epithet of Mazdāh's organ **Θwōrəštar*. Whether this epithet existed before Zarathuštra's time, or was invented by him, we cannot tell. At all events, after the vivid figure of Angra Mainyu appeared on the scene as Spənta Mainyu's opponent, the description 'Spənta Mainyu' of the Entity in question was bound to prevail over the old name **Θwōrəštar*. The survival of the latter is probably due to its transparent character of noun of agent to the base *θwarəs-* 'to fashion', which enabled Zarathuštra to use it in the meaning of 'the fashioner' κατ' ἐξοχήν, with reference to Spənta

to *tvə̄m* 'Thou', but this is syntactically intolerable. It should be noted that contrary to Maria W. Smith's opinion (§70, pp. 52 *sq.*), *Y* 51.7 *dāidī mōi yā gąm tašō apascā urvarā̊scā* 'give me (O Mazdah,) who didst fashion the cow and the waters and the plants', does not have to mean that Mazdāh is the 'Fashioner of the cow' (*gə̄uš tašan-*); this was shown by Bartholomae long ago, *op. cit.* 27 *sqq.* Being Spənta Mainyu's 'father', Ahura Mazdāh is, of course, ultimately the author of whatever Spənta Mainyu, his creative organ, creates or fashions. Hence in so far as Spənta Mainyu is the fashioner of the cow, his 'father' Ahura Mazdāh can be said to have fashioned not only the waters and plants, but also the cow.

† *Asiatische Studien*, 1–4, 1954, pp. 79 *sqq.* In his interpretation of *pāyū θwōrəštārā* (p. 83) Leumann has not taken into account the Zoroastrian tradition that the dvandva refers to Mithra, which, as far as *pāyū* is concerned, deserves preference over Leumann's purely etymological connection with the Vedic *pāyu-*s; the latter lack the individuality of the Avestan *pāyu-*, who, to judge from the context of *Y* 42.2 and *Y* 57.2, is a god of considerable importance.

Mainyu. Thus Spənta Mainyu's original name became a rarely used epithet of his.

If this much is granted, the meaning of the dvandva *pāyū θwōrəštārā* will be 'Mithra and Spənta Mainyu'. As *Y* 42 is an early post-Gāthic text, it seems that at an early stage after Zarathuštra's death his disciples, timidly and almost clandestinely, introduced into their litanies the ever-watchful (*pāyu-*) god of contract, and associated him with Spənta Mainyu. A possible reason for this innovation is that to early Zarathuštrians the Ahura (formerly *Vouruna) of the pantheistic Mazdāhians would seem a considerably poorer god than their own Ahura Mazdāh, except in one respect: he had a kind of emanated twin brother, Mithra, a tireless watcher (*pāyu-*) over the interests of those who abide by Truth and Contract. Having decided—if this is how things happened—to appropriate Mithra under the harmless title *pāyu-*, it looks as if, anxious to preserve the suggestion of partnership implied in the non-Zarathuštrian (cf. above, pp. 44, 49) dvandva *miθra ahura*, yet not daring to give to their own Ahura a twin partner, they joined the 'protector' in a dvandva with Spənta Mainyu, the creative aspect of Ahura Mazdāh.

Once it is agreed that the first element of the dvandva *pāyū θwōrəštārā* is Mithra, the relative sentence *yā vīspa θwərəsatō dāmąn* 'the two who fashioned all creatures' (*Y* 57.2, see above, p. 54) acquires special significance, as it seems to foreshadow the later, Persian conception of Mithra as a δημιουργός (see above, p. 41). The Vedic Mitra, through forming the intimate association with Varuṇa of which the dvandva *mitrāvaruṇā* is the label, came to have a share in Varuṇa's control of the waters (cf. above, p. 30, and Lüders, *op. cit.* 51, n. 10). Similarly Mithra the *pāyu-*, through his dvandva association with the 'fashioner', of whom alone it could truly be said that 'he fashioned all creatures', came to have a share in Spənta Mainyu's creative activity. Understandably enough the δημιουργός side of Mithra does not appear in the Mithra Yašt or in any other of the Avestan references to the god which derive from *genuine* Mithra worship. To genuine Mithra worshippers, that is, pantheistic Mazdāhians, the thought of combining Mithra with Spənta Mainyu instead of Ahura would seem an unattractive

INTRODUCTION

heresy. Hence they had no occasion to discover through Spənta Mainyu demiurgic qualities in Mithra, and did not pass on any demiurgic notions to the Zoroastrian author of the Mithra Yašt. Here is further indication, if any more is needed, that the so-called 'Zoroastrianization' of non-Zarathuštrian gods is very far from implying the imposition of Zarathuštrian notions on the gods of pantheistic Mazdāhism. If the author of the Mithra Yašt had had any intention of 'Zarathuštrianizing' Mithra, he could have turned to his own Zarathuštrian school of thought, and developed the association of Mithra with Spənta Mainyu into joint subordination to Mazdāh, assigned to Mithra demiurgic functions, replaced the dvandva *miθra ahura* by the dvandva *pāyū θwōrəštārā*, and so forth. Instead, he disregarded this line of thought almost completely,† and glorified Mithra on lines which he can only have found in non-Zarathuštrian sources. His aim, which he pursued with scrupulous honesty, was to portray Mithra not as Zarathuštrians might wish him to be, but as he *was* in the eyes of men who had behind them a long tradition of Mithraic worship, and knew what to expect from an authoritative hymn, such as the Mithra Yašt was intended to be. Several centuries later, however, we may assume that a Persian Zoroastrian school evolved from the dvandva *pāyū θwōrəštārā* and its use as subject of *θwərəsatō* 'they fashioned' in Y 57.2, the idea of a demiurgic Mithra, thus preparing the ground for the Persian Manichean identification of the Spiritus Vivens with Mihr-yazd (see above, p. 41).

VII. MITHRA'S COMPANIONS

Mithra's importance and sovereign status, apart from the superlative titles and epithets he bears, and his intimate relation with Mazdāh, transpire from his unusually large retinue of divinities. The following accompany him in the Mithra Yašt:

Aši 66, 68, and, *incognito*, 76, 143;
Pārəndi 66;
the Mazdayasnian Religion (68; cf. 64 and 92), and her *alter ego*, Čistā (126, see note 9[4]);

† At most a trace of it can be discerned in his notion that it was Spənta Mainyu who built Mithra's chariot (st. 143).

MITHRA'S COMPANIONS

Sraoša 41, 100;
Rašnu 41, 100, 126, cf. also 79, 81, 139;
Nairyō.saŋha 52, and, *incognito*, 66;
Ham.varəti (cf. note 5[1], end) and Θwāša, 66;
the Kavyan Fortune 66, and Fire-Fortune 127;
Vərəθraγna (70), and his *alter ego* (9, 66, 68, 127), see note 9[4];
wind (9), waters and plants (100);
the Fravašis 66, 100, cf. also 3;
Justice (Arštāt), see 139 with note;
in addition, Fire intervenes on Mithra's behalf (3), the Incremental Immortals build him an abode (51) and acknowledge him as Judge of Creation (92), the Incremental Spirit builds his chariot (143), and Haoma, in his capacity as priest to Mazdāh, officiates in the worship of Mithra (88 *sqq*., 120).

Divinities especially associated with Mithra outside the Mithra Yašt (apart from the *θwōrəštar*- discussed in the preceding chapter) are Rāman, to all appearances an attendant secondarily accrued to him in late Zoroastrianism (see note 145[2]), Ārmaiti, cf. below, p. 195, n. †, and Apąm Napāt 'the Grandson of the Waters'.

Apąm Napāt's relation to Mithra is intriguing. The two are the only 'Zoroastrian' gods, apart from Mazdāh, to bear the title *ahura*- 'lord'. Both are, as we saw (p. 27), pacifiers of countries in turmoil and supporters of governmental authority. Either takes charge of fortune ($x^v arənah$-) at times when that precious commodity is in danger of falling into wrong hands:

Apąmnapāt takes the $x^v arənah$- to the bottom of the sea in *Yt* 19.51 after an inconclusive contest for it between Fire and the dragon Aži Dahāka. It looks as if the Grandson of the Waters intended to save the $x^v arənah$ from coming to harm while it was being pursued by the two rivals. Earlier in the same Yašt 19 (sts. 34 *sqq*.) the $x^v arənah$, having been forfeited by Yima, is seized successively by Mithra, Θraētaona, and Kərəsāspa. The most satisfactory explanation of this episode is Darmesteter's (*ZA*, ii, 625, n. 52): during the reign of Yima's successor, the evil Zohāk (Aži Dahāka), Mithra keeps the $x^v arənah$ in trust; in time Farīdūn (Θraētaona) obtains the $x^v arənah$, defeats Zohāk, and reigns; after him, during Minōčihr's childhood, the $x^v arənah$ passes to Sām Narīmān (Kərəsāspa).

INTRODUCTION

Accordingly, the resemblance between Mithra and Apąmnapāt consists in that both are politically minded. Hence their concern in the fate of the $x^v ar\partial nah$, symbol of legitimate supreme authority. In the case of Apąmnapāt the political interest he evinces supports the 'Herrenwürde' which B. Geiger sought to establish as a characteristic quality of the god (*Die Amǝša Spǝntas*, 222, n. 1). Mithra's concern with proper government—arising from his political chieftainship and preoccupation with the covenant (see above, p. 42)—turned him into a maker, as well as undoer, of kings; witness the Mithra Yašt (sts. 109, 111), the monument of Antiochus I of Commagene at Nimrud Dagh (see Cumont, *TMMM*, II, 188), and the inscription of Carnuntum (*fautori imperii sui*, see Cumont, *op. cit.* I, 281; II, 146). His role of divine ruler who installs earthly rulers gave rise to the idea that legitimate kings were σύνθρονοι θεῷ Μίθρᾳ (Cumont, *op. cit.* I, 229, n. 5).† The special link we have noticed in the Avesta between Mithra and Apąm Napāt confirms the correctness of Cumont's identification of Oceanus on Mithraic reliefs with Apąm Napāt (*TMMM*, I, 142, cf. also 98 *sq.*). The only amendment Cumont's interpretation requires is that Apąm Napāt is present on the reliefs not (or not only) because he is a water-god but (also) on the strength of a share he has in Mithra's responsibility for orderly political government.

The divinities surrounding Mithra in the Yašt dedicated to him will be discussed in the Commentary where necessary. Here only the Fire-Fortune of st. 127 need be considered, an Avestan divine hypostasis whose existence has not so far been recognized, although it emerges from the mere literal translation of the text, as explained in note 127[5]. This Fire-Fortune (the compound is one of identification, not of dependence) is the ancestor of the chief Sasanian sacred fire *Ātur-farn-bag*, whose name, it now becomes clear, has been consistently misunderstood. Because the two other main Sasanian fires are normally referred to as *Ātur Gušnasp* 'the fire G.'

† Mithra's connection with 'Herrscher-Gewalt' was over-emphasized by Widengren, *Hoc.* 146–63, on the strength of two groundless assumptions: (1), that Mithra is the god of the sky (see above, pp. 36 *sqq.*), and (2), that the Parthian title βαγπūr, corresponding to Sogdian βαγpǝše, should be interpreted as meaning 'Mithra's son' (*Hoc.* 157). The precariousness of the notion introduced by Marquart, that in Old Persian the noun *baga-* 'god' was used as an alternative name of Mithra, was demonstrated by Henning, *JRAS*, 1944, 134 *sq.*

and *Ātur Burzīn Mihr* 'the fire B.M.', it was thought that *Āturfarn-bag* meant 'the fire *Farn-bag*'. Pagliaro, *Oriental Studies*... Pavry, 378, understood *Farn-bag* as meaning 'God of glory'. Bailey, *Zor. Prob.* 44, gave different translations of the two forms in which the name is quoted in Middle Iranian, *Āturfarnbag* and *Āturxᵛarrah*: the former he understood as the fire 'distributor (*bag*) of good fortune (*farn*)', the latter as 'whose good is from fire', which as name of a fire sounds unconvincing. *A priori* one expects *Āturfarnbag* and *Āturxᵛarrah* to have the same meaning. The Avestan identification of *ātarš* with *x*ᵛ*arənō* in st. 127 makes it clear that, in the two Middle Iranian forms of the name, *Ātur* is part of the name, not an appositive as in *Ātur Gušnasp*, the equivalent of which would read *Ātur Āturfarnbag*, or *Ātur Āturxᵛarrah*. *Āturfarnbag* accordingly means 'God Ātur-Farn', as Parthian *'whrmyzdbg* means 'God Ohrmizd', and *myhryzd* 'God Mihr'.

It may be further noted that as the 'Fire which is the Kavyan Fortune' travels in st. 127 in front of Mithra, its name cannot be an epithet of Mithra's, as Pagliaro had claimed, *art. cit.* 379, and the attempt made (*ibid.*) to connect the three Sasanian fires with the τριπλάσιος Μίθρας mentioned by Pseudo-Dionysius must be considered as having failed. [Read *Farnbāg* and see pp. 278 *sq.*]

VIII. THE WESTERN MITHRAS

In the preceding chapters we have had repeated occasion to refer to aspects of the Roman Mithras which are clearly derived from the Iranian Mithra as attested in the Avesta. Mithras' relation to Sol, though the details escape us, is clearly based on Avestan premises: Mithras, like Mithra, is a light-bringer, no doubt because he, too, is the first light of the day; Mithra's epithet *hvāraoxšna-* 'endowed with own light', and the context in which it is used in st. 142 (cf. above, p. 31), find their reflex in Mithras' epithet *gen(itori?) lum(inis?)* (Cumont, *TMMM*, II, 147, No. 370), which Cumont, I, 161, related to the portrayal of the god with a torch in his hand, 'emblème de la lumière qu'il apporte au monde'; the epithet *Oriens* allows of a similar interpretation, cf. *ibid.* 128, n. 6; even Mithras' position behind Sol on the reliefs can be traced to

INTRODUCTION

the same source as the Avestan notion that Mithra precedes the sun (see above, pp. 32, 39 *sq.*). We have noticed the aptness of Oceanus = Apąmnapāt's presence on Mithraic reliefs (above, p. 60). Clearly the portrayal of Fortuna and Hermes has equally ancient roots in the ties which link Aši (see note 68²) and Nairyō.saṇha (see note 52¹) to Mithra. Similarly the boar seen next to Heracles will not be unconnected with Vərəθrayna's incarnation in a boar (see note 70²),† and one may suspect that the cock which is occasionally found on Mithraic reliefs (Cumont, I, 210), has something to do with Mithra's friend Sraoša, whose bailiff (Av. *sraošāvarəz-*), according to *Vend.* 18.14 *sq.*, is the cock.

We have also seen that Mithras' role of a life-giver is anticipated by three Avestan epithets of Mithra (see above, pp. 32 *sq.*). These, however, some scholars will consider insufficient explanation of the behaviour of the Roman Mithras, whose life-giving function is most forcefully expressed in the central scene of most Mithraic reliefs: as Mithras kills the (primordial?) bull, sperm is emitted by the dying animal (Cumont, II, 209, No. 28, cf. I, 190), and ears of corn sprout from its tail (Cumont, I, 186 *sq.*). Let us first state our view on how this legend came to be connected with Mithras. It is basically Cumont's view that the legend is a Mithraic version of Ahriman's murder of the primordial bull. Cumont, however, did not commit himself as to which of the two versions was the original one, no doubt because he had not realized the implication of the three Avestan epithets we have quoted.

The legend is not attested in the Avesta, but Middle Persian Zoroastrian tradition assures us that Ahriman, the ancient Fiendish Spirit Angra Mainyu, had killed the primordial bull, from whose

---

† It may even be suggested with due reserve that the two boars, which together with four horses surround Mithras on horseback on the Rückingen relief, are Vərəθrayna and his *alter ego* Dāmōiš Upamana, who, incarnate in boars, are described in our Yašt as respectively preceding (st. 70) and following (sts. 68, 127) Mithra's chariot (cf. below, p. 168, n. †). The four horses, to which A. Alföldi devoted an interesting article in *Germania*, 30 (1952), 362 *sqq.*, may then simply represent the four coursers who draw Mithra's chariot (sts. 125, 68). The awkward fact that on the relief Mithras himself is on horseback may be due to a contamination with Mithraic hunting scenes, such as the one at Dura-Europos (cf. M. Rostovtzeff, *Dura-Europos and its Art*, plate XVIII facing p. 96), to which Alföldi, p. 366, did not fail to draw attention.

marrow grew the species of grain, and whose seed, carried to the moon, produced the species of animals. It was always known that a connection existed between Ahriman and Mithras, because in Mithraic inscriptions a *deus Arimanius* is mentioned. Recently Duchesne-Guillemin† and Zaehner‡ took the seemingly correct step§ of identifying the lion-headed deity of Western Mithraism with Ahriman. Moreover, Zaehner has attractively argued‖ that the Mithraists derive from Iranian daēva-worshippers who sought refuge in Babylonia when Xerxes prohibited the worship of daēvas (see above, p. 51, n. †).

What fifth-century Avestan authors meant by daēva-worshippers is clear from *Yt* 5.94 *sq.*, a passage which is quoted below, note 45³, in a revised translation: they are people who worshipped the daēvas (cf. above, p. 51), *as well as* the gods, or some of the gods, worshipped by Mazdayasnians. Thus, in the Avestan passage referred to, the daēva-worshippers offer libations to Anāhitā, although she evidently was not a daēvī; the Zoroastrian author, of course, portrays Anāhitā as rejecting these libations; but it is obvious that the daēva-worshippers believed the goddess to be agreeable to their offerings. What is true of Anāhitā has a good chance of applying also to Mithra. He, too, may have been worshipped not only by Zoroastrians who had taken over his cult from pantheistic Mazdāhians (formerly *Vourunians), but also, and this already in Zarathuštra's lifetime, by people who did not hesitate to combine their allegiance to Mithra with apotropaic sacrifices to the daēvas. The Mithra Yašt itself confirms this view when in st. 108 the god is said to distinguish between worshippers who consider it their duty to offer him 'good sacrifice' and those who see fit to revere him with 'bad sacrifice'. The latter are most probably the daēva-worshippers, whose sacrifice to Mithra Zoroastrians could only regard as an affront to the god. Angra Mainyu, who may be an invention of Zarathuštra's (cf. above, p. 47), is, of course, a comparative late-comer among the daēvas, but he can be relied upon to have travelled faster among daēva-worshippers than Zarathuštra's

† *Numen*, II (1955), 191 *sqq.*   ‡ *BSOAS*, XVII, 237 *sqq.*
§ Which, however, has since been challenged by M. Boyce in *BSOAS*, XIX, 314 *sq.*   ‖ *Art. cit.* 240 *sqq.*, and *Zurvan*, 14 *sqq.*, 20.

INTRODUCTION

'ahuric' ideas among daēva-despising *Vourunians. The daēva-worshippers, because of their superstition that all evil gods must be 'worshipped', that is, placated, would no doubt be quick to include the powerful new demon, the Fiendish Spirit, among the recipients of their apotropaic offerings. In the wake of Darius' introduction of the cult of Ahura Mazdāh into Western Iran (cf. above, p. 16), the arch-demon Angra Mainyu would not be slow to make his appearance among the Magi, who may well have been prepared, when commissioned to do so, to offer him, along with the daēvas of old, placatory tribute. To this practice Xerxes' prohibition officially put an end, and it is not impossible that the westward spread of Ahrimanian Mithraism began at that early period.

In a cult where Mithras was the chief god and Ahriman the chief demon, any action of the latter which had eventual beneficial effects had a fair chance of being transferred to the former. The Zoroastrian account of Ahriman's killing of the bull, though admittedly attested only in a late version, does present Ahriman as the unintentional instrument of the dissemination of plants and animals. Considering that the Mithra whom the daēva-worshippers exported to Babylonia was, on the evidence of the Avesta, a life-giver who caused plants to grow (see above, p. 32), it seems not unreasonable to suppose that Ahriman's murder of the bull was transferred to Mithra because the latter's epithet 'life-giver' had marked him for producing *intentionally* the good effects which with Ahriman were *unintended* results of criminal behaviour.

The above interpretation differs substantially from the one currently held today, which was most ably put forward and defended by Lommel (*Rel.* 182 *sqq.*; *Paideuma*, III (1949), 207 *sqq.*). In the Yajur Veda, Lommel points out, the gods try to persuade Mitra to join them in killing Soma; he at first refuses, declaring that cattle would turn away from him if he took part in such action; eventually he gives in and joins in the murder. The killing of Soma, Lommel explains, symbolizes the pressing of the sacred plant Soma, which causes rain, and consequently the growth of plants; Soma is the elixir of life, which after dropping on earth as rain, mounts to the moon and is drunk out of the moon by the gods, who use the moon as a cup. The animal representing the moon (*Mond-*

Tier) is the bull, Lommel continues. Here then, and in the resentment of cattle at Mitra's intention to kill Soma, Lommel sees the proof that Soma is not only a plant but also a bull. Thus the basis is gained for assuming that Mithra's murder of the bull is a pre-Zarathuštrian myth, which survives only on Mithraic reliefs because the Zoroastrians, when they adopted Mithra as one of their gods, finding it intolerable that he should have killed the primordial bull, purged him of this crime and imputed it to Ahriman instead. Accordingly, much has been made by Duchesne-Guillemin and other writers of Zarathuštra's alleged Mithraphobia, said to reveal itself in *Y* 32.10, 12, 14, cf. also *Y* 44.20, *Y* 46.4, *Y* 51.14. In these Gāthic passages, Zarathuštra undoubtedly disapproves of animal sacrifices, but it cannot be stressed enough that he nowhere in any way implies that Mithra, whom he never mentions, is their recipient. Zarathuštra's alleged 'hostilité à l'égard de Mithra, le dieu des offrandes cruelles et des extases d'ivresse' (Duchesne-Guillemin, *Ormazd et Ahriman*, 40) is a modern invention in all three respects: *hostilité*, *offrandes cruelles*, and *extases d'ivresse*.

The first objection to Lommel's theory is that none of its equations is more than approximate. If Soma is called a bull, so are other gods who are not bulls. As a *Mond-Tier* is not the moon herself it is hardly safe to confirm through a moon equation that Soma is a bull. The resentment of cattle at Mitra's murder of Soma is an even less convincing argument for identifying Soma with a bull; the mere fact that Soma provides 'wide pastures'† would suffice to explain the reluctance of cattle to see him slain. The equation of the Indian with the Iranian situation is also very doubtful: Mitra is only one of the Soma-killing gods, while the reliefs present Mithras as sole slayer of the bull.

More serious than the inadequacies of Lommel's equations is the absence of even the slightest reference in the Avesta to Mithra's alleged slaughter of the bull. It may be said that this objection applies equally to the theory we hold, that the original murderer of the bull is Angra Mainyu, since he, too, is nowhere so described in

† Soma has in the Rig Veda the epithet *urúgavyūti*, which is the etymological and semantic equivalent of the Avestan Mithra's constant epithet *vouru.-gaoyaoiti-*, on which cf. above, p. 43. The reason for this qualification is probably the same in both cases, for Soma, like Mithra, produces rain.

the Avesta. The answer is that the Avesta altogether has little to say about the Fiendish Spirit, except for cursing him in general terms. There is no hymn of 145 stanzas devoted to him, supplemented by sundry informative passages strewn over the scripture, as in the case of Mithra. To this one may retort that the author of the hymn may purposely have concealed from us and from his listeners (who, however, must have been well aware of it!) all traces of Mithra's tauroctonous feat. But the idea that Zoroastrian authors should have been so squeamish about admitting that Mithra had killed—very reluctantly †—the primordial bull for the most praiseworthy purpose of creating life strikes me as unrealistic. For these same Zoroastrians had no hesitation in stating that Anāhitā and Drvāspā had accepted *and* rewarded the sacrifice of hecatombs of cattle (*Yt* 5.20, 25, etc., *Yt* 9.3, 8, etc.), or, what seems an even worse heresy, that Ahura Mazdāh had worshipped Mithra and Anāhitā (cf. above, pp. 52 *sq.*). Even from a practical point of view it is difficult to see how the Zoroastrian author or authors of the Mithra Yašt, who, as everybody agrees, were not merely trying to please Zarathuštrian worshippers, but also worshippers who throughout their lives had practised the 'pagan' cult of Mithra, could have got away with such a high-handed treatment of the divine newcomer. Here was a god supposedly revelling in 'offrandes cruelles' and 'extases d'ivresse', who under the eyes of his outraged worshippers was being turned into a just and merciful judge, punisher of wrong (st. 35), which includes the offering of blood-libations (st. 113), protector of the very cow after whose blood he had forever been thirsting, always vigilant and sober, certainly no longer prepared to condone, let alone encourage, the 'extases d'ivresse' of his inveterate supporters. To top it all, what to the non-Zarathuštrian worshippers of Mithra must have seemed his chief merit —the killing of the primordial bull in order to produce life—was taken away from him and thrust as if it had been an act of spite on the archdemon Angra Mainyu. Such cavalier treatment of their awful god would scarcely have induced the Mithra-worshippers to join the fold of the Zoroastrian church.

† So the pained expression on Mithras' face informs us, see Cumont, *TMMM*, I, 193.

The above theory on the clever camouflage of Mithra perpetrated by the authors of the Avesta is part and parcel of the modern trend in Iranian religious studies always to suspect that the Avesta either conceals the truth or tells the untruth, and that it is so artificial a scripture, so little based on the realities of the religious life of ordinary men and women, that its authors could permit themselves whatever arbitrary combination, distortion, or suppression crossed their fancy. It must be said in fairness that Lommel, who by his book *Die Religion Zarathustras*, and his careful translation of the Yašts, has contributed more than any other living author to a balanced and sound understanding of the Zoroastrian religion, does not otherwise take such a sceptical view of the truthfulness of the scripture of the religion of Truth. It is an irony that little else but this theory, out of a book which is packed with penetrating and sober observations on the Zoroastrian religion, has caught the imagination of present-day religious historians; the greater part of that thoughtful book seems to have passed unnoticed.

If we resist the unrewarding temptation to look for deliberate deception in the Avesta, the silence of Avestan authors with regard to Mithra's slaughter of the bull can be taken as reliable evidence that the attribution of this slaughter to Mithras is an innovation, and not an ancient inheritance. It goes without saying that Zarathuštra's condemnation of blood sacrifices should in no way influence our opinion about Mithra. The sacrifices the prophet had in mind were very likely offered to the daēvas, and to a few other divinities such as Anāhitā and Drvāspā (cf. above, p. 66). If daēva-worshippers offered blood also to Mithra, Zarathuštra's wrath would turn against *them*, not against the god. We have tried to show above, p. 49, that Zarathuštra and his early followers had more reason to like than to detest Mithra, and the god's unofficial appearance as a *pāyu-* in an early post-Gāthic text (see above, pp. 54 *sqq.*), bears out our view. If such was the attraction exerted by 'pagan' Mithra on early Zarathuštrians, he cannot have thirsted for the blood of animals. In the Younger Avesta Mithra remains strictly averse to blood sacrifices. Even st. 119, which so far has generally been taken to contain instruction that cattle and birds

INTRODUCTION

should be sacrificed to Mithra, turns out on closer inspection to have a very different meaning.

At the beginning of this chapter we listed a number of traits of the Western Mithras, which in our view indicate that he represents a development of the Iranian Mithra as attested in the Avesta. It is true that by what has aptly been called a 'severe shock treatment' of Mithraic studies (Zaehner, *BSOAS*, XVII, 240), Wikander has denied the Iranian origin of Mithras, and suggested a Balkan origin instead.† This novel idea was disproved with conclusive archaeological arguments by A. Alföldi in the article quoted above, p. 62, n. †. On the linguistic side Zaehner, *loc. cit.*, rightly stresses the significance of the name of the grade *Persa* in Mithraic mysteries, and the Iranian origin of the word *nama* in the phrase *nama Sebesio*.‡§ As a further argument for the very close dependence of Mithras on the Iranian Mithra, the suggestion will be found below, note 1¹, that the names *Cautes* and *Cautopates* of the two torchbearers flanking Mithras, which also serve as epithets of Mithras himself, are nothing but variants of the Avestan Mithra's most common epithet *vouru.gaoyaoiti-*.

This brings us to our last point, namely, that no inference on the nature of Old Iranian Mithra should be drawn from the group of

† *Vetenskaps-Societetens i Lund, Årsbok*, 1950.
‡ Zaehner's interpretation of Mithras' epithets *Sebesio* and *Nabarzes* as representing respectively **savišyō* = Av. *saošyå* [sic] 'saviour', and Av. *nā bərəzō* 'the great male' (thus already Markwart, *Das erste Kapitel der Gāthā uštavatī*, 12), is, however, unacceptable. That Mithra should have been called a 'great male' is unbelievable, seeing that the very common Avestan word *nar-* 'man, hero, male' is, not surprisingly, never used as an attribute of gods. On *Sebesio* see below, p. 273, n. †.
§ Less reliable, but also perhaps pointing to an Iranian—this time Zarathuštrian—line of thought in Western Mithraism, is the striking resemblance, noticed by H. J. Rose (*Journal of Hellenic Studies*, XLIV, 181 sq.), between the wording of an inscription at Rome ascribed to the year 361 A.D. and Zarathuštra's noble utterance of *Y* 33.14. The dedicator of the inscription brings as an offering (τοῦτο φέρω τὸ θῦμα) 'works, thought, action, excellence of life, all the choiceness of...wisdom (ἔργα νόον πρῆξιν βίον ἔξοχον ἐσθλὰ πρόπαντα...πραπίδων)'; Zarathuštra 'will give as an offering his own life, and the first choice of good mind, good deed, and good word'. The inscription commemorates the reintroduction of the taurobolium and criobolium in honour of a divinity who is merely referred to as παλίνορσιν εὐρυβίην 'the mighty one that rose again'. Even though the god so described may be Attis, it is possible that the Gāthic reminiscence derives from a store of Mithraic liturgic formulae. Cf. F. Cumont, *Les religions orientales dans le paganisme romain*, 4th ed., p. 229.

68

the three Mithrases Mithras, Cautes, and Cautopates. The point arises from section IV, headed *Mihr, Srōš, Rašn, the Threefold Mithras and the Incarnate God*, of Zaehner's article referred to above, pp. 243 *sqq*. In his book *Zurvan*, p. 102, Zaehner had ventured the guess that certain three 'judges' (*rat*) mentioned in the *Bundahišn*, 'who are needed for the material world and will carry all evil away from it in the last days', are Mihr, Srōš, and Rašn. This guess disagrees with the statement Zaehner later noticed in the *Dātastān i dēnīk*, that 'Gayōmart, Jamšēt, and Zoroaster are the judges (*rat*) of the blessed, the means of thankfulness to many, who were created for the consummation of the *fraškart*'. Instead of concluding that his guess was wrong, Zaehner unexpectedly proceeds on the assumption that it was correct, and pretends that there are grounds for investigating a mysterious relation between the triads Mihr–Srōš–Rašn on the one hand, and Gayōmart–Jamšēt–Zoroaster on the other. Accordingly, noting that 'Hartman† has recently sought to establish the identity of Gayōmart with Mithra (Mihr)', Zaehner compares *Yt* 17.16, according to which Mithra has as father Ahura Mazdāh (see below, p. 195) and possibly as mother Ārmaiti, with a late Pahlavi text which allows the inference that Ohrmazd and Spandarmat are the parents of Gayōmart. The two passages combined show, in Zaehner's view, that Mithra and Gayōmart are brothers. 'Thus'— Zaehner concludes—'it would seem *certain* that according to one tradition Mithra and...Sraoša...and Rašnu were regarded as brothers of Man, *if not identical with him*. Gayōmart, in this tradition, *is the Sun-Man*, and *as such* he is the brother of Mithra and born of the Earth' (my italics).

Let us for argument's sake admit that Mithra and Gayōmart had the same parents. How does their *identity* follow from their brotherhood? Perhaps through Hartman's having 'sought to establish' it? But the whole of Hartman's book is one huge error, as Mary Boyce has cogently‡ shown, and Zaehner, to judge by some of his remarks, is well aware.§ Next, in what way does the tradition

---

† Sven S. Hartman, *Gayōmart*, 1953.
‡ See *BSOAS*, XVII, 174 *sqq*.
§ [Cf. now Zaehner's review of Hartman's book in *JRAS*, 1956, 224.]

INTRODUCTION

invoked by Zaehner indicate that Gayōmart is the Sun-Man? What *is* the Sun-Man? Why is it *as such* that Gayōmart is Mithra's brother and born of the Earth?

Having reached the conclusion that Mithra and Gayōmart are all but identical, Zaehner proceeds to identify Rašnu with Jamšēt (Yima), quoting for this purpose a Pahlavi passage which, to say the least, is ambiguous (see below, p. 225, n. †). But Rašnu, according to Zaehner, is not only Yima. Because in st. 100 of the Mithra Yašt he flies on Mithra's left, he is the 'sinister' aspect of Mithra, just as Sraoša is the 'dexter' or propitious. In the Dumézilian terminology Zaehner here chooses to adopt (p. 247), 'Rašnu is the "Varuṇa" aspect of the Mithra of *Yt* 10.100 while Sraoša is his "Mithra" aspect'. Such unhelpful definitions need not concern us here, as Rašnu, unnoticed by Zaehner, is also found on Mithra's right-hand side (see above, p. 39). After postulating on an erroneous premise that Rašnu is an aspect of Mithra, Zaehner is driven to seek in the obscure st. 79 '*further* evidence' (my italics) of this aspect theory. Zaehner's treatment of the stanza is discussed in an addition to note 79[1] below, pp. 225 *sq.*

The relationship between Sraoša and Zoroaster, the third respective members of the two triads whom Zaehner feels impelled to identify, has not yet been investigated, but the author foresees (p. 249) that it will reveal 'a final merging and marrying of an ancient solar religion having Mithra as its central figure with the ethical dualism of Zoroaster'.

The present Introduction, whose main points were thought out long before Zaehner's article appeared, is intended to refute and forestall precisely the kind of speculation he has advanced. The line Zaehner takes is that because Pseudo-Dionysius mentions a τριπλάσιος Μίθρας and the Mithraic reliefs show a big Mithras flanked by two small Mithrases, a threefold Mithra must perforce be found at the Old Iranian, perhaps even at the Indo-Iranian, stage. Similarly, because in Western Mithraism, and generally in the Christian era, Mithra has clear connections with the sun, Zaehner takes for granted the existence at a pre-Zarathuštrian period of a 'solar religion having Mithra as its central figure'. Our line is to take into consideration the chronological order in which the available data present themselves, and not suspect *a priori* that the data are there to deceive us. If chronologically successive data A, B, C, are such that one can understand how B developed from A, and C either from A, or from B, or from both, it would be perverse to insist that B and/or C must be earlier than A, and produce elaborate and unbelievable reasons why A shows no signs of having been preceded by B and/or C.

All the evidence goes to show that the Avestan Mithra is earlier than solar Mithra, and does not represent the last remnant, or a desolarization, of an earlier solar Mithra. The life-giving capacity of the Western Mithras is comfortably derived from the Avestan Mithra. As regards Mithras' killing of the bull, this requires a hypothesis whichever way it is to be explained, and the hypothesis which least upsets the available data should be given preference. The transfer of the murder of the bull from Ahriman to Mithras in a cult where clearly both were worshipped involves only the hypothesis of a transfer. The transfer of the same deed from Mithra to Ahriman involves not only the assumption of a transfer, but also the hypothesis of deliberate deception on the part of the authors of the Avesta, as well as the awkward need to explain whom beside themselves these authors might have hoped to deceive by the ostrich's device of ignoring what everybody around them knew to be the case.

Let us now revert to, and end with, the τριπλάσιος Μίθρας. He is attested in the fourth century A.D.; accordingly we can give him a full five hundred years† during which he may have occurred to the minds of philosophically inclined Mithraists as they were gazing at the big Mithras and his small-size replicas, the two torch-bearers Cautes and Cautopates. And the origin of the torch-bearers? Cumont long ago surmised that they owe their presence on the reliefs to mere imitation of a conventional type of figurative religious art (see below, note 1[1]). If the explanation we have proposed in the same note of the names Cautes and Cautopates is accepted, Cumont's explanation will be confirmed from the

---

† The beginnings of the representation in sculpture of Mithras tauroctonos are dated by Cumont in about the second century B.C., see *TMMM*, I, 237, with note 1.

INTRODUCTION

Iranian side. If our explanation is rejected, there still remains in Cumont's favour the fact that even Zaehner's determined attempt to find nothing less than a triple series of Threefold Mithrases, viz. Mithras–Cautes–Cautopates, Mithra–Rašnu–Sraoša, and Gayō-mart–Yima–Zoroaster, merely serves to confirm the impression that the ancient Iranian Mithra shows no inclination to fold in three.

TEXT AND TRANSLATION

MIHIR YAŠT (YAŠT X)†

KARDE I

1. mraoṯ ahurō mazdå spitamāi zaraθuštrāi:
āaṯ yaṯ miθrəm yim¹ vouru.gaoyaoitīm
frādaδąm² azəm spitama āaṯ dim³ daδąm
avå̃ntəm yesnyata⁴ avå̃ntəm vahmyata⁴
yaθa mąmčiṯ yim ahurəm mazdąm.

2. mərənčaite¹ vīspąm² dainhaom³
mairyō miθrō.druxš⁴ spitama;
yaθa satəm kayaδanąm
avavaṯ⁵ ašava.jačiṯ⁶;
miθrəm mā janyå spitama
mā⁷ yim drvataṯ pərəså̃nhe
mā yim xᵛādaēnāṯ⁸ ašaonaṯ;
vayå̃⁹ zī asti miθrō
drvataēča ašaonaēča.

3. āsu.aspīm daδāiti
miθrō yō vouru.gaoyaoitiš
yōi¹ miθrəm nōiṯ aiwi.družinti²;
razištəm pantąm daδāiti
ātarš mazdå ahurahe
yōi¹ miθrəm nōiṯ aiwi.družinti³;
ašaonąm vaṇuhīš sūrå
spəntå̃ fravašayō daδāiti⁴
āsnąm frazaintīm
yōi¹ miθrəm nōiṯ aiwi.družinti³.

4. ¹ahe raya xᵛarənaṇhača
təm yazāi surunvata yasna
miθrəm vouru.gaoyaoitīm zaoθrābyō;
miθrəm vouru.gaoyaoitīm yazamaide
rāmašayanəm hušayanəm²
airyābyō dainhubyō³.

† Superior figures in the text refer to the critical apparatus on pp. 301 *sqq*.

HYMN TO MITHRA†

SECTION I

1. Said Ahura Mazdāh to Zarathuštra the Spitamid:
'When I created grass-land magnate[1] Mithra, O Spitamid, I made him such[2] in worthiness to be worshipped and prayed to as myself, Ahura Mazdāh.

2. The knave[1] who is false to the treaty,[2] O Spitamid, wrecks the whole country, hitting as he does the Truth-owners[3] as hard as would a hundred obscurantists.[4]
Never break a contract, O Spitamid, whether you conclude it with an owner of Falsehood, or a Truth-owning follower of the good Religion;[5] for the contract applies to both, the owner of Falsehood and him who owns Truth.'

3. To those who are not false to the contract grass-land magnate Mithra grants (possession of) fast horses, while Fire (, the son) of Ahura Mazdāh, grants them the straightest path,[1] and the good, strong, incremental[2] Fravašis of the owners of Truth give[3] them noble[4] progeny.[5]

4. On account of his splendour and fortune I will audibly worship grass-land magnate Mithra with libations. Grass-land magnate Mithra we worship, since it is he who bestows peaceful and comfortable dwellings on the Iranian countries.

† Superior figures in the translation refer to the commentary on pp. 149 *sqq.*

5. āca.nō ǰamyāṯ avaiṅhe[1]
   āca.nō ǰamyāṯ ravaiṅhe[2]
   āca.nō ǰamyāṯ rafnaṇhe[3]
   āca.nō ǰamyāṯ marždikāi
   āca.nō ǰamyāṯ baēšazāi[4]
   āca.nō ǰamyāṯ vərəθraγnāi[5]
   āca.nō ǰamyāṯ havaṇhāi
   āca.nō ǰamyāṯ ašavastāi
   uγrō aiwiθūrō[6] yesnyō[7]
   vahmyō anaiwi.druxδō[8]
   vīspəmāi[9] aṇuhe astvaite[10]
   miθrō yō vouru.gaoyaoitiš.

6. təm amavantəm yazatəm
   sūrəm dāmōhu[1] səvištəm[2]
   miθrəm yazāi zaoθrābyō
   təm pairi.ǰasāi vantača[3] nəmaṇhača
   təm yazāi surunvata yasna
   miθrəm vouru.gaoyaoitīm [4]zaoθrābyō;
   miθrəm vouru.gaoyaoitīm[4] yazamaide
   haomayō[5] gava[6,7] barəsmana
   hizvō[8] daṇhaṇha[9] mąθrača
   vačača šyaoθnača zaoθrābyasča
   aršuxδaēibyasča vāγžibyō[10].
   [11]yeṅhē hātąm āaṯ yesnē paitī vaṇhō[12]
   mazdå ahurō vaēθā ašāṯ hačā
   yåṇhąmčā tąsčā tåsčā yazamaide.

KARDE 2

7. [1]miθrəm vouru.gaoyaoitīm yazamaide
   arš.vačaṇhəm vyāxanəm
   hazaṇra.gaošəm hutāštəm
   baēvarə.čašmanəm bərəzantəm
   pərəθu.vaēδayanəm sūrəm
   axᵛafnəm ǰaγaurvåṇhəm[2];

8. yim yazənte[1] daiṅhupatayō[2]
   arəzahe[3] ava.ǰasəntō

76

5. May he join us for assistance, may he join us for (the granting of) spaciousness, may he join us for support, may he join us for mercy, may he join us for therapy, may he join us for ability to defeat the opponent,[1] may he join us for a comfortable existence, may he join us for ownership of Truth,[2] he who is strong and victorious, he whom the whole[3] material world must needs worship, pray to, and refrain from deceiving, grass-land magnate Mithra.

6. This powerful strong god Mithra, strongest in the (world of) creatures,[1] I will worship with libations. I will cultivate him with praise and reverence, worship him with audible prayer, with libations, Mithra the grass-land magnate.

    We worship grass-land magnate Mithra with Haoma-containing milk[2] and Barsman twigs, with skill of tongue[3] and magic word, with speech and action and libations, and with correctly uttered words.

    We worship the male and female Entities[4] in the worship of whom Ahura Mazdāh knows (there is [*or:* consists] what is) best (*lit.* better) according to Truth.

SECTION 2

7. Grass-land magnate Mithra we worship, whose words are correct, who is challenging,[1] has a thousand ears, is well built, has ten thousand eyes, is tall, has a wide outlook,[2] is strong, sleepless, (ever-)waking,

8. whom the heads of countries worship as they go to the battlefield[1] against the blood-thirsty enemy armies,

avi⁴ haēnayå xrvišyeitīš⁵
   avi hąm.yanta rasmaoyō
   antarə daińhu⁶ pāpərətāne⁷.
9. ¹yatāra vā.dim² paurva frāyazāiti³
   fraorət̰ fraxšni⁴ avi manō
   zarazdātōit̰⁵ aṇuhyat̰⁶ hača
   ātaraθra⁷ fraorisyeiti⁸
   miθrō yō vouru.gaoyaoitiš
   haθra vāta⁹ vərəθrājanō
   haθra dāmōiš upamanō.
   ahe raya...tåsčā yazamaide (=4–6).

KARDE 3

10. miθrəm vouru.gaoyaoitīm¹...jaγaurvåṇhəm² (=7),
11. yim yazənte¹ raθaēštārō
    barəšaēšu paiti aspaṇąm²
    zāvarə jaiδyantō hitaēibyō
    drvatātəm tanubyō
    pouru.spaxštīm³ t̰bišyaṇtąm⁴
    paiti.jaitīm dušmainyunąm⁵
    haθrā.nivāitīm⁶ hamərəθanąm⁷
    aurvaθanąm t̰bišyaṇtąm⁸.
    ahe raya...tåsčā yazamaide (=4–6).

KARDE 4

12. miθrəm vouru.gaoyaoitīm...jaγaurvåṇhəm¹ (=7)
13. yō paoiryō mainyavō yazatō
    tarō harąm āsnaoiti
    paurva.naēmāt̰ aməšahe
    hū yat̰ aurvat̰.aspahe,
    yō paoiryō zaranyō.pīsō¹
    srīrå barəšnava gərəwnāiti
    aδāt̰² vīspəm āδiδāiti³
    airyō.šayanəm⁴ səvištō,

towards those who, (in the area lying) between two countries at war, join their (respective) regiments.

9. Whichever of the two countries is the first to worship him believingly[1] with fore-knowing[2] thought and trusting mind, to that one (*lit.* there) turns grass-land magnate Mithra, together with[3] the victorious wind, together with the Likeness of Ahura's creature.[4] (*Here repeat 4–6.*)

SECTION 3

10–11. Grass-land magnate Mithra we worship... (=7)..., whom the warriors worship at (=bending down close to) the manes[1] of their horses, requesting strength for their teams, health for themselves, much watchfulness[2] against antagonists, ability to strike back at enemies, ability to rout lawless,[3] hostile opponents. (*Here repeat 4–6.*)

SECTION 4

12–13. Grass-land magnate Mithra we worship... (=7)..., who is the first supernatural[1] god to approach[2] across the Harā, in front of the immortal swift-horsed[3] sun; who is the first to seize the beautiful gold-painted mountain tops; from there the most mighty surveys the whole land inhabited by Iranians,

14. yahmya¹ sāstārō aurva²
    paoiriš³ īrā̊⁴ rāzayente⁵
    yahmya garayō bərəzantō
    pouru.vāstrā̊ṇhō āfəntō
    θātairō⁶ gave frāδayente⁷†
    yahmya⁸ ǰafra varayō
    urvāpā̊ṇhō hištənte⁹
    yahmya āpō nāvayå
    pərəθwiš¹⁰ xšaoδaṇha¹¹ θwaxšənte¹²
    āiškatəm¹³ pourutəmča¹⁴
    mourum hārōyum¹⁵
    gaoṃča suxδəm¹⁶
    xᵛāirizəmča¹⁷
15. avi arəzahi¹ savahi²
    avi fradaδafšu vīdaδafšu³
    avi vouru.barəšti vouru.ǰarəšti
    avi imaṯ karšvarə
    yaṯ xᵛaniraθəm bāmīm
    gavašayanəm⁴ gavašitīmča baēšazyąm
    miθrō sūrō⁵ ādiδāiti⁶,
16. yō vīspāhu karšvōhu¹
    mainyavō yazatō
    vazaite² xᵛarənō.då³
    yō vīspāhu karšvōhu¹
    mainyavō yazatō
    vazaite² xšaθrō.då³;
    aēšąm gūnaoti⁴ vərəθraɣnəm
    yōi dim⁵ dahma⁶ vīduš.aša
    zaoθrābyō frāyazənte.
    ahe raya...tå̄sča yazamaide (=4–6).

KARDE 5

17. miθrəm vouru.gaoyaoitīm...ǰaɣaurvā̊ṇhəm¹ (=7),
    yō nōiṯ kahmāi aiwi.draoxδō²
    nōiṯ nmānahe nmānō.patōe
    nōiṯ vīsō vīspatōe

† In his Additions Geldner prefers the reading *frāδayən*.

14. where gallant rulers organize many attacks, where high, sheltering[1] mountains with ample pasture provide solicitous[2] for cattle; where deep lakes stand with surging waves;[3] where navigable[4] rivers rush wide with a swell[5] towards Parutian Iškata,[6] Haraivian Margu,[7] Sogdian Gava,[8] and Chorasmia.

15. (The seven climes of the earth, which are) Arəzahī, Savahī, Fradaδafšu, Vidaδafšu, Vouru.barəštī, Vouru.-jarəštī, and that splendid clime which is X^{v}aniraθa,[1] the land of settled dwelling and healthy village colonization,[2] (all this area) strong Mithra surveys,

16. the supernatural god who flies over all climes bestowing good fortune, the supernatural god who flies over all climes bestowing power; victoriousness he increases in those who (are) trained by their knowledge of Truth[1] (to) worship him with libations. (*Here repeat 4–6.*)

SECTION 5

17. Grass-land magnate Mithra we worship...(=7)..., to whom nobody may be false, neither the head of the house who presides over the house, nor the head of the clan who presides over the clan, nor the head of the

nōiṯ zantə̄uš zantupatə̄e
nōiṯ daiṅ́hə̄uš³ daiṅ́hupatə̄e⁴.

18. yezi vā.dim¹ aiwi.družaiti²
nmānahe vā nmanō.paitiš
vīsō vā vīspaitiš³
zantə̄uš vā zantupaitiš
daiṅ́hə̄uš⁴ vā daiṅ́hupaitiš⁴
fraša upa.sčandayeiti⁵
miθrō grantō upa.ṯbištō
uta nmānəm uta vīsəm
uta zantūm⁶ uta daȟyūm
uta nmānąm nmānō.paitiš⁷
uta vīsąm vīspaitiš⁷
uta zantunąm zantupaitiš⁷
uta daȟyunąm⁸ daiṅ́hupaitiš⁹
uta daȟyunąm fratəmaδātō¹⁰.

19. ahmāi¹ naēmāi uzjasāiti
miθrō grantō upa.ṯbištō
yahmāi naēmanąm miθrō.druxš
naēδa mainyu paiti.pāite².

20. aspačiṯ¹ yōi² miθrō.drująm³
vazyąstra⁴ bavainti⁵
tačintō nōiṯ apayeinti⁶
barəntō nōiṯ frastanvanti⁷
vazəntō⁸ nōiṯ framanyente⁹.
apaši¹⁰ vazaite¹¹ arštiš
yąm aṅhayeiti¹² avi.miθriš
frə̄na aγanąm mąθranąm
yå vərəzyeiti avi.miθriš;

21. yaṯčiṯ¹ hvastəm² aṅhayeiti³
yaṯčiṯ¹ tanūm⁴ apayeiti
aṯčiṯ⁵ dim nōiṯ rāšayente⁶
frə̄na aγanąm mąθranąm
yå vərəzyeiti avi.miθriš⁷;
vātō tąm arštīm baraiti
yąm aṅhayeiti⁸ avi.miθriš

tribe who presides over the tribe, nor the head of the country who presides over the country.

18. If the head of the house who presides over the house, or the head of the clan who presides over the clan, or the head of the tribe who presides over the tribe, or the head of the country who presides over the country, are false to him, Mithra enraged and provoked comes forth to smash[1] the house, the clan, the tribe, the country, the heads of the houses who preside over the houses, the heads of the clans who preside over the clans, the heads of the tribes who preside over the tribes, the heads of the countries who preside over the countries, and the councils of premiers[2] of the countries.

19. The direction in which Mithra enraged and provoked will sally forth is that in respect of which the contract-breaker is least (*lit.* not at all)[1] on his guard[2] in his mind.

20. Even the horses of those who are false to the contract are loath to be mounted (*lit.* loaded): they perform the motion of running (*lit.* they run), but do not stir from their places (*lit.* do not go away); they bear (their riders) yet do not convey (them) forward;[1] they (begin to) drive (the chariot), but do not persevere.[2]

    Back flies the spear which the Antimithra throws, because[3] of the evil spells which the Antimithra performs;

21. even if he (=the Antimithra) throws a good throw, even if he attains somebody (with his spear), he does (*lit.* they do) not injure him, as a result of the evil spells which the Antimithra performs: the wind carries (off) the spear which the Antimithra throws, because of the

frāna[9] aγanąm mąθranąm
yā vərəzyeiti avi.miθriš.
ahe raya...tāscā yazamaide (=4-6).

KARDE 6

22. miθrəm vouru.gaoyaoitīm...jaγaurvåṇhəm[1] (=7)
yō narəm anaiwi.druxtō
apa ązaṇhaṯ baraiti
apa iθyajaṇhaṯ[2] baraiti.

23. apa.nō hača ązaṇhaṯ
apa hača ązaṇhibyō[1]
miθra barōiš anādruxtō.
tūm ana miθrō.drująm mašyānąm
avi xᵛaēpaiθyā̊sə.tanvō[2]
θwyąm[3] ava.barahi[4];
[5]apa aēšąm bāzvā̊
aojō tūm[6] grantō
xšayamnō barahi[7]
apa pāδayā̊ zāvarə
apa čašmanā̊ sūkəm
apa gaošayā̊ sraoma.

24. nōiṯ dim arštōiš huxšnutayā̊[1]
nōiṯ išaoš[2] para.paθwatō[3]
ava.ašnaoiti[4] šanmaoyō[5]
yahmāi fraxšni[6] avi manō
miθrō jasaiti avaiṇhe[7]
yō baēvarə.spasanō sūrō
vīspō.vīδvā̊ aδaoyamnō[8].
ahe raya...tāscā yazamaide (=4-6).

KARDE 7

25. miθrəm vouru.gaoyaoitīm...jaγaurvå̄ṇhəm[1] (=7)
ahurəm gufrəm amavantəm
dātō.saokəm vyāxanəm[2]
vahmō.səndaṇhəm bərəzantəm
ašahunarəm[3] tanumąθrəm
bāzuš.aojaṇhəm[4] raθaēštąm

evil spells which the Antimithra performs. (*Here repeat 4–6.*)

SECTION 6

22. Grass-land magnate Mithra we worship...(=7)..., who (if) not treated with falsehood removes a man from anxiety, removes him from danger.[1]

23. May you, O Mithra, not having been treated with falsehood (by us), remove us from anxiety, remove us from anxieties!

    This is how[1] you induce fear for their own person(s) in men who are false to the contract: (they know that) when enraged you can[2] carry off the vigour of their arms, the strength of their feet, the light of their eyes, the hearing of their ears.

24. One does not hit[1] with thrusts[2] of well-sharpened[3] spear(s), nor with thrusts of far-flying[4] arrow(s), the one to whose assistance comes with fore-knowing mind Mithra the strong, whose spies are ten thousand,[5] who knows everything and cannot be deceived. (*Here repeat 4–6.*)

SECTION 7

25. Grass-land magnate Mithra we worship...(=7)..., the profound, strong lord, the profit-bestowing champion, the exalted gratifier of prayers,[1] the much-talented personification of the divine word,[2] the warrior endowed with strength of arm,[3]

26. kamərəδō.janəm daēvanąm
akatarəm[1] sraošyanąm
ačaētārəm[2] miθrō.drująm mašyānąm
hamaēstārəm pairikanąm
yō daińhaom[3] anādruxtō
uparāi amāi daδāiti
yō daińhaom[4] anādruxtō
uparāi vərəθrāi daδāiti;

27. yō daińhə̄uš[1]† rąxšyąiθyå[2]
para razištå baraiti
paiti x^{v}arənå vārayeiti
apa vərəθraγnəm baraiti
avarəθå hīš apivaiti[3]
baēvarə γənąnå[4] nisirinaoiti
yō baēvarə.spasanō sūrō
vīspō.vīδvå aδaoyamnō.
ahe raya...tåscā yazamaide (=4–6).

KARDE 8

28. miθrəm vouru.gaoyaoitīm...jaγaurvå̇həm (=7)
yō stunå[1] vīδārayeiti[2]
bərəzimitahe nmānahe
stawrå[3] ąiθyå[4] kərənaoiti;
āat̰ ahmāi nmānāi daδāiti
gə̄ušca vąθwa vīranąmca
yahva[5] xšnūtō[6] bavaiti[7];
upa anyå sčindayeiti[8]
yāhva t̰bištō bavaiti.

29. tūm[1] akō vahištasča
miθra ahi daińhubyō[2]
tūm akō vahištasča
miθra ahi mašyākaēibyō;
tūm āxštōiš[3] anāxštōišca[4]
miθra xšayehe[5] daḣyunąm.

† As stated by Geldner in his Additions, F1 and E1 have *daihə̄uš*.

26. the head-smasher of evil gods, punisher of men false to the contract with an even worse punishment,[1] the engager of witches; who, if not treated with falsehood, will lead a country to superior strength; who, if not treated with falsehood, will lead a country to superior vigour;

27. who carries off the straightest (paths [cf. 3]) of the defiant[1] country, diverts[2] its chances, removes its victoriousness; he pursues them (=the countries, *or rather* the inhabitants) defenceless, deals out ten thousand blows, he who has ten thousand spies, is strong, all-knowing, undeceivable. (*Here repeat 4–6.*)

SECTION 8

28. Grass-land magnate Mithra we worship...(=7)..., who arranges[1] the columns of the high-pillared house, builds the strong (*or* makes strong the) gate-posts;[2] herds of cattle and (teams of) slaves[3] he bestows on the house in which[4] he is propitiated; the others, in which he is provoked, he smashes.

29. You, Mithra, are both wicked and very good to the countries, you, Mithra, are both wicked and very good to men; you, Mithra, control peace and strife of the countries.

30. tūm[1] sraogənå sraoraθå
 ništarətō. spayå[2] niδātō. barəzištå
 nmānå masitå[3] daδāhi
 tūm sraogənəm[4] sraoraθəm
 ništarətō. aspaēm[5] niδātō. barəzištəm
 nmānəm daδāhi bərəzimitəm
 yasə.θwā aoxtō.nāmana yasna
 raθwya vača yazaite
 barō.zaoθrō ašava.

31. aoxtō.nāmana θwā yasna
 raθwya vača sūra
 miθra yazāi zaoθrābyō;
 aoxtō.nāmana θwā yasna
 raθwya vača səvišta
 miθra yazāi zaoθrābyō;
 aoxtō.nāmana θwā yasna
 raθwya vača aδaoyamna
 miθra yazāi zaoθrābyō.

32. surunuyå nō miθra yasnahe
 xšnuyå nō miθra yasnahe
 upa.nō yasnəm āhiša[1]
 paiti.nō zaoθrå[2] vīsaŋuha
 [3]paiti.hīš yaštå vīsaŋuha[3]
 hąm hīš[4] čimāne[5] baraŋuha
 nī.hīš dasva garōnmāne.

33. dazdi ahmākəm tat̰ āyaptəm
 yasə.θwā yāsāmahi[1] sūra
 urvaiti[2] dātanąm sravaŋhąm
 īštīm aməm vərəθraγnəmča[3]
 havaŋhum ašavastəmča
 haosravaŋhəm hurunīmča
 mastīm spānō vaēiδīmča[4]
 vərəθraγnəmča ahuraδātəm
 vanaintīmča uparatātəm
 yąm ašahe vahištahe
 paiti.parštīmča[5] mąθrahe spəntahe,

30. It is you who provide the great houses with bustling women and fast chariots,[1] with spread-out rugs[2] and piled-up cushion-heaps;[3] it is you who provide with bustling women and fast chariots, with spread-out rugs and piled-up cushion-heaps the high-pillared house of the Truth-owning man who regularly mentions you by name in his spoken prayer, offering libations.[4]

31. Mentioning you regularly by name in my spoken prayer, O strong one, I worship you, Mithra, with libations; mentioning you regularly by name in my spoken prayer, O strongest one, I worship you, Mithra, with libations; mentioning you regularly by name in my spoken prayer, O undeceivable one, I worship you, Mithra, with libations!

32. Listen, O Mithra, to our prayer, satisfy, O Mithra, our prayer, condescend to our prayer! Approach our libations, approach them as they are sacrificed, collect them for consumption,[1] deposit them in Paradise!

33. Give us the following boon(s) for which we ask you, O strong one, by virtue of the stipulation[1] of the given promises (*lit.* words): riches, strength, and victoriousness, comfortable existence and ownership of Truth, good reputation and peace of soul, learning, increment, and knowledge, Ahura-created victoriousness, the conquering superiority of (= deriving from) Truth which is what is best, and the interpretation[2] of the incremental divine word,

34. yaθa vaēm humanaŋhō
framanaŋhasča[1] urvāzəmna[2] haomanaŋhimna[3]
vanāma[4] vīspə harəθə̄,
yaθa vaēm humanaŋhō
framanaŋhasča[1] urvāzəmna[2] haomanaŋhimna[3]
vanāma[4] vīspə dušmainyuš,
yaθa vaēm humanaŋhō
framanaŋhasča[1] urvāzəmna[2] haomanaŋhimna[5]
vanāma
vīspå t̰baēšå taurvayama
daēvanąm mašyānąmča
yāθwąm pairikanąmča
sāθrąm kaoyąm karafnąmča[6].
ahe raya...tåsčā yazamaide (=4–6).

KARDE 9

35. miθrəm vouru.gaoyaoitīm...jaγaurvå̇ŋhəm[1] (=7),
[2]arənat̰.čaēšəm vindat̰.spāδəm[3]
hazaŋra.yaoxštīm[4] xšayantəm
xšayamnəm vīspō.vīδvå̇ŋhəm;

36. yō arəzəm frašāvayeiti[1]
[2]yō arəze paiti hištaiti[2]
yō arəze[3] paiti hištəmnō
frā rasmanō[4] sčandayeiti[5];
yaozənti vīspe[6] karanō
rasmanō arəzō.šūtahe
frā maiδyąnəm[7] xrå̇ŋhayete[8]
spāδahe xrvīšyantahe[9];

37. avi dīš aēm xšayamnō
āiθīm barāiti[1] θwyąmča[2];
para kamərəδå spayeiti[3]
miθrō.drująm mašyānąm
[4]para kamərəδå vazaite[5]
miθrō.drująm mašyānąm;

38. xrūmå šaitayō[1] frazinte[2]
anašitå maēθanyå

34. so that we, being in good spirit, cheerful,[1] joyful, and optimistic, may overcome all opponents,[2] so that we, being in good spirit, cheerful, joyful, and optimistic, may overcome all enemies, so that we, being in good spirit, cheerful, joyful, and optimistic, may overcome[3] all hostilities of evil gods and men, sorcerers and witches, tyrants,[4] hymn-mongers,[5] and mumblers.[6] (*Here repeat 4–6.*)

SECTION 9

35. Grass-land magnate Mithra we worship…(=7)…, the punisher of wrong,[1] the levier of armies, whose perception is thousandfold,[2] who rules as an all-knowing potentate;

36. who sets the battle in motion, who takes his stand in the battle, who, having taken his stand in the battle, smashes the regiments: all the flanks are surging[1] of the battle-tossed[2] regiment(s), the centre of the blood-thirsty army is quaking;[3]

37. well may he (*lit.* he will be able to) bring them terror[1] and fear: off he throws the evil heads of the men that are false to the treaty, off fly the evil heads of the men that are false to the treaty;

38. he sweeps away the crumbling[1] dwellings,[2] the no longer inhabitable abodes in which (used to) live the

yāhva miθrō. drujō šyete³
haiθīm. ašava. ǰanasča⁴ drvantō;
xrūmīm gāuš yā čaᵯraᵯhāxš
varaiθīm⁵ pantąm azaite⁶
yā darənāhu miθrō. drująm mašyānąm
frazaršta⁷ aēšąm raiθya⁸
asrū azānō hištənte⁹
anu. zafanō¹⁰ takahe.

39. išavasčiṯ¹ aēšąm ərəzifyō. parəna
huθaxtaṯ² hača θanvanāṯ
ǰya. ǰatåᵯhō vazəmna
ašəmnō. vīδō³ bavaiti⁴
yaθa grantō upa. ṯbištō
apaiti. zantō⁵ miθnāiti
miθrō yō vouru. gaoyaoitiš;
arštayasčiṯ⁶ aēšąm huxšnuta
tiɣra darəɣa. arštaya
vazəmna hača bāzubyō
ašəmnō. vīδō⁷ bavaiti⁸
yaθa grantō upa. ṯbištō
apaiti. zantō⁵ miθnāiti
miθrō yō vouru. gaoyaoitiš;
zarštvačiṯ⁹ aēšąm fradaxšanya
vazəmna hača bāzubyō
ašəmnō. vīδō¹⁰ bavaiti
yaθa grantō upa. ṯbištō
apaiti. zantō miθnāiti
miθrō yō vouru. gaoyaoitiš;

40. ¹karətačiṯ aēšąm
hufrāyuxta yōi² niɣrāire
sarahu³ mašyākanąm
ašəmnō. ǰanō⁴ bavaiti
yaθa grantō upa. ṯbištō
apaiti. zantō miθnāiti
miθrō yō vouru. gaoyaoitiš;
vazračiṯ⁵ aēšąm
hunivixta yōi⁶ niɣrāire

owners of Falsehood who are false to the treaty and strike at what virtually owns Truth:[3] the cow, accustomed to pastures, is driven along the dusty[4] road of captivity, dragged forward in the clutches of treaty-infringing men as their draught-animal;[5] choking[6] with tears they (=cows) stand,[7] slobbering at the mouth (*lit.* a flow (being) along the mouth).

39. Their eagle-winged arrows, though propelled in their flight from a well-stretched bow by the bow-string, pierce no wounds,[1] since grass-land magnate Mithra, enraged and provoked, is hostile,[2] not having been acknowledged.[3]

 Their well-sharpened, pointed, long-shafted spears take off in flight from their arms, but pierce no wounds, since grass-land magnate Mithra, enraged and provoked, is hostile, not having been acknowledged.

 Their sling-stones,[4] too, take off in flight from their arms, but pierce no wounds, since grass-land magnate Mithra, enraged and provoked, is hostile, not having been acknowledged.

40. Well-discharged are their knives when thrown at[1] the heads of men; yet they strike no wounds, since grass-land magnate Mithra, enraged and provoked, is hostile, not having been acknowledged.

 So, too, their maces are well brandished when swung

sarahu⁷ mašyākanąm
ašəmnō.jano⁸ bavaiti
yaθa grantō upa.t̰bištō
apaiti.zantō miθnāiti
miθrō yō vouru.gaoyaoitiš.

41. miθrō avi.θrą̇ṇhayete¹
rašnuš paiti.θrą̇ṇhayete¹
sraošō ašyō vīspaēibyō
naēmaēibyō hąm.vāiti²
paiti θrātāra yazata,
tē³ rasmanō raēčayeinti⁴
yaθa grantō upa.t̰bištō
apaiti.zantō miθnāiti
miθrō yō vouru.gaoyaoitiš

42. uityaojanå̄¹ miθrāi vouru.gaoyaoitə̄e:
²āi miθra vouru.gaoyaoite³
ime nō aurvantō aspa
para miθrāδa⁴ nayente⁵
ime nō uγra.bāzava⁶
karəta miθra sčindayeinti.⁷

43. pasčaēta dīš fraspayeiti¹
miθrō yō vouru.gaoyaoitiš
pančasaγnāi sataγnāišča
sataγnāi hazaṇraγnāišča
hazaṇraγnāi baēvarəγnāišča
baēvarəγnāi² ahąxštaγnāišča
yaθa grantō upa.t̰bištō
miθrō yō vouru.gaoyaoitiš.
ahe raya...tå̄sčā yazamaide (=4–6).

KARDE 10

44. miθrəm vouru.gaoyaoitīm...jaγaurvą̇ṇhəm¹ (=7)
yeṅhe zəm.fraθō maēθanəm
vīδātəm astvanti aṇhvō
mazat̰² anązō³ bāmīm
pərəθu aipi vouru.aštəm⁴;

on the heads of men; yet they strike no wounds, since grass-land magnate Mithra, enraged and provoked, is hostile, not having been acknowledged.

41. Mithra drives the frightened regiments hither, Rašnu drives them thither, Sraoša, the friend of Aši,[1] chases them everywhere: the(ir) protective gods desert them,[2] as grass-land magnate Mithra, enraged and provoked, is hostile, not having been acknowledged.

42. Thus they (= regiments) cry (*lit.* speak) to grass-land magnate Mithra: 'Ah grass-land magnate Mithra! It is to (*lit.* before) Mithra these fast horses are taking us![1] These strong-armed (*viz.* Mithra, Rašnu, Sraoša), O Mithra, are destroying us with the knife!'[2]

43. Thereupon he knocks them down, Mithra the grass-land magnate, with one hundred killings for each blow (aimed) at fifty, with one thousand killings for each blow aimed at a hundred, with ten thousand killings for each blow aimed at a thousand, with countless killings for each blow aimed at ten thousand;[1] for he is enraged and provoked, Mithra the grass-land magnate. (*Here repeat 4–6.*)

SECTION 10

44. Grass-land magnate Mithra we worship...(=7)..., whose abode[1] is set[2] in the material world as far as the earth extends, unrestricted in size,[3] shining, reaching widely[4] abroad;[5]

45. yeṅhe ašta¹ rātayō
²vīspāhu paiti barəzāhu²
vīspāhu vaēδayanāhu
spasō³ åṇhāire miθrahe
miθrō. drujəm⁴ hišpōsəmna⁵†
ave⁶ aipi daiδyantō
ave⁶ aipi hišmarəntō
yōi paurva miθrəm družinti⁷
avaēšąmča paθō⁸ påntō
yim isənti⁹ miθrō. drujō
haiθīm. ašava. janasča drvantō

46. avå pavå pasča pavå
parō pavå spaš¹ vīδaēta²
aδaoyamnō frā aṇhe³ vīsaiti
miθrō yō vouru.gaoyaoitiš
yahmāi frašni⁴ avi manō
miθrō jasaiti avaiṅhe⁵
yō baēvarə. spasanō sūrō
vīspō.vīδvå aδaoyamnō.
ahe raya...tåsčā yazamaide (=4–6).

KARDE II

47. miθrəm vouru.gaoyaoitīm...jayaurvåṇhəm¹ (=7),
yim frasrūtəm² zaranimnəm³
pərəθu.safåṇhō vazənti
avi haēnayå xrvišyeitiš⁴
avi hąm.yanta rasmaoyō
antarə daiṅhu⁵ pāpərətāne.

48. āat yat miθrō fravazaite¹
avi haēnayå xrvišyeitiš²
avi hąm.yanta rasmaoyō
antarə daiṅhu² pāpərətāne
aθra narąm miθrō. drująm
³apąš gavō darəzayeiti
para⁴ daēma vārayeiti

† In his Additions Geldner prefers the reading *hispōsəmna*.

45. for whom on every height, in every watchpost, eight servants sit as watchers of the contract,[1] watching the contract-breaker(s); they see them, they notice them, as soon as[2] they begin to be false to the contract, and they guard the paths of those whom seek out[3] the owners of Falsehood who are false to the contract and strike at what virtually owns Truth.[4]

46. To whose assistance comes Mithra with fore-knowing thought, the strong, all-knowing, undeceivable master of ten thousand spies, him Mithra, the undeceivable grass-land magnate, is ready to help[1] and protect,[2] protecting behind, protecting in front, a watcher and observer all round. (*Here repeat 4–6.*)

SECTION II

47. Grass-land magnate Mithra we worship...(=7)..., whom notorious in his anger[1] the broad-hooved (horses) drive against the blood-thirsty enemy armies, towards those who (in the area lying) between two countries at war, join their (respective) regiments.

48. When Mithra comes driving against the blood-thirsty enemy armies, towards those who (in the area lying) between two countries at war join their (respective) regiments, (and, having arrived) there,[1] fetters behind the evil hands of men false to the treaty, switches off their eyesight, deafens their ears: (then) one[2] no longer

apa gaoša taošayeiti[5]
nōit̰ pāδa vīδārayeiti
nōit̰ paiti.tavå bavaiti
tā[6] daiṁhāvō tē haməraθō[7]
yaθa dužbərəntō[8] baraiti
miθrō yō vouru.gaoyaoitiš.
ahe raya...tåsčā yazamaide (=4–6).

KARDE 12

49. miθrəm vouru.gaoyaoitīm...jaγaurvåṁhəm[1] (=7),
50. yahmāi maēθanəm frāθwərəsat̰
    yō daδvå ahurō mazdå
    upairi harąm bərəzaitīm
    pouru.fraorvaēsyąm[1] bāmyąm
    yaθra nōit̰ xšapa nōit̰ təmå
    nōit̰ aotō vātō nōit̰ garəmō
    nōit̰ axtiš pouru.mahrkō
    nōit̰ āhitiš daēvō.dāta
    naēδa dunmąn[2] uzjasaiti[3]
    haraiθyō[4] paiti[5] barəzayå
51. yat̰ kərənāun[1] aməšå[2] spənta
    vīspe hvarə.hazaoša[3]
    fraorət̰ fraxšni[4] avi manō
    zrazdātōit̰[5] aṅuhyat̰[6] hača
    yō vīspəm ahūm astvantəm ādiδāiti[7]
    haraiθyāt̰ paiti barəzaṁhat̰.
52. āat̰ yat̰ duždå fradvaraiti
    yō[1] aγāvarəš θwāša gāma
    θwāšəm yujyeiti[2] vāšəm
    miθrō yō vouru.gaoyaoitiš
    sraošasča ašyō sūrō
    nairyō.saṅhasča yō māyuš[3],
    rasmō.jatəm vā.dim[4] jainti
    amō.jatəm vā.
    ahe raya...tåsčā yazamaide (=4–6).

disjoins³ the feet, one has no strength to counter:⁴ the lands and the men (*lit.* opponents) he treats as (he treats) those who treat (him) badly,⁵ Mithra the grass-land magnate. (*Here repeat 4–6.*)

SECTION 12

49–50. Grass-land magnate Mithra we worship...(=7)..., for whom Ahura Mazdāh, the creator, fashioned an abode above¹ the much-twisting,² shining Harā the high, where is no night or darkness, no wind cold or hot, no deadly illness, no defilement produced by evil gods— neither do mists rise from Harā the high:

51. which (abode) the incremental Immortals¹ built, all in harmony with the sun, believingly, with fore-knowing thought and trusting mind; who (= Mithra) from Harā the high surveys the whole material world.

52. When fast-stepped the evil-doing maligner hastens to the fore, grass-land magnate Mithra yokes his fast chariot—likewise strong Sraoša, the friend of Aši, and delightful Nairyō.saŋha¹—: he slays him (in any case, so that he is) slain, be it that he stays with his regiment, be it that he ventures out to attack.² (*Here repeat 4–6.*)

KARDE 13

53. miθrəm vouru.gaoyaoitīm...jaɣaurvånhəm (=7)
    yō bāδa ustānazastō
    ɣərəzaite¹ ahurāi mazdāi
    uiti aojanō:
54. azəm vīspanąm dāmanąm
    nipāta¹ ahmi hvapō²
    azəm vīspanąm dāmanąm
    nišharəta ahmi hvapō;
    āat̰ mā nōit̰ mašyāka
    aoxtō.nāmana yasna yazənte³
    yaθa anye yazatąȧhō
    aoxtō.nāmana yasna⁴ yazənti⁵
55. ¹yeiδi.zī.mā mašyāka
    aoxtō.nāmana yasna yazayanta²
    yaθa anye yazatąȧhō
    aoxtō.nāmana yasna yazənti³
    frā nuruyō ašavaoyō⁴
    θwarštahe zrū āyu⁵ šūšuyąm⁶
    xᵛahe gayehe xᵛanvatō aməšahe
    upa θwarštahe⁷ jaɣmyąm.
56. aoxtō.nāmana θwā yasna
    raθwya vača yazaite
    barō.zaoθrō ašava;
    aoxtō.nāmana θwā yasna
    raθwya vača sūra
    miθra yazāi zaoθrābyō;
    aoxtō.nāmana...yazāi zaoθrābyō (=31).
57-59. surunuyå nō miθra yasnahe...karafnąmča (=32-34).
    ahe raya...tåsčā yazamaide (=4-6).

KARDE 14

60. miθrəm vouru.gaoyaoitīm...jaɣaurvånhəm (=7)
    yeṅhe vohu haosravaṅhəm
    vaṅuhi kərəfš vaṅuhi frasasti¹;

SECTION 13

53. Grass-land magnate Mithra we worship...(=7)..., who at times[1] complains to Ahura Mazdāh with outstretched hands, as follows:

54. 'I am the beneficent[1] protector of all creatures, I am the beneficent guardian of all creatures; yet men do not worship me by mentioning my name in their prayer(s), as other gods are worshipped with prayer(s) in which their names are mentioned.

55. If indeed men were to worship me by mentioning my name in their prayer(s), as other gods are worshipped with prayer(s) in which their names are mentioned, I should go forth to men who own Truth, for the duration of a limited time; interrupting my own radiant immortal life I should come.'[1]

56. Mentioning you regularly by name in his spoken prayer, the owner of Truth worships you offering libations. Mentioning you regularly by name in my spoken prayer, O strong one, I worship you, Mithra, with libations; mentioning...(=31)...libations.

57–59. (=32–34+4–6)

SECTION 14

60. Grass-land magnate Mithra we worship...(=7)..., who has good reputation, good appearance, good renown; who grants favours at will, who grants grass-land

vasō.yānəm vasō.gaoyaoitīm[2]
ataurvayō[3] iδa fšuyantəm vāstrīm
vasō.yaonāi intąm[4] huδā̊ŋhəm
yō baēvarə.spasānō[5] sūrō
vīspō.vīδvā̊ aδaoyamnō.
ahe raya...tā̊sčā yazamaide (=4–6).

KARDE 15

61. miθrəm vouru.gaoyaoitīm...jaγaurvā̊ŋhəm (=7)
ərəδwō.zəngəm[1] zaēnaŋhuntəm[2]
spasəm[3] taxməm vyāxanəm[4]
frat̰.āpəm[5] zavanō.srūtəm
tat̰.āpəm[6] uxšyat̰.urvarəm
karšō.rāzaŋhəm vyāxanəm[7]
yaoxštivantəm[8] aδaoyamnəm
pouru.yaoxštīm dāmi.δātəm[9];

62. yō nōit̰ kahmāi miθrō.drująm mašyānąm
aojō daδāiti[1] nōit̰ zāvarə,
yō nōit̰ kahmāi miθrō.drująm mašyānąm
x^varənō daδāiti nōit̰ miždəm[2].

63. apa aēšąm bāzvā̊
aojō tūm[1] grantō
xšayamnō...aδaoyamnō (=23–4).
ahe raya...tā̊sčā yazamaide (=4–6).

KARDE 16

64. miθrəm vouru.gaoyaoitīm...jaγaurvā̊ŋhəm (=7)
yahmi[1] vyāne[2] daēnayāi
srīrayāi pərəθu.frākayāi[3]
maza amava niδātəm
yahi[4] paiti čiθrəm vīδātəm
vīspāiš avi[5] karšvąn yāiš hapta;

65. yō āsunąm āsuš
yō arədranąm arədrō
yō taxmanąm taxmō

at will, *ataurvayō iδa fšuyantəm vāstrīm vasō.yaonāi intąm*,[1] the benign; who has ten thousand spies, is strong, all-knowing, undeceivable. (*Here repeat 4–6.*)

SECTION 15

61. Grass-land magnate Mithra we worship...(=7)..., who stands watchful[1] with upright shanks,[2] the strong challenging watcher, the replenisher of waters[3] who listens to the call, thanks to whom water (=rain) falls and plants grow, the challenging director of (boundary) lines,[4] the creator's perceptive, undeceivable creature,[5] endowed with much perception;

62. who to no man false to the contract gives vigour or strength, who allows no fortune or payment to any man false to the contract.

63. When enraged you can carry off...(=23–4)...cannot be deceived. (*Here repeat 4–6.*)

SECTION 16

64. Grass-land magnate Mithra we worship...(=7)..., in[1] whose soul[2] (there is) a great, powerful[3] pledge[4] to the beautiful, far-spreading Religion, according to which (pledge)[5] its (=the Religion's) seed (*or* family)[6] is distributed over all seven climes;[7]

65. who is fast among the fast, loyal[1] among the loyal, strong among the strong, a champion among champions; who

yō vyāxananąm¹ vyāxanō²
yō fraxšti.då³ yō āzuiti.då⁴
yō vąθwō.då yō xšaθrō.då
yō puθrō.då yō gayō.då
yō havaṇhō.då yō ašavastō.då;

66. yim hačaite¹ ašiš vaṇuhi
pārəndiča² raoraθa
uγrača naire hąm.varəitiš³
uγrəmča kavaēm x^varənō
uγrəmča θwāšəm x^vaδātəm
uγrasča dāmōiš upamanō
uγråsča ašaonąm⁴ fravašayō
yasča pourunąm haθrākō
ašaonąm mazdayasnanąm.
ahe raya...tåsčā yazamaide (=4-6).

KARDE 17

67. miθrəm vouru.gaoyaoitīm...jaγaurvåṇhəm (=7)
yō vāša mainyu.hąm.tāšta¹
bərəzi.čaxra² fravazaite
hača karšvarə ³yaṯ arəzahi⁴
upa karšvarə³ yaṯ⁵ x^vaniraθəm bāmīm
raθwya čiθra⁶† hačimnō
x^varənaṇhača ⁷mazdaδāta
vərəθraγnača⁷ ahuraδāta;

68. yeṅhe vāšəm hangrəwnāiti¹
ašiš vaṇuhi yā bərəzaiti
yeṅhe daēna māzdayasniš²
x^vīte³ paθō rāδaiti⁴;
yim aurvantō mainyavåṇhō
auruša⁵ raoxšna⁶ frādərəsra
spənta vīδvåṇhō asaya⁷
manivasaṇhō⁸ vazənti
yaṯ dim⁹ dāmōiš upamanō
hu.irixtəm¹⁰ bāδa irinaxti;

† In his Additions Geldner prefers the reading *čaxra*.

grants the entreaty,² who dispenses fat³ and herds, who gives power and sons, who bestows life and comfortable existence, who ensures ownership of Truth;

66. whose escort are good Aši, Bounty¹ in her fast carriage,² strong manly Valour, the strong Kavyan Fortune, the strong Firmament³ which obeys its own law, the strong Likeness of Ahura's creature, the strong Fravašis of the owners of Truth, and he (*viz.* Nairyō.saŋha) who shares place and time with many Truth-owning worshippers of Mazdāh.⁴ (*Here repeat 4–6.*)

SECTION 17

67. Grass-land magnate Mithra we worship...(=7)..., who comes driving in a supernaturally fashioned, high-wheeled chariot, from the (eastern) continent Arəzahī¹ to (us in) the splendid continent Xᵛaniraθa, equipped with prompt energy,² Mazdāh-created fortune, and Ahura-created victoriousness;

68. tall good Aši guides¹ his chariot; the Mazdayasnian Religion² paves its paths for good travel;³ it is driven by supernatural, white, radiant, transparent,⁴ incremental, intelligent, shadeless coursers that hail from supernature, as often as⁵ the Likeness of Ahura's creature launches it well-launched.⁶

yahmaṯ[11] hača fratərəsənti
vīspe mainyava daēva
yaēča varənya drvantō;

69. mōi.tū[1] iθra ahurahe
grantahe vaēɣāi jasaēma
yeṅhe hazaṉrəm vaēɣanąm
paiti hamərəθāi[2] jasaiti[3]
yō baēvarə.spasānō[4] sūrō
vīspō.vīδvå aδaoyamnō.
ahe raya...tåsčā yazamaide (= 4–6).

KARDE 18

70. miθrəm vouru.gaoyaoitīm...jaɣaurvåṉhəm (= 7)
yeṅhe paurva.naēmāṯ vazaite[1]
vərəθraɣnō ahuraδātō
hū kəhrpa[2] varāzahe
paiti.ərənō[3] tiži.dąstrahe
aršnō tiži.asūrahe
hakərəṯ.janō varāzahe
anu.pōiθwahe grantahe
paršvanikahe taxmahe
ayaṉhō pāδahe[4] ayaṉhō.zastahe
ayaṉhō.jyehe[5] ayaṉhō.dumahe
ayaṉhō.paitišxᵛarənahe

71. yō frąštačō hamərəθāδa[1]
upa.haxtō[2] ā.manaṉha[3]
haθra nairya[4] hąm.varəta[5]
stija nijainti hamərəθə̄[6]
naēδa manyete jaɣnvå[7]
naēδa čim ɣənąm sadayeiti
yavata aēm nijainti
mərəzuča stūnō gayehe
mərəzuča xå uštānahe;

72. hakaṯ vīspå aipi.kərəntaiti
yō hakaṯ astəsča varəsəsča
mastarəɣnasča[1] vohunišča[2]

Wherefore all supernatural evil gods, as well as the concupiscent[7] owners of Falsehood, are moved by fear:[8]

69. 'Let us not meet here with the charge of the wrathful lord who comes with a thousand batterings (*lit.* whose thousand batterings come) to the opponent, the strong, all-knowing, undeceivable master of ten thousand spies.' (*Here repeat 4–6.*)

SECTION 18

70. Grass-land magnate Mithra we worship...(=7)..., in front of whom flies Ahura-created Vərəθrayna[1] in the shape of a wild, aggressive, male boar[2] with sharp fangs[3] and sharp tusks,[4] a boar that kills at one blow, is unapproachable, grim, speckle-faced,[5] and strong, has iron hind feet, iron fore-feet, iron tendons, an iron tail, and iron jaws;

71. as he (= Vərəθrayna = boar) catches up with the opponent(s),[1] beset by passion—simultaneously by manly valour—, he knocks them (*lit.* the opponents) down with a toss (of his head):[2] he does not even[3] think he has struck, nor has he the impression he is hitting[4] anybody,[5] until he has smashed even the vertebrae,[6] the pillars of life, even the vertebrae, the springs of vitality;[7]

72. he cuts to pieces everything at once, mingling (*lit.* he who mingles) together on the ground the bones, the

zəmāδa³ hąm.raēθwayeiti⁴
miθrō.drująm⁵ mašyānąm.
ahe raya...tåsčā yazamaide (=4–6).

KARDE 19

73. miθrəm vouru.gaoyaoitīm...jaγaurvåṇhəm (=7)
yō bāδa ustānazastō
urvāzəmnō¹ avarōit̰ vāčim²
uityaojanō:
ahura mazda mainyō³ spəništa
dātarə gaēθanąm astvaitinąm ašāum

74. ¹yeδi.zī.mā mašyāka
aoxtō.nāmana yasna yazayanta²
yaθa anye yazatåṇhō
aoxtō.nāmana yasna yazənti³
frā nuruyō⁴ ašavaoyō⁵
θwarštahe zrū āyū⁶ šūšuyąm⁷
xᵛahe gayehe xᵛanvatō aməšahe
upa θwarštahe⁸ jaγmyąm.

75. buyama tē šōiθrō.pānō
mā buyama¹ šōiθrō.iričō
mā nmānō.iričō
mā vīsō.iričō²
mā zantu.iričō
mā daiṅhu.iričō³
māδa yat̰ nō uγra.bāzāuš⁴
nivānāt̰⁵ parō t̰bišyanbyō⁶.

76. tūm aēšąm t̰bišyatąm¹
tūm aēšąm t̰baēšaṇuhatąm²
t̰baēšå sčindayehi³;
sčindaya⁴ ašavajanō
hvaspō ahi hurāθyō⁵
zavanō.sva ahi⁶ sūrō.

77. āča.θwā zbayāi avaiṅhe¹,
āča.nō jamyāt̰ avaiṅhe¹
aš.frāyaštiča zaoθranąm hufrāyaštiča

hair, the brains, and the blood of men false to the contract. (*Here repeat 4–6.*)

SECTION 19

73. Grass-land magnate Mithra we worship...(=7)..., who at (other [*cf. 53*]) times joyfully raises his voice to Ahura Mazdāh, speaking with outstretched hands as follows: 'O Truth-owning Ahura Mazdāh, most incremental spirit, creator of the material world!

74. If indeed men were to worship me by mentioning my name in their prayer(s), as other gods are worshipped with prayer(s) in which their names are mentioned, I should go forth to men who own Truth, for the duration of a limited time; interrupting my own radiant immortal life I should come.'

75. May we become such as are capable of protecting the homesteads,[1] may we become such as do not need to abandon either homestead, or house, or clan, or tribe, or country, or what(ever else) the strong-armed (=Mithra) shall guard for us from the enemies![2]

76. It is you who destroy the enmities of the enemies, it is you who destroy the enmities of the inimical: do destroy the slayers of Truth-owning men!
    You have good horses and a good chariot-driver:[1] strong, you bring profit when invoked.

77. I invoke you for assistance:[1] 'May he join us for assistance, (moved) by abundant and good sacrifice of liba-

aš.frabərəitiča zaoθranąm hufrabərəitiča
yaθa θwā aiwišayamna[2]
darəγa aiwišayana
hušitəm bərəγmya.šaētəm.

78. tūm tā[1] daiṅhāvō[2] nipāhi
yā[3] hubərəitīm[4] yātayeiti[5]
miθrahe vouru.gaoyaoitōiš;
tūm tā[1] frasčindayehi[6]
yā[3] rąxšyeitīš daiṅhāvō;
āča.θwā zbayāi avaiṅhe[7],
āča.nō ǰamyāṱ[8] avaḣyāi
uγrō aiwiθūrō yesnyō[9] vahmyō
miθrō raēvå daiṅhupaitiš.
ahe raya...tåscā yazamaide (=4–6).

KARDE 20

79. miθrəm vouru.gaoyaoitīm...ǰaγaurvåṅhəm (=7)
yō rašnuš[1] daiδe[2] maēθanəm
yahmāi rašnuš[3] darəγāi haxəδrāi
frabavara manavaintīm[4].

80. tūm[1] maēθanahe pāta nipāta
ahi adružąm[2], tūm[3] varəzānahe
paiti niš.harəta ahi[4] adružąm[2],
θwā paiti zī haxəδrəm daiδe vahištəm
vərəθraγnəmča ahuraδātəm
yahmi sōire[5] miθrō.drujō
aipi vīθiši[6] ǰata
paurva mašyākå̇ṅhō.
ahe raya...tåsca yazamaide (=4–6).

KARDE 21

81. miθrəm vouru.gaoyaoitīm...ǰaγaurvåṅhəm (=7)
yō rašnuš daiδe[1] maēθanəm
yahmāi rašnuš darəγāi haxəδrāi
bavara[2] manavaintīm[3];

tions, by abundant and good offering of libations!', so that through you we may inhabit with long dwelling a welcome dwelling good to dwell in.

78. You protect the countries in the same measure in which[1] they strive (*lit.* it strives) to take care[2] of grassland magnate Mithra; you destroy the countries to the same extent to which they are defiant.

    I invoke you for assistance: 'May he join us for assistance, Mithra the strong, victorious, splendid master of countries, worthy to be worshipped, worthy to be prayed to!' (*Here repeat 4–6.*)

SECTION 20

79. Grass-land magnate Mithra we worship...(=7)..., the judge who makes the abode gain prominence,[1] (for that man) to whom Rašnu has given (it) for long succession.[2]

80. You are the protector, the defender, of the abode of those who reject falsehood; you are the guardian[1] of the community of those who reject falsehood: with you as master it (=the community)[2] obtains the most excellent succession and Ahura-created victoriousness, (because) in it (*lit.* in which [community]) the many men false to the contract are floored (*lit.* lie), struck at the divinatory trial. (*Here repeat 4–6.*)

SECTION 21

81. Grass-land magnate Mithra we worship...(=7)..., the judge who makes the abode gain prominence, (for that man) to whom Rašnu has given (it) for long succession;

82. yeŋhe hazaŋrəm yaoxštinąm
fradaθaṯ ahurō mazdå
baēvarə dōiθranąm vīdōiθre[1];
āaṯ ābyō dōiθrābyō
aiwyasča yaoxštibyō
spasyeiti[2] miθrō.zyąm
miθrō.druǰəmča[3];
āaṯ ābyō dōiθrābyō
aiwyasča[4] yaoxštibyō
aδaoyō asti miθrō
yō baēvarə.spasanō[5] sūrō
vīspō.vīδvå aδaoyamnō.
ahe raya...tåsčā yazamaide (=4–6).

KARDE 22

83. miθrəm vouru.gaoyaoitīm...ǰaγaurvåṇhəm (=7)
yim dainhəuš[1] dainhupaitiš[1]
bāδa ustānazastō
zbayeiti avainhe[2],
yim zantə̄uš zantupaitiš
bāδa ustānazastō
zbayeiti avainhe[2],

84. yim vīsō vīspaitiš
bāδa ustānazastō
zbayeiti avainhe,
yim nmānahe nmānō.paitiš
bāδa ustānazastō
zbayeiti avainhe,
yim dvāčina piθe[1] hačimna[2]
bāδa ustānazastō
zbayeiti avainhe,
yim driγuščiṯ[3] ašōṯkaēšō
apayatō havāiš[4] dātāiš
bāδa ustānazastō
zbayeiti avainhe,

82. on whom Ahura Mazdāh has conferred a thousand perceptions, (and) ten thousand eyes for seeing all-round; thanks to these eyes and perceptions he spots the infringer of the contract and the man false to the contract; thanks to these eyes and perceptions Mithra cannot be deceived, he who has ten thousand spies, is strong, all-knowing, undeceivable. (*Here repeat 4–6.*)

SECTION 22

83. Grass-land magnate Mithra we worship...(=7)..., whom invoke for assistance, with outstretched hands, sometimes the head of the country who presides over the country, sometimes the head of the tribe who presides over the tribe,

84. sometimes the head of the clan who presides over the clan, sometimes the head of the house who presides over the house, sometimes she (*viz.* the cow) who longs to be milked as she feels (*lit.* is with) the pain of swelling,[1] sometimes also the pauper[2] who follows the doctrine of Truth but is deprived of his rights;

85. yeṅhe[1] vāxš[2] gərəzānahe
us ava raočā̊ ašnaoiti
ava pairi imąm ząm ǰasaiti
vī[3] hapta karšvąn ǰasaiti
yaṱčiṱ nəmaṅha vāčim baraiti
yaṱ gaoščiṱ[4],

86. yā varəta[1] azimna[2]
bāδa ustānazastō
zbayeiti avaiṅhe
gavaiθīm paitišmarəmna[3]:
kaδa.nō[4] arša gavaiθīm
apayāṱ paskāṱ vazəmnō
miθrō yō vouru.gaoyaoitiš
kaδa.nō fraourvaēsayāiti[5]
ašahe paiti pantąm
druǰō vaēsmənda azəmnąm[6].

87. āaṱ[1] yahmāi xšnūtō[2] bavaiti
miθrō yō vouru.gaoyaoitiš
ahmāi ǰasaiti avaiṅhe;
āaṱ yahmāi ṱbištō bavaiti
miθrō yō vouru.gaoyaoitiš
ahmāi fraščindayeiti
nmānəmča vīsəmča
zantūmča daḣyūmča
daiṅhusastīmča.
ahe raya...tā̊sčā yazamaide (=4-6).

KARDE 23

88. miθrəm vouru.gaoyaoitīm...ǰaγaurvā̊ṅhəm (=7)
yim yazata haomō
frāšmiš baēšazyō srīrō
xšaθryō [1]zairidōiθrō
barəzište paiti barəzahi[2]
haraiθyō paiti barəzayā̊
yaṱ vaočē[3] hūkairīm[4] nąma
anāhitəm anāhitō

85. the lamenting voice of the latter, even though he raises his voice reverently,[1] reaches up to the (heavenly) lights, makes the round of the earth, pervades the seven climes; so does the voice of the cow (*lit.* as (that) of the cow, too, (does)),[2]

86. who, being led away captive, calls at times for assistance with outstretched hands, longing for the herd: 'When will grass-land magnate Mithra, the hero,[1] driving from behind, make us reach[2] the herd, when will he divert to the path of Truth us who are (*lit.* is)[3] being driven to the estate[4] of Falsehood?'

87. Then grass-land magnate Mithra comes to the assistance of that (invoker) by whom he has been propitiated; but of him who has antagonized grass-land magnate Mithra he destroys (respectively [*cf. 83 sq.*]) the house, the clan, the tribe, the country, and the empire (*lit.* command of countries).[1] (*Here repeat 4–6.*)

SECTION 23

88. Grass-land magnate Mithra we worship...(=7)..., whom glowing[1] Haoma the healer, beautiful, majestic, and golden-eyed, worshipped on the highest peak of Harā the high,[2] which is called Hukairya by name; the immaculate (worshipped) the immaculate with im-

anāhitāṯ parō barəsman⁵
anāhitayāṯ parō zaoθrayāṯ
anāhitaēibyō parō vaγžibyō⁶;

89. yim zaotārəm stayata
ahurō mazdå ašava
āsu.yasnəm¹ bərəzi.gāθrəm;
yazata zaota āsu.yasnō
bərəzi.gāθrō² bərəzata vača
³zaota ahurāi mazdāi³
zaota aməšanąm spəntanąm;
hō vāxš us ava raočå ašnaoṯ
ava pairi imąm ząm jasaṯ
vījasāṯ vīspāiš avi karšvąn yāiš hapta;

90. yō paoiryō hāvana
haomą¹ uzdasta
stəhrpaēsaṇha mainyutāšta²
haraiθyō paiti barəzayå;
bərəjayaṯ³ ahurō mazdå
bərəjayən⁴ aməšå spənta
yeṇhå kəhrpō huraoδayå
yahmāi hvarə aurvaṯ.aspəm
dūrāṯ⁵ nəmō baoδayeiti;

91. nəmō miθrāi vouru.gaoyaoitēe
hazaṇrō.gaošāi baēvarə.čašmaine¹;
yesnyō² ahi vahmyō
yesnyō buyå vahmyō
nmānāhu mašyākanąm;
ušta buyāṯ ahmāi naire
yasə.θwā bāδa frāyazāite
aēsmō.zastō barəsmō.zastō
gaozastō hāvanō.zastō
frasnātaēibya zastaēibya
frasnātaēibya hāvanaēibya
frastərətāṯ³ paiti barəsman⁴
uzdātāṯ paiti haomāṯ
srāvayamnāṯ paiti ahunāṯ vairyāṯ.

maculate Barsman twigs, immaculate libation, immaculate words;

89. whom (= Haoma) Truth-owning Ahura Mazdāh installed as promptly-sacrificing, loud-chanting priest: as Ahura Mazdāh's promptly-sacrificing, loud-chanting priest, as the priest of the incremental Immortals, he, the priest, sacrificed (chanting) with loud voice; his voice reached up to the (heavenly) lights, made the round of the earth, pervaded all seven climes;

90. who (= Haoma)[1] was the first mortar-priest to elevate the star-decked, supernaturally fashioned Haoma-stalks on the high Harā; (even) Ahura Mazdāh praised (his) well-grown body which the incremental Immortals were praising;[2] whom (= Haoma) from afar the swift-horsed sun causes to perceive his reverence.

91. Homage to grass-land magnate Mithra who has a thousand ears (and) ten thousand eyes! Worthy of worship and prayer you are; may you go on being (held) worthy of worship and prayer in the houses of men! Hail to that man who at various times will worship you, firewood, Barsman-twigs, milk, pestle and mortar in hand, having washed his hands, washed pestle and mortar, spread out the Barsman-twigs, elevated the Haoma-stalk, recited the Ahuna Vairya prayer!

92. ¹aya daēnaya fraorənta²
ahurō mazdå ašava
frā vohu manō
frā ašəm vahištəm
frā xšaθrəm vairīm
frā spənta ārmaiti³
frā haurvata amərətāta,
frā.hē aməšå spənta
bərəja⁴ vərənta daēnayāi,
frā.hē⁵ mazdå hvāpå⁶
ratuθwəm barāṯ gaēθanąm
yōi.θwā⁷ vaēnən dāmōhu
ahūm⁸ ratūmča gaēθanąm
yaoždātārəm åŋhąm dāmanąm vahištąm⁹.

93. ¹aδa vaēibya ahubya
vaēibya nō ahubya nipayå
āi miθra vouru.gaoyaoite
aheča aŋhə̄uš yō astvatō
yasča asti manahyō
pairi drvataṯ mahrkāṯ
pairi drvataṯ aēšmāṯ
pairi drvataēibyō² haēnə̄byō³
yå us xrūrəm⁴ drafšəm gərəwnąn⁵
aēšmahe parō draomə̄byō⁶
yå aēšmō duždå drāvayāṯ
maṯ vīδataoṯ⁷ daēvō.dātāṯ;

94. aδa.nō.tūm miθra vouru.gaoyaoite
zāvarə dayå¹ hitaēibyō
drvatātəm tanubyō
pouru.spaxštīm² ṯbišyantąm³
paiti.jaitīm dušmainyunąm⁴
haθrānivāitīm⁵ hamərəθanąm⁶
aurvaθanąm ṯbišyantąm⁷.
ahe raya...tåsčā yazamaide (=4-6).

92. This Religion Truth-owning Ahura Mazdā professed;[1] out of desire[2] for the Religion (also) Good Mind, Truth which is what is best,[3] Power to be chosen, incremental Devotion, Wholeness and Life professed (it); to it generous[4] Mazdāh gave[5] the jurisdiction over the living beings, so did (*lit.* to it) the incremental Immortals, who consider you (=Mithra)[6] the temporal and religious judge of living beings in the (world of) creatures,[7] the one who purifies the best (Religion)[8] for the creatures.

93. Now then, in both lives, O grass-land magnate Mithra, in both—this material existence, and the one which is spiritual—do protect us from Death and Wrath, the two owners of Falsehood, from the evil armies of the owners of Falsehood who raise a gruesome banner, from the onslaughts of Wrath, which are run by Wrath the malignant with (the co-operation of) the Disintegrator (of the body) whom the evil gods created!

94. Now then, grass-land magnate Mithra, give strength to our teams, health to ourselves, much watchfulness against antagonists, ability to strike back at enemies, ability to rout lawless, hostile, opponents! (*Here repeat 4–6.*)

KARDE 24

95. miθrəm vouru.gaoyaoitīm...jaɣaurvå̆ŋhəm (=7)
    yō zəm.fraθå̄¹ aiwyāiti
    pasča hū frāšmō.dāitīm
    marəzaiti va karana²
    aiṅhå̄³ zəmō yat paθanayå̄
    skarənayå̄ dūraēpārayå̄;
    vīspəm imat ādiδāiti
    yat antarə ząm asmanəmča,

96. vazrəm zastaya dražəmnō¹
    satafštānəm satō.dārəm
    fravaēɣəm vīrō.nyå̄nčim²
    zarōiš ayaṅhō³ frahixtəm⁴
    amavatō zaranyehe
    amavastəməm zaēnąm⁵
    vərəθravastəməm zaēnąm⁵;

97. yahmat¹ hača fratərəsaiti
    aṅrō mainyuš pouru.mahrkō
    yahmat hača fratərəsaiti
    aēšmō duždå̄ pəšō.tanuš
    yahmat hača fratərəsaiti
    būšyąsta darəɣō.gava
    yahmat hača fratərəsənti
    vīspe² mainyava daēva
    yaēča varənya drvantō,

98. mā miθrahe vouru.gaoyaotōiš
    grantahe vaēɣāi jasaēma,
    mā.nō grantō aipi.janyå̄¹
    miθra² yō³ vouru.gaoyaoitiš⁴
    yō aojištō yazatanąm
    ⁵yō tančištō yazatanąm⁵
    yō θwaxšištō yazatanąm
    yō āsištō yazatanąm
    yō as vərəθrająstəmō yazatanąm
    fraxštaite⁶ paiti āya⁷ zəmā

SECTION 24

95. Grass-land magnate Mithra we worship...(=7)..., who goes along the whole width of the earth after the setting of the glow of the sun, sweeping across both edges[1] of this wide, round earth whose limits are far apart: everything he surveys between heaven and earth,

96. holding his mace in his hand; with its hundred bosses and hundred blades[1] (it is) a feller of men as it swings forward; strongest of weapons,[2] most valiant of weapons, it is cast in strong, yellow, gilded[3] iron;

97. from whom the Fiendish Spirit, very deadly, recoils in fear, from whom malignant Wrath, his body forfeited,[1] recoils in fear, from whom long-handed Procrastination[2] recoils in fear, from whom all supernatural evil gods, and the concupiscent owners of Falsehood, recoil in fear (, saying):

98. 'May we not meet with the onslaught of grass-land magnate Mithra in his rage (*lit.* enraged)! May you not strike us in your rage (*lit.* enraged), grass-land magnate Mithra!'

(Thus [*cf.* 95]) he who is[1] the mightiest of gods, the strongest of gods, the most mobile of gods, the fastest of gods, the most victorious of gods, comes forth on

miθrō yō vouru.gaoyaoitiš.
ahe raya...tåsčā yazamaide (=4–6).

KARDE 25

99. miθrǝm vouru.gaoyaoitīm...jaγaurvǻŋhǝm (=7),
yahmaṱ hača fratǝrǝsǝnta[1]
vīspe mainyava daēva
yaēča varǝnya drvantō;
fravazaite[2] daiṅhupaitiš
miθrō yō vouru.gaoyaoitiš
dašinǝm upa karanǝm
aiṅhǻ zǝmō yaṱ paθanayǻ
skarǝnayǻ dūraēpārayǻ;

100. dašinǝm hē upa arǝδǝm
vazaite[1] yō vaṇhuš sraošō ašyō,
vairya.stārǝm[2] hē upa arǝδǝm
vazaite[3] rašnuš bǝrǝzō yō amavǻ,
vīspō hē upa arǝδǝm[4]
vazǝnti yǻ āpō yǻsča urvarǻ
yǻsča ašaonąm[5] fravašayō;

101. avi.dīš[1] aēm xšayamnō hamaθa
baraiti išavō[2] ǝrǝzifyō.parǝna[3];
āaṱ yaṱ aθra para.jasaiti vazǝmnō
yaθra daiṅhāvō avi.miθranyǻ[4]
hō paoiryō gaδąm nijainti
aspaēča paiti vīraēča,
haθra taršta θrǻṇhayete[5]
vaya[6] aspa.vīraja[7].
ahe raya...tåsčā yazamaide (=4–6).

KARDE 26

102. miθrǝm vouru.gaoyaoitīm...jaγaurvǻŋhǝm (=7)
aurušāspǝm
tiži.arštīm darǝγa.arǝštaēm[1]
xšviwi.išūm[2] parō.kǝvīdǝm
hunairyǻnčim raθaēštąm;

this earth, grass-land magnate Mithra. (*Here repeat 4–6.*)

SECTION 25

99. Grass-land magnate Mithra we worship...(=7)..., from whom all supernatural evil gods, and the concupiscent owners of Falsehood, recoil in fear. Along flies grass-land magnate Mithra, master of countries, over the right-hand (=southern) border[1] of this wide, round earth whose limits are far apart;

100. on his right flies good Sraoša, the friend of Aši; on his left[1] flies tall Rašnu the strong; all round him fly the waters and plants, and the Fravašis of the owners of Truth;[2]

101. he[1] knows how (*lit.* he is able) to supply *them* regularly with eagle-winged arrows; but when in his flight he arrives where the countries are anti-Mithrian, it is *he* who first strikes his club at horse and man,[2] (who) striking at horse and man frightens both with sudden fright. (*Here repeat 4–6.*)

SECTION 26

102. Grass-land magnate Mithra we worship...(=7)..., the skilful warrior who has white horses and pointed spears with long shafts, who shoots[1] afar with swift arrows;

103. yim harətārəmča aiwyāxštārəmča
 frada𝜃at ahurō mazdå
 vīspayå fravōiš gaē𝜃ayå,
 [1]yō harətača aiwyāxštača
 vīspayå fravōiš gaē𝜃ayå,
 yō anavaṇuhabdəmnō[2] zaēnaṇha
 nipāiti mazdå dāmąn
 yō anavaṇuhabdəmnō zaēnaṇha
 nišhaurvaiti mazdå dāmąn.
 ahe raya...tåscā yazamaide (= 4–6).

KARDE 27

104. mi𝜃rəm vouru.gaoyaoitīm...jaγaurvåṇhəm (= 7)
 yeṇhe darəγāčit bāzava
 fragrəwənti[1] mi𝜃rō.aojaṇhō†
 yatčit ušastaire[2] hindvō[3] āgəurvayeite[4]
 yatčit daošataire niγne[5]
 yatčit sanake[6] raṇhayå
 yatčit vīmaiðīm[7] aiṇhå zəmō,

105. təmčit[1] mi𝜃rō hangrəfšəmnō[2]
 pairi apaya bāzuwe[3];
 dušxᵛarənå naštō[4] razišta
 ašātō[5] asti aṇuhaya[6];
 i𝜃a[7] mainyete dušxᵛarənå,
 nōit imat vīspəm dužvarštəm
 nōit vīspəm aiwi.druxtə̄e[8]
 mi𝜃rō vaēnaiti apišma[9];

106. āat azəm manya[1] manaṇhō:
 nōit mašyō gaē𝜃yō stē[2]
 aojō manyete[3] dušmatəm
 ya𝜃a mi𝜃rasčit mainyavō
 aojō manyete humatəm;
 nōit mašyō gaē𝜃yō stē[4]
 aojō mraoiti dužuxtəm[5]
 ya𝜃a mi𝜃rasčit mainyavō

† In his Additions Geldner emends to *mi𝜃ō.aojaṇhō*.

103. whom Ahura Mazdāh appointed inspector and supervisor of the promotion[1] of the whole world, who is the inspector and supervisor of the promotion of the whole world, the caretaker who without falling asleep, watchfully protects the creatures of Mazdāh, the caretaker who without falling asleep, watchfully observes the creatures of Mazdāh. (*Here repeat 4–6.*)

SECTION 27

104. Grass-land magnate Mithra we worship...(=7)..., whose[1] long arms reach out to catch the violators of the contract:[2] if (the violator is) by the eastern[3] river[4] he is caught,[5] if (he is) by the western (river) he is struck down;[6] whether (he is) at the source[7] of the Raṇhā, whether (he is) in the middle of the earth,

105. Mithra (will be) seizing him still, reaching round (him)[1] with his two arms. The ill-fated, having forfeited the straightest (path), is miserable in mind:
   'So'[2]—thinks the ill-fated—'(it is) not (true that) all this ill-doing, Mithra does not see *all*, when his face is not turned to[3] (man's) trickery!'

106. But I (=the worshipper) think in my mind: 'There is no material man in existence[1] who thinks evil thoughts to (so) great an extent as supernatural Mithra thinks good thoughts; there is no material man in existence who speaks evil words to (so) great an extent as super-

aojō mraoiti hūxtəm;
nōiṯ mašyō gaēθyō stē⁶
aojō vərəzyeiti dužvarštəm
yaθa miθrasčiṯ mainyavō
aojō vərəzyeiti hvarštəm;
107. nōiṯ mašīm gaēθīm stē¹
masyå hačaite² āsnō xratuš
yaθa miθrəmčiṯ mainyaom³
hačaite² āsnasčiṯ xratuš;
nōiṯ mašyō gaēθyō stē⁴
aojō surunaoiti gaošaiwe
yaθa miθrasčiṯ mainyavō
sruṯ.gaošō hazaᵑra.yaoxštiš⁵
vīspəm vaēnaiti družintəm⁶;
amava miθrō fraxštāite⁷
uγra vazaite⁸ xšaθrahe
srīra daδāiti⁹ daēmāna
dūrāṯ.sūka dōiθrābya¹⁰:
108. kō mąm yazāite kō družāṯ
kō huyešti kō dužyešti
mąm zī mainyete yazatəm;
kahmāi raēšča¹ xᵛarənasča
kahmāi tanvō drvatātəm
azəm baxšāni² xšayamnō;
kahmāi īštīm pouruš.xᵛāθrąm
azəm baxšāni³ xšayamnō
kahmāi āsnąmčiṯ frazaintīm
us apara⁴ barəzayeni⁵;
109. kahmāi azəm uγrəm xšaθrəm
xᵛainisaxtəm¹ pouru.spāδəm
amainimnahe manaᵑhō
paiti.daθāni vahištəm
sāθrasčiṯ hamō.xšaθrahe
kamərəδō.janō aurvahe
vanatō avanəmnahe
yō ništayeiti kərətēe sraošyąm,
išarə hā ništāta² kiryeiti³

natural Mithra speaks good words; there is no material man in existence who commits evil deeds to (so) great an extent as supernatural Mithra performs good deeds;

107. no material man in existence is endowed with (*lit.* followed by) greater insight than that which supernatural Mithra is endowed with; no material man in existence hears with his two ears as much as supernatural Mithra, who has listening ears and a thousand perceptions, sees,[1] (namely) *all* [*cf. 105*] perpetrators of falsehood.'

Forceful(ly) Mithra comes forth, strong in (*lit.* of) power[2] he flies, with a beautiful far-shining glance he looks (round) with his eyes:

108. 'Who is he that worships me, who is he that is false to me? Who is he that thinks I am to be worshipped with good sacrifice, who is he that thinks I am to be worshipped with bad sacrifice?[1] On whom may I bestow riches and fortune, on whom health of body, on whom property that affords much comfort? For whom shall I raise noble progeny hereafter?

109. On whom shall I bestow against his expectation an excellent (*vahištəm*) powerful kingdom, beautifully strong[1] thanks to[2] a numerous army?[3] (Once he rules) he appeases through Mithra, by honouring the treaty,[4] even the mind of an antagonized, unreconciled[5] conqueror (*vanatō*) unconquerable, who gallant(ly) strikes the evil head of even an equally powerful tyrant,[6] who

yezi grantō ništayeiti⁴
ṭbištahečiṭ axšnuštahe⁵
miθra⁶ manō rāmayeiti
huxšnūitīm⁷ paiti miθrahe;

110. kahmāi yaskəmča mahrkəmča
kahmāi ainištīm dučiθrəm¹
azəm baxšāni xšayamnō;
kahmāi āsnąmčiṭ frazaintīm
haθra.jata² nijanāni;

111. kahmāi azəm uγrəm xšaθrəm
xᵛainisaxtəm¹ pouru.spāδəm
amainimnahe manaṉhō
apabarāni vahištəm
sāθrasčiṭ hamō.xšaθrahe
kamərəδō.janō aurvahe
vanatō avanəmnahe
yō ništayeiti kərətə̄e sraošyąm,
išarə hā ništāta² kiryeti³
yezi grantō ništayeiti²
xšnūtahečiṭ aṭbištahe
miθra⁴ manō yaozayeiti
axšnūitīm paiti miθrahe.
ahe raya...tā̊sčā yazamaide (=4-6).

KARDE 28

112. miθrəm vouru.gaoyaoitīm...jaγaurvā̊ṉhəm (=7),
ərəzatō.frašnəm zaranyō.vārəθmanəm
aštraṉhāδəm¹ amavantəm
taxməm vīspaitīm² raθaēštąm;
čiθrå miθrahe frayanå³
yasə.tąm⁴ dahyūm⁵ āčaraiti⁶
yaθa hubərətō baraiti
paθanå jafrå gaoyaotə̄e⁷
āaṭ hva⁸ pasu vīra
vasō.xšaθrō fračaraite⁹.

orders the execution of punishment, (and) as soon as it is ordered it is executed at his angry bidding (*lit.* if he, angry, orders).

110. On whom may I bestow illness and death,[1] on whom poverty that brings misery?[2] Whose noble progeny shall I slay with one blow?

111. From whom shall I carry off against his expectation an excellent powerful kingdom, beautifully strong thanks to a numerous army? (While he rules) he incenses[1] through Mithra, by not honouring the treaty, even the mind of a reconciled, not antagonized conqueror unconquerable, who gallant(ly) strikes the evil head of even an equally powerful tyrant, who orders the execution of punishment, (and) as soon as it is ordered it is executed at his angry bidding.' (*Here repeat 4–6.*)

SECTION 28

112. Grass-land magnate Mithra we worship...(=7)..., whose pike[1] is of silver, whose armour of gold, who drives with the whip, the powerful, strong, broad-shouldered[2] warrior. The clans dear[3] to Mithra—when he visits the(ir) country—he treats as (he treats) those who treat (him) well;[4] the(ir) valleys (are) wide[5] for pasture, and their own cattle and slaves go about at will;[6]

113. taδa[1] nō[2] ǰamyāṯ avaiṅhe[3]
 miθra ahura bərəzanta[4]
 yaṯ bərəzəm barāṯ aštra
 vāčim[5] aspanąmča[6] srifa
 xšufsąn aštrå kahvąn ǰyå
 naviθyąn[7] tiγrąṇhō aštayō[8]
 taδa[9] hunavō gouru.zaoθranąm
 ǰata paiθyånte[10] frā.vərəsa[11].
114. [1]aδa.nō.tūm miθra vouru.gaoyaoite
 zāvarə dayå hitaēibyō
 drvatātəm tanubyō
 pouru.spaxštīm[2] ṯbišyantąm[3]
 paiti.ǰaitīm dušmainyunąm[4]
 haθrānivāitīm[5] hamərəθanąm[6]
 aurvaθanąm ṯbišyantąm[3].
 ahe raya...tåsčā yazamaide (=4–6).

KARDE 29

115. miθrəm vouru.gaoyaoitīm...ǰaγaurvåṇhəm (=7).
 āi miθra vouru.gaoyaoite
 nmānya ratvō[1] vīsya zantuma
 daḣyuma zaraθuštrōtəma.

116. vīsaitivå asti miθrō
 antarə haša suptiδarənga[1]
 θrisaθwå antarə varəzāna[2]
 čaθwarəsaθwå antarə haδō.gaēθa
 panča.saθwå antarə huyāγna[3]
 xšvaštivå antarə hāvišta
 haptaiθivå antarə aēθrya aēθra.paiti
 aštaiθivå antarə zāmātara[4] x^vasura
 navaitivå antarə brāθra,
117. satāyuš antarə pitarə[1] puθrəmča
 hazaṇrāiš[2] antarə daḣyu
 baēvarōiš asti miθrō

113. may he therefore come to our assistance,¹ O exalted Mithra and Ahura! When loudly resound the whip and the neighing² of horses, when the whips are tossing,³ the bow-strings twanging(?), the sharp arrows⁴ darting, then the evil sons of those who have offered viscid (*lit.* heavy) libations (=libations of blood), having been struck, will go down writhing.⁵

114. And so, grass-land magnate Mithra, may you give us strength for our teams, health for ourselves, much watchfulness against antagonists, ability to strike back at enemies, ability to rout lawless, hostile opponents! (*Here repeat 4–6.*)

SECTION 29

115. Grass-land magnate Mithra we worship…(=7).…
O you, grass-land magnate Mithra, (who are simultaneously) the religious chief (of the house, called) Nmānya, the religious chief (of the clan, called) Vīsya, the religious chief (of the tribe, called) Zantuma, the religious chief (of the country, called) Dahyuma, the religious (supreme) chief (called) Zaraθuštrōtəma!¹

116. Twentyfold is the contract between two friends shouldering (mutual) obligations,¹ thirtyfold between two fellow-citizens, fortyfold between two partners,² fiftyfold between husband and wife,³ sixtyfold between two fellow-students,⁴ seventyfold between disciple and teacher, eightyfold between son-in-law and father-in-law, ninetyfold between two brothers,

117. hundredfold between father and son, thousandfold between two countries; ten-thousandfold is the contract

yō daēnayā̊ māzdayasnōiš
ava² hačaite³ amahe⁴ ayąn⁵
aθa aṅhāiti⁶ vərəθraɣnahe⁷.

118. nəmaṅha aδara dāta
āǰasāni upara dāta;
yaθa avat̰ hvarəxšaētəm
tarasča harąm bərəzaitīm
frača āiti¹ aiwiča vazaite²
avaθa azəmčit̰ spitama
nəmaṅha aδara dāta
³āǰasāni upara dāta³
tarasča aṅrahe mainyə̄uš
drvatō zaošą⁴.
ahe raya... tā̊sčā yazamaide (=4–6).

KARDE 30

119. miθrəm vouru.gaoyaoitīm... ǰaɣaurvā̊ṅhəm (=7);
miθrəm yazaēša¹ spitama
framrviša² aēθryanąm;
yazayanta³ θwąm mazdayasna⁴
pasubya staoraēibya
vayaēibya patarətaēibya⁵
yō⁶ parənīnō⁷ fravazå̄nte⁸.

120. miθrō vīspə̄¹ mazdayasnanąm²
yąm ašaonąm ərəδwāča kərəθwāča
haomō āvistō³ aiwi.vistō yā̊ zaota
aiwiča vaēδayå̄nte⁴ frača yazå̄nte⁵
nā ašava yaoždātąm
zaoθrąm fraṅuharāt̰⁶
yō kərənavāt̰ yim yazaite
miθrəm yim vouru.gaoyaoitīm
xšnūtō⁷ at̰bištō hyāt̰⁸.

121. paiti.dim pərəsat̰ zaraθuštrō:
kuθa ahura mazda
nā ašava yaoždātąm
zaoθrąm¹ fraṅuharāt̰

of the Mazdayasnian Religion:[1] thereby follow days of strength, thus there will be (days) of victory![2]

118. 'I shall come (—said Mithra—) amidst the homage of the lowly and the exalted;[1] as yonder[2] sun goes forth to cross Harā the high in his flight, so I, too, O Spitamid,[3] shall come forward to cross the pleasure of the Falsehood-owning Fiendish Spirit, amidst the homage of the lowly and the exalted.' (*Here repeat 4–6.*)

SECTION 30

119. Grass-land magnate Mithra we worship...(=7).... 'Worship Mithra, O Spitamid (—said Ahura Mazdāh—)[1] reveal him to the disciples! You (, Mithra,) the worshippers of Mazdāh shall worship, together with the small and large cattle, together with the birds and fowls[2] that fly[3] on wings!

120. Mithra (is) the furtherer and guardian of all the Truth-owning worshippers of Mazdāh: these shall (therefore) dedicate and sacrifice (the libations of) which consecrated and dedicated Haoma is the pourer,[1] (and) the Truth-owning man shall drink[2] a purified libation, to bring it about that grass-land magnate Mithra whom he worships shall be propitiated, not antagonized!'

121. Zarathuštra asked him: '(Tell me,) Ahura Mazdāh, how shall a Truth-owning man drink the purified libation,

       yō kərənavāt̰ yim yazaite
       miθrəm yim vouru.gaoyaoitīm
       xšnūtō at̰bištō hyāt̰².
122. āat̰ mraot̰ ahurō mazdå:
     θri.ayarəm θri.xšaparəm tanūm frasnayayanta¹
     θrisatəm upāzananąm pairi.ākayayanta²
     miθrahe vouru.gaoyaoitōiš yasnāiča vahmāiča;
     bi.ayarəm bi.xšaparəm tanūm³ frasnayayanta⁴
     vīsaiti⁵ upāzananąm pairi.ākayayanta⁶
     miθrahe vouru.gaoyaoitōiš yasnāiča vahmāiča;
     mā.čiš mē å̇ŋhąm zaoθranąm franuharāt̰
     yā nōit̰⁷ staotanąm yesnyanąm āmātō vīspe ratavō.
     ahe raya...tåsčā yazamaide (=4-6).

KARDE 31

123. miθrəm vouru.gaoyaoitīm...jaγaurvå̇ŋhəm (=7),
     yim yazata ahurō mazdå
     raoxšnāt̰ paiti garō.nmānāt̰,
124. uzbāzāuš¹ paiti amərəxtīm
     fravazaite² miθrō yō vouru.gaoyaoitiš
     hača raoxšnāt̰ garō.nmānāt̰
     vāšəm srīrəm³ vavazānəm
     hāmō.taxməm vīspō.paēsəm⁴ zaranaēnəm;
125. ahmya vāše vazå̇nte¹
     čaθwārō aurvantō
     spaētita hama.gaonå̇ŋhō²
     mainyuš.xᵛarəθa³ anaošå̇ŋhō;
     tē para.safå̇ŋhō
     zaranaēna⁴ paiti.šmuxta⁵
     āat̰ hē apara ərəzataēna;
     āat̰ tē vīspa⁶ frā.yuxta⁷
     hąm ivąmča⁸ simąmča simōiθrąmča
     dərəta hukərəta upairispātā⁹
     aka bastąm xšaθrəm vairīm.

to bring it about that grass-land magnate Mithra whom he worships shall be propitiated, not antagonized?'

122. Ahura Mazdāh said: 'Let them wash their bodies for three days and three nights, let them undergo a penance of thirty inflictions (=strokes of the whip), (be)for(e) worshipping and praying to grass-land magnate Mithra; let them wash their bodies for two days and two nights, let them undergo a penance of twenty inflictions, (be)for(e) worshipping and praying to grass-land magnate Mithra; let no one drink of these libations unless he is (*lit.* in so far as (he is) not) experienced in the (section called) "All Chiefs" of the (liturgy called) "Prayers of Praise".'[1] (*Here repeat 4–6.*)

SECTION 31

123. Grass-land magnate Mithra we worship...(=7)..., whom (even) Ahurah Mazdāh worshipped in Paradise,

124. raising his arms towards the indestructible:[1] from Paradise grass-land magnate Mithra drives out his beautiful, golden, all-adorned chariot, which is easy to drive[2] and runs evenly.[3]

125. Four coursers pull at (*lit.* on) his chariot: all of the same whiteness, they are immortal, having been reared on supernatural food; their front-hooves are shod with gold, their hind-hooves with silver; all are harnessed to the yoke[1]—as well as to the yoke pin(s)[2] and yoke strap(s)[3]—, which (=yoke) is connected to a solid,[4] well-made shaft[5] by means of a metal hook.[6]

126. dašinəm hē arəδe
vazaite¹ rašnvō razištō
spə̄ništō upa.raoδištō;
āaṱ hē hāvōya² arəδe
vazaite razištąm čistąm³
baraṱ.zaoθrąm ašaonīm
spaēta vastrā̊ vaŋhaiti⁴ spaēta
daēnayā̊ māzdayasnōiš⁵ upamanəm;
127. upavazata taxmō
dāmōiš upamanō
hū¹ kəhrpa varāzahe
paiti.ərənō² tiži.dąstrahe
aršnō tiži.asūrahe³
hakərəṱ.ǰanō varāzahe
anu.pōiθwahe grantahe
paršvanikahe taxmahe
yūxδahe⁴ pāirivāzahe⁵;
nixšata⁶ ahmāṱ vazata
ātarš yō upa.suxtō
uγrəm yō⁷ kavaēm xᵛarənō.
128. hištaite¹ aom vāšahe
miθrahe vouru.gaoyaoitōiš
hazaŋrəm θanvarəitinąm² [asti yō³
gavasnahe⁴ snāvya⁵ ǰya⁶] hukərətanąm;
mainyavasā̊⁷ vazənti
mainyavasā̊ patənti
kamərəδe paiti daēvanąm.
129. hištaite¹ aom vāšahe
miθrahe vouru.gaoyaoitōiš
hazaŋrəm išunąm² kahrkāsō.parnanąm³
zaranyō.zafrąm⁴ srvī.stayąm [asti⁵ yā
aŋhaēna⁶ sparəγa] hukərətanąm;
mainyavasā̊ vazənti
mainyavasā̊ patənti
kamərəδe paiti daēvanąm.
130. hištaite aom vāšahe
miθrahe vouru.gaoyaoitōiš

126. On his right flies Rašnu, very straight, most incremental, and extremely tall;[1] on his left flies the libation-bearing, Truth-owning (goddess) Razištā Čistā, white, dressed in white garments,[2] the Likeness of the Mazdayasnian Religion;[3]

127. flying behind (*lit.* up to) (him)[1] comes the strong Likeness of Ahura's creature, in the shape of a wild, aggressive, male boar with sharp fangs and sharp tusks, a boar that kills at one blow, is unapproachable, grim, speckle-faced, and strong, dexterous[2] as it leaps about;[3] in front of[4] him (= Mithra) flies the blazing Fire which (is) the strong Kavyan Fortune.[5]

128. There, on grass-land magnate Mithra's chariot, are in readiness a thousand well-made 'bow-stretchers'[1]— that is a bow-string made of the sinews of deer[2]—; hailing from supernature they fly, hailing from supernature they fall, on to the evil head(s) of the evil gods.[3]

129. There, on grass-land magnate Mithra's chariot, are in readiness a thousand well-made vulture-feathered arrows, golden-mouthed,[1] each having as barbs two (small) horns—that is two sprouts made of bone[2]—; hailing from supernature they fly, hailing from supernature they fall, on to the evil head(s) of the evil gods.

130. There, on grass-land magnate Mithra's chariot, are in readiness a thousand well-made spears, sharp at the

   hazaŋrəm arštinąm
   brōiθrō.taēžanąm¹ hukərətanąm;
   mainyavasā̊ vazənti
   mainyavasā̊ patənti
   kamərəδe paiti daēvanąm²;
   hištaite aom vāšahe
   miθrahe vouru.gaoyaoitōiš
   hazaŋrəm čakušanąm³
   haosafnaēnąm⁴ bitaēγanąm hukərətanąm;
   mainyavasā̊ vazənti
   mainyavasā̊ patənti
   kamərəδe paiti daēvanąm.

131. hištaite aom vāšahe
   miθrahe vouru.gaoyaoitōiš
   hazaŋrəm karətanąm
   vayō.dāranąm¹ hukərətanąm;
   mainyavasā̊ vazənti
   mainyavasā̊ patənti˙
   kamərəδe paiti daēvanąm;
   hištaite aom vāšahe
   miθrahe vouru.gaoyaoitōiš
   hazaŋrəm gaδanąm
   ayaŋhaēnanąm hukərətanąm;
   mainyavasā̊ vazənti
   mainyavasā̊ patənti
   kamərəδe paiti daēvanąm.

132. hištaite aom vāšahe
   miθrahe vouru.gaoyaoitōiš
   vazrəm¹ srīrəm hunivixtəm
   satafštānəm² satō.dārəm
   fravaēγəm³ vīrō.nyā̊nčim⁴
   zarōiš ayaŋhō⁵ frahixtəm
   amavatō zaranyehe⁶
   amavastəməm zayanąm⁷
   vərəθravastəməm zayanąm⁷;
   mainyavasā̊ vazənti⁸

blade; hailing from supernature they fly, hailing from supernature they fall, on to the evil head(s) of the evil gods.

There, on grass-land magnate Mithra's chariot, are in readiness a thousand well-made, two-wedged hatchets of steel; hailing from supernature they fly, hailing from supernature they fall, on to the evil head(s) of the evil gods.

131. There, on grass-land magnate Mithra's chariot, are in readiness a thousand well-made two-edged knives; hailing from supernature they fly, hailing from supernature they fall, on to the evil head(s) of the evil gods.

There, on grass-land magnate Mithra's chariot, are in readiness a thousand well-made iron clubs; hailing from supernature they fly, hailing from supernature they fall, on to the evil head(s) of the evil gods.

132. There, on grass-land magnate Mithra's chariot, is in readiness his beautiful, easily brandished mace; with its hundred bosses and hundred blades (it is) a feller of men as it swings forward; strongest of weapons, most valiant of weapons, it is cast in strong, yellow, gilded iron;[1]

mainyavasā̊ patənti
kamərəδe paiti daēvanąm.

133. pasča jainti daēvanąm
pasča niγninti¹ miθrō. drująm
mašyānąm fravazaite²
miθrō yō vouru.gaoyaoitiš
tarō arəzahi³ savahi⁴
tarō fradaδafšu vīdaδafšu
tarō vouru.barəšti⁵ vouru.jarəšti⁶
tarō imaṯ karšvarə
yaṯ xᵛaniraθəm bāmīm.

134. avi bāδa fratərəsaiti
aṇrō mainyuš pouru.mahrkō;
avi bāδa fratərəsaiti
aēšmō¹ duždā̊ pəšō.tanuš²;
avi bāδa fratərəsaiti
būšyąsta darəγō.gava;
avi bāδa fratərəsənti
vīspe mainyava daēva
yaēča varənya drvantō,

135. ¹mā miθrahe vouru.gaoyaoitōiš
grantahe vaēγāi jasaēma,
mā.nō grantō aipi.janyā̊²
miθrō² yō² vouru.gaoyaoitiš
yō aojištō yazatanąm
³yō tančištō yazatanąm³
yō θwaxšištō yazatanąm
yō āsištō yazatanąm
yō as⁴ vərəθrająstəmō yazatanąm
fraxštaite⁵ paiti āya⁶ zəmā
miθrō yō vouru.gaoyaoitiš.
ahe raya...tā̊sčā yazamaide (=4-6).

KARDE 32

136. miθrəm vouru.gaoyaoitīm...jaγaurvā̊ṇhəm (=7)
yahmāi auruša aurvanta¹
yūxta² vāša θanjasā̊nte³

hailing from supernature it flies (*lit.* they fly), hailing from supernature it (*lit.* they) fall(s), on to the evil head(s) of evil gods.

133. After the smiting of evil gods, after the slaying of men false to the contract, grass-land magnate Mithra comes flying across Arəzahī, Savahī, Fradaδafšu, Vīdaδafšu, Vouru.barəštī, Vouru.jarəštī, and that splendid clime which is Xᵛaniraθa:

134. now it is the Fiendish Spirit, very deadly, who recoils in fear, now malignant Wrath, his body forfeited, now long-handed Procrastination; now recoil in fear all supernatural evil gods and the concupiscent owners of Falsehood (, all of them crying):[1]

135. 'May we not meet with the onslaught of grass-land magnate Mithra in his rage (*lit.* enraged)! May you not strike us in your rage (*lit.* enraged), grass-land magnate Mithra!'

(Thus [*cf. 133*]) he who is the mightiest of gods, the strongest of gods, the most mobile of gods, the fastest of gods, the most victorious of gods, comes forth on this earth, grass-land magnate Mithra.[1] (*Here repeat 4–6.*)

SECTION 32

136. Grass-land magnate Mithra we worship...(=7)..., for whom white coursers, yoked to his one-wheeled,

aēva⁴ čaxra⁵ zaranaēna
asānasča⁶ vīspō.bāma
yezi.šē⁷ zaoθrą̇ baraiti
avi.šē⁸ maēθanəm.

137. ušta¹ ahmāi naire mainyāi
uiti² mraot̰ ahurō mazdå
āi ašāum zaraθuštra
yahmāi zaota³† ašava
aṇhəuš dahmō tanu.mą̇θrō
frastərətāt̰⁴ paiti barəsmən⁵
miθrahe vača yazāite⁶
rāštəm⁷ ahmāi naire mainyāi⁸
miθrō maēθanəm āčaraiti,
yezi.šē⁹ yānāδa¹⁰ bavaiti
saṇhəmčit̰ anu sastrāi¹¹
saṇhəmčit̰ anu mainyāi¹².

138. sādrəm ahmāi naire mainyāi
uiti¹ mraot̰ ahurō mazdå
āi ašāum zaraθuštra
yahmāi zaota²† anašava
adahmō atanu.mą̇θrō
pasča barəsma frahišta
pərənəmča barəsma staranō
darəγəmča yasnəm yazānō.

139. nōit̰ xšnāvayeiti ahurəm mazdąm
nōit̰ anye aməšå¹ spənta
nōit̰ miθrəm yim vouru.gaoyaoitīm
yō mazdąm tarō manyete
tarō anye aməšå spənta
tarō miθrəm yim vouru.gaoyaoitīm
tarō dātəmča rašnūmča² arštātəmča
frādat̰.gaēθąm varədat̰.gaēθąm.
ahe raya...tå̇sčā yazamaide (= 4–6).

† In his *Prolegomena*, XLIV, n. 2, Geldner prefers the reading *zaotō*.

golden chariot¹ which is all-glittering² with (precious) stones,³ pull⁴ (it), when he takes his libations to his abode [*cf. 32*].

137. 'Hail to the authoritative¹ man'—said Ahura Mazdāh—, 'O Truth-owning Zarathuštra, on whose behalf a priest who is an owner of Truth, has experience of the world,² and personifies the divine word, having spread out the Barsman-twigs, offers sacrifice with utterance of (the name of) Mithra!³ Straightway Mithra visits the residence of this authoritative man, if as a result of his (=the man's) favour (shown to the priest), it (=the utterance of Mithra's name) is in accordance with the prescription for recitation, in accordance with the prescription for thinking⁴ (=the prescriptions for praying orally and mentally).'

138. 'Woe to the authoritative man'—said Ahura Mazdāh—, 'O Truth-owning Zarathuštra, on whose behalf a priest who is no owner of Truth, has no experience, and does not personify the divine word, takes his stand behind the Barsman-twigs, even if he spreads them out fully,¹ even if he performs a long sacrifice!'

139. Neither Ahura Mazdāh, nor the other¹ incremental Immortals, nor grass-land magnate Mithra, are propitiated by him who thinks overbearingly of Mazdāh, the other incremental Immortals, grass-land magnate Mithra, the Law, Rašnu, or world-furthering, world-promoting Justice.² (*Here repeat 4–6.*)

KARDE 33

140. miθrəm vouru.gaoyaoitīm...jaɣaurvå̇ṇhəm (=7);
yazāi miθrəm spitama
vaṇhuš[1] taxməm mainyaom
aɣrīm[2] hvāmarždikəm
amiθwəm[3] uparō.nmānəm
aojaṇhəm taxməm raθaēštąm;

141. vərəθravå̇ zaēna hačimnō hutāšta
təmaṇhāδa[1] jiɣāurum[2] aδaoyamnəm;
aojištanąm asti aojištəm[3]
tančištanąm asti tančištəm[4]
baɣanąm asti aš.xraθwastəmō[5]
vərəθravå̇ x^varəna[6] hačimnō
hazaṇrā.gaošō[7] baēvarə.čašmanō
yō baēvarə.spasānō sūrō
vīspō.vīδvå̇ aδaoyamnō.
ahe raya...tå̇scā yazamaide (=4-6).

KARDE 34

142. miθrəm vouru.gaoyaoitīm...jaɣaurvå̇ṇhəm (=7)
yō paoiriš[1] vaēiδiš[2] sūrəm[3] frāδaiti
spəntahe mainyə̄uš dāmąn[4]
huδātō mazištō yazatō
yaθa tanūm raočayeiti
yaθa må̇ṇhō[5] hvāraoxšnō[6];

143. yeṅhå̇ ainikō brāzaiti[1]
yaθa tištryō.stārahe
yeṅhe vāšəm hangrəwnāiti[2]
aδaviš[3] paoirīš[4] spitama
yaθa dāmąn sraēštāiš
hū.bāmya[5] xšaētāi[6]
yazāi hąm.taštəm
yō daδvå̇ spəntō mainyuš
stəhrpaēsaṇhəm mainyu.tāštəm[7]

SECTION 33

140. Grass-land magnate Mithra we worship...(=7)....
'I will worship[1] Mithra, O Spitamid' (—said Ahura Mazdāh—), 'who is good, strong, supernatural, foremost, merciful,[2] incomparable,[3] high-dwelling,[4] a mighty strong warrior.

141. Valiant, he is equipped with a well-fashioned weapon [cf. 96], he who watches in darkness, the undeceivable. He is what (is) mightiest among the very mighty, he is what (is) strongest among the very strong; he has by far the greatest insight among the gods.[1] Fortune attends him, the valiant, who with his thousand ears and ten thousand eyes is the strong, all-knowing, undeceivable master of ten thousand spies.' (*Here repeat 4–6.*)

SECTION 34

142. Grass-land magnate Mithra we worship...(=7)..., the well-created, very great god who in the morning brings into evidence the many shapes, the creatures of the Incremental Spirit, as he lights up his body,[1] being endowed with own light like the moon;[2]

143. whose[1] face blazes like (that) of the star Sirius.[2]
'(Him)[3] I will worship, O Spitamid' (—said Ahura Mazdāh—) 'of whom frequently[4] she,[5] the undeceiving— who shines[6] like the majestic sun's[7] most beautiful creature (*viz.* daylight)[8]—guides the star-decked, supernaturally fashioned[9] chariot built (by him) who [10] is the creative Incremental Spirit! (Him I will worship,) the

yō baēvarə.spasānō sūrō
vīspō.vīδvå aδaoyamnō.
ahe raya...tåsčā yazamaide (=4-6).

KARDE 35

144. miθrəm vouru.gaoyaoitīm...jaγaurvåṁhəm[1] (=7);
[2]miθrəm aiwi.daȟyūm[3] yazamaide,
miθrəm antarə.daȟyūm yazamaide,
miθrəm ā.daȟyūm yazamaide,
miθrəm upairi.daȟyūm yazamaide,
miθrəm aδairi.daȟyūm yazamaide,
miθrəm pairi.daȟyūm yazamaide,
miθrəm aipi.daȟyūm yazamaide.

145. miθra ahura bərəzanta
aiθyejaṁha[1] ašavana yazamaide;
strə̄ušča må̄ṁhəmča hvarəča
urvarå[2] paiti barəsmanyå[2]
miθrəm vīspanąm daȟyunąm
daiṅhupaitīm yazamaide.
ahe raya...tåsčā yazamaide (=4-6).

strong, all-knowing, undeceivable master of ten thousand spies!'¹¹ (*Here repeat 4–6.*)

SECTION 35

144. Grass-land magnate Mithra we worship…(=7)…. We worship Mithra when he faces the country, we worship Mithra when he is between (two) countries, we worship Mithra when he is inside the country, we worship Mithra when he is above the country, we worship Mithra when he is below the country, we worship Mithra when he makes the round of the country, we worship Mithra when he is behind the country.¹

145. (Standing) by the Barsman plant we worship Mithra and Ahura—the two exalted owners of Truth that are removed from danger—, as well as the stars, the moon, and the sun. We worship Mithra, who in (*lit.* of) all countries is the head of the country.¹ (*Here repeat 4–6.*)²

COMMENTARY

COMMENTARY

1¹. *vouru.gaoyaoiti*-. Lit. 'having, or providing, wide cattle-pastures'. Above, p. 43, we have surmised that this epithet was not given to Mithra in consequence of his care for the cow, but is on the contrary the source of the Avestan conception of Mithra as protector of the cow. On the etymology of *gaoyaoiti*- see H. W. Bailey, *TPS*, 1954, 138.

The somewhat uncouth translation 'grass-land magnate' has been chosen mainly because of its comparative brevity. That the unusual length and heaviness of this constant epithet of Mithra became in time inconvenient even to ardent devotees of the god can be inferred from the epithet *Cautes* of the Roman Mithras, which in my opinion is the result of a haplological reduction of OIr. *gauyauti- to *gauyti, with subsequent disappearance of *y*, and with initial *k* as in Gr. καννάκη(ς) beside γαυνάκης < OIr. *gaunaka- (cf. Lüders, *APAW*, 1936, No. 3, 7 *sqq.*).

Cautes and *Cautopates* are the names of the two torch-bearers who flank Mithras tauroctonos on nearly all reliefs on which the immolation is portrayed. The torch *Cautes* holds is turned upwards, that of *Cautopates* downwards. Since the two torch-bearers are exact small-size replicas of the tauroctonous god, and the names of either occur in inscriptions as attributes of Mithras (see Cumont, *TMMM*, I, 203 *sq.*, 208), we are justified in tracing the two names to ancient epithets of the god. In Cautopates I see the reflex of an Iranian *gauyauti-pati- 'grass-land chief', synonymous with *varu-gauyauti-.

Non-Iranian worshippers of Mithra in taking over the two epithets would be aware, even without understanding their meaning, that they consisted of an essential element **gau(yau)ti*, to which two different elements had been respectively prefixed and added. Being indifferent to meaning, they may well have reduced **varu-gau(yau)ti*-, which had virtually become a name of the god, to its main element, following a common hypocoristic procedure.† They would be less inclined to alter the structure of **gau(yau)ti-pati*-, which agreed with the pattern of Iranian names ending, in their Greek garb, in -πάτης; but they did grecize its compound-vowel, probably on the analogy of names beginning with Μιθρο.‡

---

† Greek examples of so-called 'hypocoristica *a posteriori*' are Νοῦς (cf. Εὔ-νους), Δῶρος (cf. Ἀντί-δωρος), Δῆμος (cf. Φιλό-δημος); among divine names cf. Αὐλίς: Ἀγρ-αυλίς, Ἀφρόδιτος: Ἑρμ-αφρόδιτος, Νόστος: Εὔ-νοστος; see F. Bechtel and A. Fick, *Die Griechischen Personennamen*, pp. 16, 445, and *passim*. For Iranian examples cf. Justi, *Iranisches Namenbuch*, p. viii, top.

‡ Cf. ο instead of ι in Σμερδομένης, corresponding to Av. **bərəzi.manah*-.

1.1–2.1]   COMMENTARY

That the function of the two torch-bearers on the reliefs need not be related to their names was already seen by Cumont, *op. cit.* 206 *sqq.*, who pointed out, on the one hand, that two δᾳδοφόροι flanking a central scene constitute a motif known from Greek and Etruscan art, and, on the other hand, that the representation of two paredroi accompanying a divinity is probably of Babylonian origin.

What then is likely to have happened is that, once in accordance with conventional figurative art two torch-bearing paredroi had been placed to the right and left of bull-killing Mithras, the desire was felt to give them names. If these were chosen by non-Iranian worshippers of Mithra we must assume that not only *(varu-)gauyauti-* (Av. *vouru.gaoyaoiti-*), but also the synonymous **gauyauti-pati-* reconstructed by us, were common epithets of the god in non-Iranian Mithraic liturgy, even before they were put to use as proper names of the δᾳδοφόροι. If, however, the name-givers of the δᾳδοφόροι were Iranians, it would not be surprising if the name **gauyauti-pati-* had been invented by them for the special purpose of providing a correlative (and synonymous) name to **varu-gauyauti-*, the name they had assigned to one of the two paredroi.†

1². *avā̊ntəm*, which everybody translates by 'so great, such', is according to Bth., *Wb.* 952, from **ā-bānt-* 'shining (like) = similar, such'. But the word looks so much like what it ought to be, viz. the acc. of *avant-* '*tantus*', that it is inadvisable to look for a different explanation, even though we should have expected to find **avantəm*, to judge by the accusatives *avavantəm* and *čvantəm*. With an adjective of size it is permissible to think of the analogical influence of *mazā̊ntəm* 'large'.

2¹. It is clear by now that in Av. *mairya-* two words have converged: (1) an adj. (fem. *mairyā-*) meaning 'deceitful', which belongs to Arm. *mel* (etc.), as Bth. claimed; and (2) the 'daēvic' noun which corresponds to 'ahuric' *nar-* 'man', and is parallel to the *jahikā-*, *jahī-* (the 'daēvic' opposite of *nāirikā-* 'married woman' and *kainyā-* 'virgin'), with whom the *mairya-* is coupled in *Yt* 8.59 and elsewhere. To (1) belongs, as Morg. has recognized (*NTS*, XII, 262), Pš. *əzmār* 'cheater' < **uz-mara-*. (2) is simply the Vedic *márya* 'young man', turned to 'daēvic' use in Avestan, as Geldner pointed out in 1904 (*GIP*, II, 53).

† Other suggestions on the origin of the names *Cautes* and *Cautopates*, to which H. W. Bailey kindly drew my attention, have no basis in reality: Louis H. Gray, *Muséon*, 1915, 189 *sqq.*, connected *Caut-* with Gr. καίω, and understood the names as meaning 'the burning' and 'he who drops the burning (torch)'; H. H. Schaeder, *ZDMG*, 1928, XCVI, imagined that Turk. *qut*, which translates MIr. derivatives of OIr. *x*ᵛ*arənah-* (cf. Bailey, *Zor. Prob.* 54 *sq.*, 226 *sq.*), goes back to a MIr. word derived from OIr. **kauta-*, an unexplained synonym of *x*ᵛ*arənah-*; hence *Cautes* < **kauta-dā-* was for him identical with Mithra's epithet *x*ᵛ*arənō.dā-* in *Yt* 10.16. I. Scheftelowitz, *Acta Orientalia*, XI, 330, n., interpreted *Cautes* as 'plenty' (comparing Arm. *koit* (etc.) 'heap'), and *Cautopates* as 'decline of plenty'.

152

The related OPers. *marīka-* in *NRb* was understood as 'servant' by the Akkadian translator, while the Elamite version was said by Hzf., *Ap. I.* 252, to have the word for 'son' in its place. For the semantics a parallel is found in French *garçon*, as Benv., *TPS*, 1945, 44, aptly remarked. In Old Persian no 'daēvic' meaning is discernible, only a depreciation (probably connected with the suffix), which eventually earned the Ormuri descendant of the word, *mrīg*, the meaning 'slave'.

In Pahl. *mērak*, Yazdī *mīrā*, etc., 'husband', OIr. *marya(ka)-* survives as ordinary, non-daēvic synonym of *nar-*; cf. Andreas, *Iranische Dialektaufzeichnungen*, 97; Christensen, *Contributions*, I, 239; Bth., *Zum sasanidischen Recht*, I, 37. [See Addenda.]

2^2. *miθrō.drug-*. Cf., with Henning, Parth. *drwxtmyhr* 'committing a breach of contract', *BSOS*, IX, 82. In Pahlavi, *miθrāndružān* 'covenant-breakers' occurs; see Bailey, *BSOS*, VI, 55, §3.

2^3. According to the definition of the function of *-van-* as given by Wn.-Debr. II, 2, 900, *ašavan-* ought to mean 'provided with Truth', or 'exercising Truth'. In recent years, however, the idea has gained ground that *ašavan-* and its OPers. (*artāvan-*) and OInd. (*r̥távan*) equivalents mean 'owning, possessing, Truth'. Thus Hzf., *Ap. I.* 290, translates OPers. *artāvan-* by 'das *'rtam* erlangt habend', though Gāthic *ašavan-* to him meant 'Anhänger des *'rtam*'.† Bailey, *Zor. Prob.* 87, renders Av. *nā ašavan-* by 'the man holding right'. For Ved. *r̥távan* Lüders, *Varuṇa*, I, 15, has 'der Wahrheit besitzende'.

The main reason for adopting this translation is Xerxes' oft-quoted statement in his *Daiva* Inscription, 51 *sqq.*: 'the man who conforms to the law which Ahuramazdāh has commanded, and worships Ahuramazdāh in proper style in accord with *Arta*, becomes happy while living, and *artāvan* when dead'. Here obviously *Arta*, viz. Truth,‡ is something to cultivate in life, as well as a reward after death, and the *artāvan* is one who has obtained, or owns, this reward. Both Hzf. and Bailey (see above) have noticed that in MIr. Zoroastrian literature the term *ahrav* (< *ašavan-*) is sometimes specially connected with the souls of the dead.

As pointed out in *BSOAS*, XVII, 483, the later Avesta contains a passage (*Y* 16.7) which is almost a commentary to Xerxes' statement:

x^vanvaitīš ašahe vərəzō yazamaide
yāhu iristanąm urvąnō šāyenti

† In *Zoroaster*, II, 433, Hzf. translated *miθra ahura...ašavana* (st. 145 of our hymn) by 'M.-A., the two...just ones'.

‡ That OPers. *arta-*, Av. *aša-*, Ved. *r̥tá* mean 'Truth' is happily beyond doubt. For Av. *aša-* this has long been known from the testimony of Plutarch and Al-Beruni; cf. Lo., *Rel.* 42 *sq.*, Hzf., *Ap. I.* 288. Conclusive proof that 'Truth' is also the meaning of Ved. *r̥tá* was offered by Lüders in *Varuṇa*, I, cf. above, Introduction, p. 6. The meaning 'oath' of Oss. *ard* < (*a*)*rta-* provides valuable confirmation of the ancient connotations of *r̥tá* as established by Lüders, see Bailey, *ap.* E. B. Ceadel, *Literatures of the East*, 101.

yā̊ ašāunąm fravašayō;
vahištəm ahūm ašaonąm yazamaide
raočaŋhəm vīspō.x^vāθrəm.

'The radiant quarters of *Aša* we worship, where dwell the souls of the dead, the Fravašis of the *ašavan*-s; the Best Existence of the *ašavan*-s we worship, (which is) light (and) affording all comforts.' *vahišta- ahu-*, the 'Best Existence', is one of the Av. idioms for 'Paradise'; to it go back Sogd. *wyštm'x*, MPers. *whyšt('w)*, NPers. *bihišt*, all meaning 'Paradise'. The idea that the Paradise is the seat of *Aša* must be of Indo-Iranian origin, since in the Vedas the 'seat of *R̥tá*' is the abode of the gods, occupying the highest position in the third sky, beyond the visible world (see Lüders, *Varuṇa*, I, 25 *sqq.*).

The inhabitants of the 'quarters of *Aša*', the souls of the dead, are called *artāvan*-s by Xerxes, but Fravašis (i.e., approximately, 'souls') of *ašavan*-s by the author of *Y* 16.7. This is at first sight surprising, because there are, according to *Visp.* 11.7, Fravašis 'of deceased *ašavan*-s, of living *ašavan*-s, and of men not yet born'. All these Fravašis are themselves in *Visp.* 11.7 said to be *ašavan* (*ašaonibyō fravašibyō*). If there are Fravašis of living *ašavan*-s the wording *yā̊†* *ašāunąm fravašayō* in *Y* 16.7 cannot be meant as an alternative description of *iristanąm urvąnō*, but must be a restrictive qualification: only such souls of the dead as are Fravašis of *ašavan*-s. Actually there are no Fravašis except of *ašavan*-s, cf. Lo., *Yäšt's*, 104 *sq.*, and we may take it for granted that every *ašavan*, alive or dead, has his Fravaši. The Avesta normally speaks of *ašāunąm fravašayō* without distinguishing between live and dead *ašavan*-s. So it happens that we are not told where the home (*vis-*, cf. *Yt* 13.49) of the Fravašis of live *ašavan*-s is to be found. It does not seem far-fetched to assume that it, too, is in the 'radiant quarters of *Aša*'.

We may now say that the apparent difference between Xerxes' use of *artāvan* to denote a blessed dead, and the Avestan practice of applying the term *ašavan* also (and mainly) to righteous men alive, corresponds to the Avestan distinction between Fravašis of dead and live *ašavan*-s. The latter distinction is real enough, for in *Yt* 13.17 we are told that the Fravašis of men alive that are *ašavan* are stronger than the Fravašis of the dead. In what must have mattered most to a Zoroastrian, the share a man's soul has of Truth, the Fravaši of an *ašavan* alive does not appear to be worse off than when he is dead. At all stages, we may take it, she dwells in, or has access to, the 'radiant quarters of Truth'. The Zoroastrian point of view may then be put as follows:

There are, and have been from the beginning, countless Fravašis in the other world, who are all, by definition, Fravašis of potential (because not yet born) or existing (dead or alive) *ašavan*-s. A man who in the great

† *yā̊* here functions as article, cf. Bailey, *TPS*, 1945, 18 *sq.*, Bth., *Wb.* 1225 *sq.*

choice between Truth (*Aša*) and Falsehood (*Drug*) has opted for Falsehood remains without a Fravaši. He who chooses Truth acquires a Fravaši; through her he becomes *ašavan*, 'owner of Truth', even while he is alive, because she, being herself *ašavan* (as stated in *Visp.* 11.7), partakes all along of Aša in the world above. His 'owning Truth', or having a Fravaši who 'owns Truth' (the two amount to the same thing), is the reward of having chosen Truth and adhering to it. Yet while he is alive the blessings of this reward are only to be discovered in the happiness derived from confidence in the future: he knows that once he sheds his body he will be largely reduced to his Fravaši, and as such will gain immediate experience of the 'radiant dwellings of Aša', the 'best existence' for an *ašavan*.

If we are right in saying that the writer of *Y* 16.7 would have admitted to the 'radiant quarters of Aša' also the Fravašis of live *ašavan*-s, then we need not doubt that for Xerxes, too, the man who is 'happy' as a result of his respect for Truth and Ahura Mazdāh, is *artāvan* even in his lifetime. He will not, however, reap the full advantage of 'ownership of Truth' until he is dead; for in life he is liable to suffer illness and misfortune, when his 'happiness' will be merely anticipatory. Accordingly the 'ownership of Truth', though a fact already in life, is fittingly held out as something which will be fully enjoyed in after-life, and is to be happily anticipated in the mean time. Xerxes does not contradict the usual Zoroastrian view of the meaning of *ašavan*, he only fails to state it completely.

There is thus no need to postulate two words *artāvan*, as Hzf. did, *loc. cit.*, one from *arta-* 'Truth', the other from **arta-* 'struggle'. Nor is there reason to doubt with Duch., *Zoroastre*, 130 *sq.*, whether the meaning 'bienheureux', which *artāvan* has in the *Daiva* Inscription 'relève de la prédication gâthique'. Zoroaster, who calls his living followers *ašavan*, would certainly not deprive them of this title when dead. On the other hand, even a *restriction* to the dead would be no more than a logical development arising from Zoroaster's doctrine, on the lines explained above. Such restriction is attested for a later period in Hesychius' gloss ἀρταῖοι· ἥρωες παρὰ Πέρσαις (see Hartmann, *OLZ*, 1937, 154), and the statement found in Stephanos of Byzantium: ἀρταίους δὲ Πέρσαι, ὥσπερ οἱ Ἕλληνες τοὺς παλαιοὺς ἀνθρώπους, ἥρωας καλοῦσι (cf. Hzf., *Ap. I.* 290). Moreover, in *BSOAS*, XVII, 483 *sqq.*, we have advanced the theory that the Ossetic spirits called (*i*)*dauæg* are descendants of ancient *artāvan*-s. Xerxes' statement is a presage of such a development. But to infer from it that *artāvan* was restricted to the dead in Xerxes' own terminology would be as risky as it would be wrong to infer from *Y* 16.7 that there are only Fravašis of *dead* 'owners of Truth' in the religion of the Younger Avesta. [See Addenda.]

Below, in note 103[1], a suggested restoration of the end of the *Beh.*

inscription will be found, according to which Darius affirms that 'whoso worships Ahuramazdāh, Truth will always be his, both (while he is) alive, and (after he is) dead'. Although of the word for Truth, *artam*, only the final -*m* is visible, there is everything to be said for this restoration, from epigraphical, syntactical, and contextual considerations. In view of Xerxes' general dependence in wording and ideas on the Darius inscriptions, it is only to be expected that the son's intentions in *Daiva* Inscr. 51 were very similar to the father's in *Beh.* v, 35. The father's position is one we would not hesitate to attribute also to Zarathuštra: the worshipper of Ahura Mazdāh is *ašavan*, 'owner of Truth', in both lives. The son's position is the same, but his statement is framed with an eye to reality: the 'ownership of Truth', being intangible, had better not be mentioned as something tangible, was presumably the advice Xerxes took, or gave, in this matter.

If Xerxes' wording compels us to translate *artāvan-* by 'owning Truth' when the dead are meant, Darius' wording, in the restoration suggested, indicates that in both lives 'being *ašavan*' implies 'possession of Truth'. Even if the *Beh.* passage is disregarded, the Avestan evidence we have examined points to the likelihood that to the Zoroastrians who wrote the Avesta, as well as to Xerxes himself, *ašavan-/artāvan-* meant what our restoration makes it mean to Darius.

It is noteworthy that while Av. *ašavan-*, OPers. *artāvan-*, go back to Indo-Iran. *$r̥tāvan-$, witness Ved. *r̥távan*, the Av. term *dragvant-* seems to be an Iranian invention. It may well be that it was Zarathuštra who created the term to describe the man who, in contrast with the 'owner of Truth', 'owned, had come into possession of, Falsehood' through making the wrong choice. The new word would be apt to transgress the bounds of any such definition imposed on it by its inventor, because of an inherent contradiction: in the eyes of the *ašavan* 'possession of Falsehood' was a sin and disadvantage, but he could not overlook that the *dragvant* himself would consider it a virtue and advantage. Accordingly, unsophisticated Zoroastrians, rather than interpret the word in a way that would flatter the *dragvant*, will soon have begun to attribute to it the simple meaning 'lying, false, morally wrong', which its Pašto derivative *daryal* still has (see Morg., *EVP*, 22). In its turn *dragvant-* would now influence the meaning of *ašavan-*, to which it stood in constant opposition, and turn it into an adjective generally denoting what is morally right. By making this detour we can understand why later Iranian derivatives of *$(a)rtāvan-$ mean 'just, right, righteous', as B. Geiger pointed out in *WZKM*, 41 (1934), 110 *sqq.*, without having to assume with Geiger that *aša-/arta-/r̥tá* itself meant 'justice, right'. The Av. word for 'justice' is *arštāt-*, see below, note 139².

2⁴. *kayaδa-*, translated approximately, in the light of Henning's remarks in *BSOAS*, IX, 91 *sq.* The related Av. *kaēta-*, MPers. *kēδ*

COMMENTARY   [2.4–3.3

'astrologer', has been compared with Sogd. *cytk* 'spirit, genius', see Gauthiot, *Grammaire Sogdienne*, 97, Benv., *TSP*, 174.401; it is found in Chinese transcription (*ki-to*) as the name of the heretics who adored the god Žun in Zabulistan, see Bailey, *TPS*, 1955, 64. A. Hillebrandt, *Vedische Mythologie*, 2nd ed., II, 430 *sq.*, suspected that OInd. *kāyādhavo* in TBr. I, 5, 9, 1 (commentary *kayadhu*) is connected with *kayaδa-*.

2⁵. Bth. summed up the general opinion on *xᵛādaēna-* by explaining the compound as 'who has the own (viz. the same as you and I) religion, co-religionist', and Duch., *Comp.* 169, followed him. However, if *xᵛā-* here had the same value as in *xᵛā.aoθra-*, *xᵛā.zaēna-*, etc., the compound could only mean 'who has his own religion', the point of *xᵛā-* 'own' being naturally that the religion in question is not someone else's.† Clearly *xᵛā-* here means 'good', as in Av. *hvā-yaona-*, *xᵛā-paθana-* 'offering good ways', which together with *hvō.yžaθa-* 'where it is good to swim', etc., have not been satisfactorily explained by Duch., *Comp.* 26 *sq.* The forms with *hvō-* (< **hva-*), *xᵛa-*, are parallel to Gr. ἔκηλος, Ved. *svá-yukta* 'well-yoked', etc., see E. Fraenkel, *Lexis*, III, 66, F. B. J. Kuiper, Νώροπι χαλκῷ (*Mededelingen der Koninklijke Nederlandse Akademie van Wetenschappen*, 14, No. 5 (1951)), p. 23, n. 2. Whenever the second member of the compound began with *u̯-*, *hva-* was liable to be lengthened to *hvā-*, see Wn., I, 46 *sq.*, II, 1, 130, Duch., *Comp.* 11;‡ examples are *hvā-vaēya-* 'whose impact is good', and, with *hva-* 'own', *hvā-vastra-* 'who has his own dress'. From such compounds, where the long vowel was justified, it was occasionally transferred analogically to a position before consonants other than *u̯*, and even vowels.

3¹. *razištəm pantąm*. On the 'straightest path (of Truth)' in Indo-Iranian religious thought, cf. B. Geiger, *Die Aməša Spəntas*, 190, 195 *sq.*§

3². The translation of *spənta-* by 'incremental' is an attempt to reproduce the Pahl. transl. *aβzōnīk* 'whose activity results in *aβzūt*, increase'; see Bailey, *BSOS*, VII, 294, who has settled, *ibid.* 284 *sqq.* and *BSOS*, VIII, 142, the etymology and meaning of *spənta-* and cognates. In *aməšā̊ spənta*, the collective name of the Entities (*hātąm*, see below, note 6⁴) Good Mind, Truth, Power, etc., *spənta* is the postponed adjective, the order of words being in accordance with the Iranian practice to which attention was drawn in *GMS*, p. 252; they are the 'incremental Immortals', as Bth. had seen, not the 'immortal Incrementals'.

3³. *daδāiti*, sing. instead of plur., under the influence of the two preceding *daδāiti*. Htl.'s solution, to take Mithra as subject and *fravašayō* as object of *daδāiti*, is not to be recommended.

---

† 'Co-religionist' in Avestan is *hāmō.daēna-*.

‡ To Av. *haiθyā varəz-*, etc., there quoted, one may add Sogd. *qš'-wrzyy*, NPers. *kašā-varz* 'tilling, cultivator', *TPS*, 1945, 141, whose *ā* is best explained as due to OIr. lengthening of the thematic vowel of *karša-*. Cf. also *hāvišta-*, note 116⁴ below.   § [See Addenda.]

157

3⁴. *āsna-* 'noble', see Bailey, *BSOS*, VI, 95, *Zor. Prob.* 10, n., to which add Aram. *'zt* 'free', *JRAS*, 1954, 126.

3⁵. *frazainti-*, cf. Sogd. βzyn, *GMS*, p. 249.

5¹. *vərəθrayna-*, which as a common noun we shall henceforward render by 'victoriousness', seems to mean 'ability to defeat *vərəθra-*', rather than merely 'defeat of *vərəθra-*', since in st. 16 it is something that can be 'increased'. Whether this *vərəθra-* was originally a form of 'opposition', 'resistance', or 'bulwark', or a dragon, or a dragon named 'warder off', does not here concern us; cf. Benv. and Renou, *Vṛtra*, *passim*, Lo.'s discussion of their views in *Der arische Kriegsgott*, 46 *sqq.*, and Bailey, *JRAS*, 1953, 116. From 'smashing (whatever form of) opposition' to 'victory' there is only one small step, which was taken by the Pahlavi translators, if not already by Avestan writers, and again by Bth. For *vərəθrăgan-* (of which *vərəθra.taurvan-* is a calque), *vərəθrəm.-gan-*, ²*vərəθraynya-*, *vārəθrayni-*, and **vārəθraynya-*, the translation 'victorious' satisfies all requirements of context.

Much ingenuity has been spent in trying to clarify the relation between the first component of *vərəθra-yna-* (etc.) and the simple ¹*vərəθra-*, for both of which the Pahl. translators used the same word *pērōzkarīh* 'victory'. But the difficulties remain. Bth. had to attribute to ¹*vərəθra-* the meaning 'attack' in order to be able to analyse *vərəθrayna-* as 'smashing the attack', and for the sake of one passage (*Yt* 13.38, quoted in note 38¹ below); in the other passages, however, he used for ¹*vərəθra-* 'victorious attack, victory'. He did not attempt to explain how the 'attack', permanently beaten in *vərəθrayna-*, came to be 'victorious'. Benv. assumed that ¹*vərəθra-* meant 'defence, resistance', yet he also admitted a derivative meaning, that of 'hostility' (*op. cit.* 10 *sq.*). Again the semantic development arouses misgivings: one does not see why the ancient Iranians should have considered the defence against an aggressor, rather than the latter's attack, an act of hostility.

But the chief objection to Benv.'s theory is that considerable persuasion is required to make credible his interpretation of ¹*vərəθra-* and its derivatives *vərəθravan-* and *vərəθravastəma-*. ¹*vərəθra-* is said to be *amaēniγnəm* in *Yt* 19.54. Bth. quite convincingly translated 'beim Angriff niederschlagend', presumably meaning 'en livrant l'assaut' (Benv., *op. cit.* 11). Benv. is forced by his interpretation of ¹*vərəθra-* to add an unusual limitation: 'qui brise (l'ennemi) dans *son* (= the enemy's) assaut' (my italics). Which other Avestan compound requires such a specification? On p. 12 the translation is simplified: 'qui brise l'assaut'; this is the meaning Benv. requires, but it would surely have been expressed by **amō.niγnəm*.

Nowhere is the defensive ability of Mithra stressed in the Avesta; this is not surprising, since there are no forces that have even the remotest chance of prevailing against the 'strongest of gods', the 'strongest of the strong' (cf. sts. 135, 141). Yet vərəθravant- applied to Mithra (st. 141) is said by Benv. to mean 'armé pour la défense' (*op. cit.* 14). Θraētaona = Firēdūn, a most enterprising hero, is at this rate described as 'the most defensive of defensive men' (*mašyąnąm vərəθravanąm vərəθravastəmō*, *Yt* 19.36).

The faithful who is most given to praying is called *vərəθra vərəθravastəmō* 'most defensive in defence', because *mąθrō spəntō mainyəvīm drujəm nižbairištō* (*Yt* 11.3), according to Benv. 'le Manθra Spanta repousse le mieux la Druj de l'air: *nižbara-* "qui repousse" reprend l'idée de défense contenue dans *vərəθra-*'. In fact *nižbar-*, which is frequently attested, only means 'to take out, remove, carry away', is so translated in Pahlavi, and continues to this day with the same meaning in Yaɣn. *živar-* 'herausbringen'.†

It cannot be said, either, that *vərəθra.baoδō* in *Yt* 13.46 offers a good case for 'defence':

Yt 13.46, (*a*)   *yaṯ hīš antarə vātō fravāiti*
                   barō.baoδō mašyąnąm
                   tē narō paiti.zānənti
                   yāhva vərəθra.baoδō;

         (*b*)   *tē ābyō frərətå frərənvainti*
                   ašaonąm vaŋuhibyō sūrābyō
                   spəntābyō fravašibyō
                   θaxtayāṯ parō aŋhuyāṯ
                   uzgərəwyāṯ parō bāzuwe.
Yt 13.47   *yatāra vā.dīš paurva frāyazənte*

                   . . . .
                   ātaraθra fraorisinti
                   uɣrå ašāunąm fravašayō
                   haθra miθrāča rašnuča
                   uɣrača dāmōiš upamana
                   haθra vāta vərəθrājana.
Yt 13.48   *tå dainhāvō haθra jatå nijaɣnənte*
                   etc.

According to M. Benv. (p. 19) st. 46 means: 'Quand parmi elles souffle un vent qui porte conscience d'elles aux hommes, c'est à elles que les guerriers font fête, ⟨elles en qui ils sentent la défense,⟩ à elles qu'ils offrent des offrandes, avant que l'épée soit tirée, avant que le bras se lève.'

There are three objections to this translation of (*a*). First, although Benv. on p. 16 rightly assumes that the stem is *vərəθra.baoδa-* (Bth., and Duch., *Comp.* 136, have °*baoδah-*), parallel to *barō.baoδa-*, he translates both compounds as if they contained *baoδah-* 'perception', a word which in the Avesta is carefully distinguished from *baoδa-* 'smell'. Gram-

† Quoted without explanation by W. Geiger, *GIP*, 337, §79*d*.

matically, since the text has *vātō barō.baoδō*, and not °*baoδå*, the obvious translation is 'wind carrying smell'; logically this is only to be expected. *vərəθra.baoδō*, as a bahuvrīhi, may then also agree with *vātō*.

Secondly, to translate *patizan-* by 'faire fête' means an unnecessary extension of the departure from the original meaning which Bth. had permitted himself. In all the passages quoted *Wb.* 1659 under *paiti* (1), the meaning 'acknowledge(d)' does full justice to the context; it represents a technical, religious, use of a verb which, as shown by Sogd. *ptz'n-*, Khot. *paysān-*, Pš. *pežandəl*, otherwise means 'to recognize'. It is possible, but not essential, to find the non-technical meaning in the present passage.

Lastly, tangible proof is required before we can accept M. Benv.'s dismissal of *yāhva vərəθrō.baoδō* as an interpolation (*op. cit.* 19). Now that we realize, thanks to Henning, *TPS*, 1942, 52 *sqq.*, that the metre of the Yašts is accentual, we see that metrically the two words fit in extremely well. As to the 'suite des idées', neither the previous translations to which Benv. took exception, nor, after what has been said, his own, carry much conviction, and this despite their all being based on one or more alterations of the text, devised to make them acceptable. In such circumstances it is safer to adhere strictly to the text and accept whatever meaning results from it, even if this implies a somewhat harsh construction.

As I see the situation, there is a contrast between the general smell of humanity (*mašyānąm*), and the smell of warriors (*narō*) which is characterized by *vərəθra-*. The Fravašis are distributed everywhere, the wind is sometimes here, sometimes there, blowing from various directions. Those Fravašis among whom a whiff of wind carries a smell of *vərəθra-* recognize by it the warriors, because these are *vərəθravan(t)-*. Hence, instead of saying 'when among them there blows a wind bearing the smell of men,† then they (the Fravašis) recognize those who are warriors if it (the wind) smells of (the warriors') *vərəθra-*', the author could say '...then (those) among whom it (is) *vərəθra-*smelling recognize (or acknowledge) the‡ warriors'.

The harshness of the construction lies in the absence of a dem. pron.

† For the gen. *mašyānam* depending on °*baoδō*, the second member of a verbal compound, cf. Wn., II, 1, 34(*f*). The gen. depends on the first member of synthetic compounds in *kamərəδō janəm daēvanąm* (st. 26, cf. sts. 128 *sqq.*), *sāθrasčit̰...kamərəδō.janō* (sts. 109, 111, cf. note 109[6]), and *vahišta.nasəm vahištahe aṅhəuš, Vend.* 18.6, 'causing to attain the best thing(s) of Paradise'; of determinative compounds in *marəyahe yat̰ parō darəšahe tanu.mazō, Vend.* 18.29, 'size of a cock's body', *hū frašmō.dāiti-* (see below, note 88[1]), and *dānunąm baēvarə paitinąm* (on which see note 38[1] at the end). The last type is common in Old Indian, see Wn., II, 1, 31 *sq.* This interesting behaviour of Avestan compounds is ignored by Duch. in his book *Comp.*

‡ For the use of *tē* as an article see Bth., *Wb.* 621, lines 7–26.

COMMENTARY [5.1

tå which could serve as antecedent to *yāhva*. In the oblique, such demonstrative antecedents are often omitted in Avestan;† the present extreme case of a suppressed nominative is mitigated by the fact that the subject of the plural *paiti.zānənti* could easily be understood from the preceding sentence in Yt 13.45 (*yå [fravašayō] pərətənte*).‡ The relative, instead of the hypothetic clause, may be due to the wish to avoid introducing another conjunction beside *yaṯ*. It is true that Benv. thought of forestalling interpretations in which the Fravašis are taken as subject of *paiti.zānənti*, by arguing that being goddesses they had no need of wind to identify humans (*op. cit.* 17 *sq.*). The point, however, is that because they are goddesses they are able to analyse the smells carried by wind to an extent denied to human beings.

A word must be added on the (*b*) sentence of Yt 13.46, which M. Benv. is scarcely right in connecting closely with what precedes. This sentence to my mind introduces a new trend of thought, which is developed in Yt 13.47, where the subject *yatāra* refers to the warriors of each of two countries (as can be seen from the parallel st. 9 of the Mithra Yašt, and from the reference to 'countries' at the beginning of Yt 13.48): warriors normally propitiate the Fravašis before engaging in combat; if, however, they belong to two countries at war with each other, then the Fravašis lend their support to that side which is first in performing the propitiation.

If the above is the correct interpretation of Yt 13.46, its importance for the definition of ¹*vərəθra*- will not be missed. Here *vərəθra*- is no longer an undefined item in an enumeration of qualities: it is singled out as the one quality of warriors whose smell suffices to make them recognizable to the Fravašis. Surely the quality selected will be the paramount requisite of a warrior. Is it then likely to be the passive attitude of 'defence, resistance', or even the somewhat more active one of 'warding off, restraining, repelling'? If one surveys the evidence arrayed by Benv., one cannot help noticing that the translation 'force de résistance' which he adopts for ¹*vərəθra*- would be much more convincing if the words 'de résistance' were dropped. There would then be no need to postulate a derivative meaning 'hostility', no need to assume that an otherwise not noticeable semi-passive attitude, a 'victoire négative' (*op. cit.* p. 12) was considered to be a great virtue of Mithra, Θraētaona, Vərəθraγna, the Fravašis, the Manθra Spənta, and Mithra's mace (*op. cit.* 14 *sq.*), no need

† Cf. st. 30: *tūm..nmānəm daδāhi..yasə.θwā..yazaite* 'you provide...the... house (of him) who worships you...'. Many such cases are quoted *Wb.* 1213–20; to these should be added Yt 13.18 (discussed in note 109⁶ below), Yt 6.5 (see below, p. 227, n. †), Yt 10.120 (see note 120¹), and Yt 10.143 (see note 143¹⁰).

‡ A somewhat doubtful alternative, which would remove the syntactical difficulty, is to follow Benv. in assuming that the text instead of the nom. and acc. pl. masc. *tē* originally had the nom. and acc. pl. fem. *tå* referring to the Fravašis. Such a replacement would have to be imputed to anticipatory analogical influence of the initial masc. *tē* of the (*b*) sentence.

to subject the epithet *amaē.niyna-* to undue pressure, no need to charge the Fravašis with a lopsided view of what should constitute a warrior's chief requisite.

M. Benv. admits that 'il arrive que le *vərəθra-* confère une force suffisante pour anéantir l'effort de l'ennemi. Celui qui en est pourvu doit alors triompher' (*op. cit.* 11). This, actually, does not only 'happen' here and there: it clearly is the case wherever the context gives an indication of the likely meaning of *vərəθra-*. 'Mais'—M. Benv. continues—'on doit se garder de croire que le sens du mot en soit modifié. Il s'agit toujours de la force de résistance...'. What is the reason for this warning? The more one scrutinizes M. Benv.'s most stimulating discussion, the more one gains the impression that the reason essentially lies in his derivation of ¹*vərəθra-* from the base *var-* 'to resist'. In these circumstances the temptation becomes irresistible to propose a different etymology which will relieve us of the necessity of imposing on *vərəθra(van[t])-* an uncongenial definition. A connection with Lat. *valeo*, etc., will not only sanction the meaning 'strength, strong', which eminently suits the context in each of the cases so far discussed, but will provide us with the last ingredient we need to complete the description of *vərəθra-*: *ahmāi tanvō drvatātəm, ahmāi tanvō vazdvarə, ahmāi tanvō vərəθrəm*, says *Y* 68.11, 'to him (you give) soundness of body, to him health† of body, to him *vərəθra-* of body'. What could here fit better than 'physical fitness', *valetudo*?

Considering that the words *valour, valiant*, belong to Lat. *valeo* 'to be strong, healthy, physically fit' we can contemplate for *vərəθra-* a range of meanings, 'physical fitness–strength–valour', with at least as much confidence as Benv.'s 'defence–hostility', or Bth.'s 'attack–victory'. It is true that ¹*vərəθra-* would thus be separated from Ved. *vṛtrá* and the first component of Av. *vərəθra-γna-*. But if the two words, though formally identical, do not agree in meaning, there is no need to derive them from the same base. The suffix *-θra-* is common enough to allow for the possibility of its having been added to more than one of the numerous IE *$\underline{u}$er/l-* bases.

The conclusion is that Av. ¹*vərəθra-* '*valetudo*', and the *vərəθra-* (=Ved. *vṛtrá*) contained in *vərəθraγna-*, are two different words. On the meaning of the latter cf. the beginning of this note; if the kind of opposition it denotes consists in mere 'resistance', then the relation of *vərəθra-* (=Ved. *vṛtrá*) to Bth.'s ²*vərəθra-* 'shield' will be that of a noun of action to a noun of instrument of one and the same base *var-*, as Benv. assumed. To the base of Lat. *valeo* we should further assign Av. *hąm.varəti-* (*hąm.vərəti-*) 'valour', as well as probably the Median masc. name Φραόρτης, OPers. *Fravarti-*, which should be kept distinct from the Av.

† Thus rightly Benv., *op. cit.* 7, n.

fem. nouns *fraorəti-* (to *var-* 'to choose') and *fravaši-* (either to *var-* 'to choose', cf. Lo.'s 'Heilküre', *Rel.* 160 *sq.*, or, with Bailey, *JRAS*, 1953, 110, to *var-* 'to cover').

5². Although I have more or less adhered to Bth.'s translation of *ašavasta-*, it looks as if this abstract has sometimes (or always?) the meaning of a collective noun (cf. the *-tāt-* abstracts mentioned in note 145¹, p. 298), like its Sogd. descendant (δyn) '*rṯwspy*' 'community (of the [Manichean] Church)' (cf. *GMS*, §470; Henning, *BBB*, pp. 120, 125). At least in the ancient Av. text (cf. *OIr. Lit.* §§28, 35) *Y* 68.4, *fradaθāi gaēθanąm havaŋhāi ašavastanąm* is best translated 'for the furthering of the material world, for the comfort of the communities of *ašavant-*s'.

5³. I accept Tedesco's explanation of *vīspəmāi* as a corruption of *vīspahmāi, pronominal dative, see *ZII*, 2, 42. Cf. the pronominal inflexion of Sogd. *wysp-*, *GMS*, §1214.

6¹. On *dāmōhu* see note 92⁷.

6². The text of the last seven lines of this stanza is identical with *Ny* 1.16, lines 7 *sqq*. *haomayō gava* is best explained as instr. of a compound *haomayō.gav-*, in which the first part is the compound form of **haomaya-* (from **haomya-* in accordance with Bth., *GIP*, I, 155.11) '*haoma-*ish'. The compound accordingly means '*haoma-*ish milk = milk with an admixture of *haoma-*'. More simply one might have said **haomō.- gav-* '*haoma-*milk'. A close parallel is provided by *aspya.payah-* 'mare-ish milk = mare's milk', which we find instead of the expected **aspa.- payah-*. These are cases of 'remplacement d'un composé par une locution à 1ᵉʳ t. en *-ya-*, et fixation de cette locution en un nouveau composé', Duch., *Comp.* 139. There is thus no need to fall back with Bth. on a strange adj. *haomayu-*, of which *haomayō* would be the loc. used instead of the instr. [See Addenda.]

6³. On the Av. forms of the word for 'tongue' see Benv.'s remarks in *Asiatica* (*Festschrift Weller*), 30 *sq.*

6⁴. There is no reason to doubt the explicit statement at the beginning of the Homily on the *yeŋhē hātąm* prayer (*Y* 21.1–2), that this prayer was composed by Zarathuštra himself:

(*a*) Y 21.1 *yesnīm vačō ašaonō zaraθuštrahe:*
 yeŋhē hātąm āaṯ yesnē paitī.

'(Homily on) the devotional utterance of Truth-owning Zarathuštra: *yeŋhē*, etc.'

In the Homily itself we read:

(*b*) Y 21.2 *yąŋhąm iδa ašaoninąm ārmaiti.paoiryanąm yasnəm para.- činasti yaθa vahməm aməšaēibyō; θrāyō ṯkaēša; vīspəm vačō yesnīm; čīm aoi yasnō? aməšə̄ spəntə̄ paiti yasnahe.*

6.4] COMMENTARY

'*yā̊ŋhąm*: here he (*scil*. Zarathuštra) teaches the worship of the female Truth-owners headed by Ārmaiti, because it is the prayer to the Immortals. Three points of doctrine (are involved).† The whole is a devotional utterance. To whom (is) the prayer (addressed)? To the incremental Immortals of (= mentioned in) the prayer.'

From this commentary we learn that in the prayer the words *hātąm*... *yā̊ŋhąm* 'entium... quarum', hence also *tåsčā* 'eas(que)', refer to the female Aməša Spəntas Ārmaiti, Amərətāt, and Haurvatāt. Inevitably, then, both *hātąm* and *tąsčā tåsčā* 'eosque easque' each refer to all the Aməša Spəntas together. From the question and answer at the end of (*b*) we learn that the prayer is dedicated to the Aməša Spəntas, who in the prayer are called 'Entities' (*hātąm*).‡ This is unmistakably confirmed by the last sentence of *Y* 4.25, which, as Geldner in his edition made clear both in a note and by the spacing of the printed lines, introduces the *yeŋhē hātąm* prayer of which *Y* 4.26 consists:

(*c*) *Y* 4.25 . . . .
     aməšā spəntā huxšaθrā
     huδā̊ŋhō yazamaide:
Y 4.26 *yeŋhē hātąm... tąsčā tåsčā yazamaide.*

It seems that because in Wo.'s translation the introductory sentence 'we worship the well-ruling beneficent Aməša Spəntas' appears to be the closing statement of *Y* 4.25, its significance for the identification of *hātąm* was overlooked. Hence even an Avestan expert like Tavadia (*Indo-Iranian Studies*, II, 122) could approve of Hertel's view that the *hātąm* of the prayer are 'human beings'.

So far we have gathered from (*b*) and (*c*) that the prayer is addressed to the Aməša Spəntas, from (*c*) that these are referred to as 'Entities, male and female' (*hātąm... tąsčā tåsčā*), and from (*a*) that this is Zarathuštra's own way of referring to the Aməša Spəntas. Before going on we may note that the gender distinction expressed by *tąsčā tåsčā* is confirmed by an early post-Zarathuštrian reference to the Aməša Spəntas, viz. *Y* 39.3: *spəntə̄ng aməšə̄ng... yōi vaŋhə̄uš ā manaŋhō šyeintī yå̄sčā ūitī* 'the incremental Immortals who consort with Good Mind, male and female ones alike'.

† Namely those commented upon in the Homily under the three headings *yeŋhē*, *hātąm*, and *yā̊ŋhąm*.
‡ This was correctly understood by Lo., *ZII*, 1, 16 *sqq*., whose interpretation of the *yeŋhe hātąm* prayer (cf. also *Yäšt's*, p. 12) I have largely adopted. Where I differ from Lo. is in his opinion that the prayer is a late 'Überarbeitung' of *Y* 51.22. What the two stanzas have in common is a close similarity of formulation and an affinity of thought that are at least as consistent with common authorship as with the kind of unintelligent plagiarism of which Lo. suspected the allegedly late compiler of the prayer. [See Addenda.]

Now, already Geldner in a note to Y 4.26 in his edition connected the *yeṅhē hātąm* prayer with the following Gāthic stanza:

(d) Y 51.22 *yehyā mōi ašāṯ hacā vahištəm yesnē paitī
vaēdā mazdå ahurō yōi åṅharəčā həntičā
tą̄ yazāi x᾽āiš nāmənīš pairičā yasāi vantā*

of which Lo., *ZII*, 1, 16 *sqq.*, offered what is evidently the correct interpretation: 'I know in whose (objective gen.) worship (consists) what in accordance with Truth is best for me: (it is) Ahura Mazdāh, and those who have been and are (viz. the 'Entities', the Aməša Spəntas). It is them I shall worship by their own names and approach with praise.' By an alternative wording (d) confirms that Zarathuštra thought of the Aməša Spəntas as 'Entities'. In addition, a new fact emerges: the Entities are no longer referred to as 'male' and 'female', but only as 'male' (*yōi, tą̄*). The significance of Zarathuštra's use of the masc. pronoun in this connection has not hitherto been realized. Since of the six Aməša Spəntas the last three are fem., the first three (Vohu Manah, Aša, Xšaθra) neuter, the obvious explanation of Zarathuštra's *yōi* is that the prophet was familiar with the masc. collective name of the Entities, *aməša spəntā*.

That the founder of Zoroastrianism should have known the term *aməša spəntā*, which to modern observers is one of the main features of the Zoroastrian religion, may seem natural enough, yet is far from being the general opinion. Thus Lo., *Rel.* 34, because the term *aməša spəntā* happens not to occur in the Gāthās, considers that a 'zusammenfassende Bezeichnung für diese sechs obersten Geister' was created only in YAv. times. (The significance as 'zusammenfassende Bezeichnung' of *hātąm*, and of *yōi åṅharəčā həntičā* in (d), the meaning of both of which Lo. was the first to recognize (in his earlier article quoted above), is thereby ignored.) More recently Duch., *Orm.* 61, wrote: 'Ceux-ci, dès le Yasna-aux-sept-chapitres, ont un terme qui les désigne: ce sont les "Bénéfiques Immortels" (*Amṛta Spanta*); en même temps, ils se matérialisent, au point de se répartir en mâles et femelles. Le polythéisme reprend ses droits.' In the light of the arguments we are here advancing, a different view should be taken of the origin of both the name Aməša Spəntā, and the 'materialistic' or 'polytheistic' character of the Entities, see Introduction, pp. 10 *sq.*, 46.

If now we inquire how it came to pass that the gender of the term *aməša spəntā* is masculine, when none of the six Entities of the group belongs to this gender, the obvious answer is that the YAv. reference to the group as 'the *seven* incremental Immortals' (*hapta aməšå spəntā*) represents the original state of affairs; beside the three neuter and three fem. Entities the group must have originally included one masc. Entity, and since this Entity determined the gender of the collective name *amaša spəntā*, it must have stood at the head of the group. One might be tempted to conclude that the masc. Entity in question was Mazdāh himself. However, Zarathuštra distinguishes Mazdāh from the Entities both in (*d*) and the *yeṅhē hātąm* prayer. This leaves only one candidate for the post of chief Incremental Immortal, viz. Mazdāh's deputy Spənta Mainyu, the Incremental Spirit. We have attempted in the Introduction, pp. 11 *sq*., to show how this conclusion may be fitted into the 'elemental' pattern which the Aməša Spəntas represent.

7¹. *vyāxana-*, translated in the light of Benv.'s remarks, *Vṛtra*, 44 *sq*.

7². °*vaēδayana-* is a vṛddhi form to **vi-dayana-*, see Bailey, *BSOS*, VII, 75, n. In st. 45 *vaēδayana-* is the place of outlook, the watch-post.

8¹. 'Battlefield' seems to be the meaning of *arəzah-*, as against *arəza-* 'battle'. *gatō arəzahe* in *Yt* 19.42 is probably gen. of a compound *gatō.arəza-*.

9¹. *fraorəṯ* 'avowedly, making a profession of faith in Mithra', with Gdn., *Lesebuch*, p. 19.51 ('sich öffentlich zu ihm bekennend'); cf. the meaning of ²*var- +frā*, and of *fraorəti-*.

9². *fraxšnin-*, according to Bth., following Gdn., '*providus*, solicitous, anxious'. I have adopted Htl.'s 'vorauserkennend', bearing in mind the function of *fra* in Parth. *frwyn-* 'to foresee'.

9³. To judge by the ending of *vərəθrajanō* one might translate: '... turns g.m. Mithra, at the same time (turn to it) the victorious winds, at the same time the Likeness of Ahura's creature'. Such interpretation, however, would run counter to the parallel passage *Yt* 13.47 (quoted above, p. 159), where the instr. sg. ending is assured.

9⁴. For *dāmōiš upamana-* Bth. offered no identification. Wi. translated 'Fluch des Weisen', Spi. 'Schwur des W.', Gdn. 'wächter der ordnung', Da. 'Imprécation du sage', Htl. 'Bestand (= die Erhaltung) der Spendung', Lo. 'Bannfluch des Schöpfers'. On the other hand, for *daēnayå māzdayasnōiš upamanəm* in st. 126 Wi. (followed by Spi.) has 'ein Gleichniß der mazdayasnischen Lehre', comparing Skt. *upamāna*, while the other translators make no difference between the two *upamana-*s, cf. note 126³. A step towards the identification of *dāmōiš upamana-* was taken by Nyberg, *Rel*. 76, who translates the name by '*dāmi* gleichend' or 'bei *dāmi* wohnend', and thinks that the divinity in question is the πάρεδρος of *dāmi-*. As to *dāmi-*, Nyberg suggests two alternative interpretations: either 'the creator', which epithet, he considers, may apply to various gods, according to the sect to which the worshipper belonged; or 'the blowing = wind god', a meaning not otherwise attested for *dāmi-*.

The meaning of *upamana-* can be approximated more closely than Nyberg suspected, by taking *daēnayå māzdayasnōiš upamanəm* 'the *u*. of the Mazdayasnian Religion' in st. 126 strictly at its face value, that is, as apposition of *razištąm čistąm*. After Benv.'s remarks on Čistā (*Vṛtra*, pp. 56–64) we may take it for granted that she is the goddess of the way.

COMMENTARY [9.4

The only point which invites disagreement is Benv.'s attempt to remove as a gloss the last four words of *Yt* 16.1, by which Razištā Čistā and Daēnā Māzdayasni are *identified*:

Yt 16.1   razištąm čistąm mazdaδātąm
            ašaonīm yazamaide,
            hupaθmainyąm hvaiwitačinąm
            . . . . . .
            mošu.kairyąm hvāyaonąm
            hvāyaozdąm yąm vaŋuhīm
            daēnąm māzdayasnīm.

'We worship Mazdāh-created Truth-owning Razištā Čistā, who offers good paths and good running, ...acting without delay, offering good course and good striving,† (who is) the good Mazdayasnian Religion.' M. Benv. himself notes several connections of Čistā and Daēnā on p. 63. In addition the Religion, just as Čistā, is said to take care of ways: *yeŋhe daēna māzdayasniš x^vīte paθō rāδaiti*, st. 68 of our hymn. Accordingly, there is every reason to accept as genuine the identification, clearly stated in *Yt* 16.1, of Čistā and Daēnā. Of course this identification can only be partial: Čistā is identical with the Religion only in so far as she is, according to st. 126, her *upamana-*.

What then is *upamana-*? The comparison of *Yt* 16.1 (*čistā-* = *daēnā-*) and *Yt* 10.126 (*čistā-* = *daēnayå upamana-*) invites the conclusion that *upamana-* means 'double, alter ego'. In *daēnayå upamana-* 'the one who is like the Religion', *upamana-* may be a masculine hypostasis of a neuter noun **upamana-* 'likeness', which would be, as Wi. guessed, the Av. correspondent of Skt. *upamāna*, with *ā* shortened before antevocalic *n* (cf. *GMS*, §122) as in *dəmana-, paitištana-, spanah-, (uz)uštana-, baēvarə.spasana-* (note 24⁵), *frayana-* (note 112³), etc. Such a hypostasis would be parallel to that which took place in Ved. *vṛtrá* and *mitrá*, cf. Benv. and Renou, *Vṛtra*, 28, 95. The use of the masc. *upamana-* in apposition to the fem. *čistā-* can be compared with the identical use of the Russian word for 'a double', двойник.

If *daenayå upamana-* can be said to be the Double or Likeness of the Religion, who is also known under the name of Čistā, one naturally asks whose double is *dāmōiš upamana-*, and what, if any, is his other name. We have seen Nyberg's answer to the first question. His reply to the

---

† I do not see how M. Benv. would justify phonologically his derivation of *yaoz-* in *hvāyaozda-* from the base *yuz-* (*op. cit.* 59, n. 2). Bth.'s derivation of °*yaozda-* from IE *ieudh-to-* is the only possible one. The meaning does not have to be the one to which Benv. objected, 'au bon combat'. We know from Parth. *ywdy-* and Oss. *udīn, uodun* that in Iranian *yud-* also means 'to exert oneself', see *BSOAS*, XIV, 487, n. 2. Lo.'s translation of *hvāyaozdąm* 'die...die gute mazdayasnische Religion wohl bereitet' goes against the normal behaviour of *-ta-* nouns.

second was to infer from the incarnation of both *dāmōiš upamana-* and Vərəθrayna in ferocious boars described in identical terms (*Yt* 10.70, *Yt* 14.15 on the one hand, *Yt* 10.127 on the other) that the two gods are themselves identical. This ill agrees with the data of our Hymn, according to which *dāmōiš upamana-* flies behind Mithra in sts. 68 and 127, whereas Vərəθrayna is said in st. 70 to precede him.† I should therefore give preference to a different solution, although it will leave our second question unanswered:

dāmi-, according to Bth., has two meanings, (1) 'creation', and (2) 'creator'. In the passages which Bth. quotes for meaning (2), the word is always in apposition to Ahura Mazdāh (*ašahyā °mīm* Y 31.8; *vaŋhəuš...°miš manaŋhō* Y 44.4; *°miš ašəm* Y 31.7), except in Y 34.10, where Ārmaiti is thus described (*ārmaitīm °mīm*); in this passage *dāmi-* is used absolutely, while as an apposition of Ahura Mazdāh it takes the acc. or objective gen. ('creator of Aša, of Vohu Manah'). Now, we know from Y 45.4 that Ārmaiti is Mazdāh's daughter:

ašāṯ hacā mazdā vaēdā yə̄ īm dāṯ
patarə̄m vaŋhəuš varəzayantō manaŋhō
aṯ hōi dugədā hušyaoθanā ārmaitiš

'I know, O Mazdāh, in accordance with Truth (him) who created it (viz. the afore-mentioned "best thing of this existence"), the father of activating Good Mind; and Ārmaiti whose deeds are good is his daughter.'‡ It is therefore possible that *ārmaitīm dāmīm* in Y 34.10 means 'Ārmaiti who originates from, or belongs to the Creator',§ *dāmi-* acting in this case as a kind of patronymic of *dāmi-* 'creator'. For an Av. vṛddhied patronymic of an *-i-* stem cf. *xštāvi-* from *xštavi-*; for the 'invisible vṛddhi' cf. *dāzgrāspi-*, *syāvaspi-*, see Horn, *KZ*, 38 (1905), 291, n. 1.

† It cannot be argued in favour of Vərəθrayna's and Dāmōiš Upamana's identity that the god who precedes Mithra's chariot in his westward drive will be found behind the chariot when Mithra and his escort turn round after sunset for their eastward journey, just as Rašnu changes position with regard to Mithra on this occasion (see the Introduction, p. 39). For there is no reason to doubt that *karde* 18 (sts. 70–2) refers to the westward drive whose inception is described in *karde* 17 (sts. 67–9); it therefore appears that in the *same drive* Dāmōiš Upamana flies behind (st. 68) and Vərəθrayna in front of Mithra (st. 70). In st. 127, which belongs to the westward drive that begins in st. 124, Dāmōiš Upamana again seems to come up from behind.

‡ Duch. translates as if she were his grand-daughter: 'C'est le père de l'active Bonne Pensée; et celle-ci a pour fille la bienfaisante Dévotion', *Zor.* p. 228, cf. also pp. 147, 182, line 3. The traditional interpretation (see Lo., *Rel.* 12, 31, and *NGGW*, 1935, 124) is, however, supported by Y 44.7 and *Vend.* 19.13; cf. also *Phl. Riv.* 8.2–4, quoted by Zaehner, *Zurvan*, 152. Moreover, Ahura Mazdāh's fathership of Ārmaiti is implied in the Av. statement that Ahura Mazdāh's daughter Aši (cf. below, p. 195) is the sister of the Aməša Spəntas (*Yt* 17.2), of whom Ārmaiti is one.

§ Cf. also Y 31.9 *θwōi as ārmaitiš* 'thine was Ārmaiti'.

COMMENTARY [9.4–11.1

In the light of this possibility we should, I think, revise Bth.'s interpretation of the two *dāmi*-s: the basic word is his ²*dāmi*- 'creator', from which by invisible vṛddhi an adj. *dāmi*- 'originating from the creator', hence 'created by the creator', was derived, the neuter of which, meaning 'creation', is Bth.'s ¹*dāmi*- (wrongly (see *infra*) described by him as fem.). Since the only *dāmi*- = 'creator' mentioned in the Avesta is Ahura Mazdāh, it should be understood that the adj. *dāmi*-, including its neuter, always refers to the creation of Ahura Mazdāh. This is contradicted by the usual translation, but not by the text, of *Y* 51.10:

*aṯ yə̄ mā nā marəxšaitē anyāθā ahmāṯ mazdā
hvō dāmōiš drūjō hunuš*

Duch., who here follows traditional lines, translates (*Zar.* 145): 'He who seeks to destroy me for whatever other cause than this, he, O Wise One, is the son of the creation of Evil.' The oddity of this translation jumps to the eye: on the one hand it implies that there are potential destroyers of Zarathuštra of whom the prophet would approve (cf. Maria W. Smith's commentary *ad loc.*); on the other hand it makes the Drug the *grandmother* of Zarathuštra's enemy. It is therefore clear that *dāmōiš* is not a fem. gen. but the neuter abl.† with which *ahmāṯ* agrees. Translate: 'The man who seeks to destroy me, he (is) outside (*lit.* elsewhere than) this (=your) creation, O Mazdāh, (being) a son of Falsehood' (*anyāθā* with the meaning of Ved. *anyátra*, as Bth. assumed; cf. Av. *anyadā*).

Thus the adj. *dāmi*-, which as such is applied only to Ārmaiti, is seen to be synonymous with *ahuraδāta*-, an epithet which is exclusive to *zam*- 'earth' (=Ārmaiti) and to Vərəθraγna.‡ If then we take *dāmōiš upamana*- as a synonym of **ahuraδātahe upamana*-, and remember that both this *upamana* and Vərəθraγna are in the habit of changing themselves to boars, we shall come to the conclusion that the *dāmi*- in question is Vərəθraγna, and *Dāmōiš Upamana* Vərəθraγna's *alter ego*.

11¹. *barəšaēšu paiti aspanąm*, Wo. 'auf den Rücken der Pferde', Lo. 'auf dem R.', Da. 'sur le dos'. Htl. has 'bei den Hälsen ihrer Rosse', imagining the warriors are standing *beside* the horses; similarly Hzf., 434, '(standing) at the neck of their h.' Pagliaro, *Oriental Studies...Pavry*,

† According to Bth., *GIP*, p. 227, the abl. of *-i*- stems, which in YAv. has the ending *-ōiṯ*, does not occur in the Gāthās. Seeing that *-ōiṯ* is an innovation (cf. ibid. pp. 119, 213), it would be in order to find *-ōiš* used for the abl. in the Gāthic dialect. Attention was drawn in *JRAS*, 1952, 178, to the possibility that *āzūtōiš* in *Y* 29.7 is to be understood thus. [See Addenda.]

‡ I doubt Benv.'s contention that *ahuraδāta*- means 'created by the ahuras' (*Vṛtra*, 47 *sqq.*). The later Zoroastrian tradition, by making Vərəθraγna a creature of Ahura Mazdāh's, confirms the traditional interpretation of *ahuraδāta*- as meaning 'created by Ahura (Mazdāh)'; cf. the Pahl. text published by Father J. de Menasce in *Rev. hist. relig.* 133 (1948), where Ohrmazd says (p. 14, §76): 'Oh Vahrām Yazat, toi que j'ai créé au premier jour comme victorieux...'; see also ibid. p. 7. A different line from Benv.'s is taken above, Introduction, p. 50.

11.1–13.1]   COMMENTARY

379, translates 'on the neck of their horses', but understands that 'the divinity is invoked to exercise his beneficent influence on the animal that should bear the warriors to victory'. Htl. insists that *barəša-* does not mean 'back' but 'neck'. In fact it means 'mane', as shown by its Iranian cognates, cf. Arm. *baš* and *barš*, Sogd. βnš, Bal. *bušk*, Pš. *wraž*, Oss. *barc*, etc. As to *raθaēštar-*, which Htl. wants to restrict to its etymological meaning 'Wagenkämpfer', our stanza, in which the *r.* seem to be on horseback, can be cited beside *ātarš spəntaraθaēštāra* (*Ny* 5.6, etc.), in support of the view that the word had been generalized in the sense of 'warrior'.

11[2]. Bth.'s translation of *pouru.spaxšti-*, 'far-reaching espying (of enemies), espying (the e.) from afar', although based on the usual, well-attested meaning of *spas-* 'to see, observe, espy', is unconvincing, because *pouru-* means 'much, many', not 'far-reaching'; who would ask to see much of his enemy? Gdn.'s 'häufiges gefangennehmen der feinde' stands on weak ground, while Lo.'s and Htl.'s rendering 'seeing the enemies first' implies an unwarranted emendation to **paourvō.sp°*. The only acceptable interpretation is Da.'s, who translated 'bonne garde contre ceux qui font du mal'. This at least agrees with the Pahl. transl. *pur pāspanīh* (to *Y* 57.26) and with the meaning 'to watch over' which *spas-* appears to have in *Yt* 11.14 (quoted below, note 41[1], p. 195). Da. also managed to make this meaning fit the awkward ('worthless' in Bth.'s view, s.v. *piθana-*) passage *Yt* 9.1: *drvāspąm...yazamaide drvō.pasvąm drvō.staorąm...pouru.spaxšti dūrāṯ piθana x^vāθravana darəyō.haxəδrayana* 'nous sacrifions à...Drvāspa...qui tient le petit bétail en santé, qui tient le gros bétail en santé,...qui veille sur eux au loin et au large (reading *paθana*) avec une longue amitié qui apporte le bien-être'. For Da. *pouru.spaxšti* was thus an instr. indicating the means by which Drvāspā is *drvō.pasvā*, etc.: 'keeping the cattle fit thanks to her *bonne garde*'. Alternatively one might adopt the variant reading *pouru.-spaxštīm*, and take this as a bahuvrīhi compound in apposition to *drvāspąm* 'Drvāspā who exerts much watchfulness afar (*dūrāṯ*)'.†

11[3]. *aurvaθa-*, translated in accordance with the meaning of Ved. *avratá, ávṛta*, see Lo., *ZII*, 5, 37, n. 1. *aurvaθanąm* and *ṯbišyantąm* are both epithets of *haməraθanąm*.

13[1]. For *mainyava-* and its MPers. derivative *mēnōk* various translations have been tried out, each with good reason: spiritual, immaterial, invisible, aerial; cf. *Wb.*, s.v., and Benv., *Vṛtra*, p. 14, n. On the title of the MPers. book *Mēnōk ī xrat* cf. Henning, *GGA*, 1935, 8, n., and Bailey, *Zor. Prob.* 2, n. 3. I have adopted 'supernatural', and for *mainivasah-* (sts. 68, 129 *sqq.*) 'hailing from supernature (*lit.* whose proper place, home, is supernature, or supernatural)', supernature being defined as

† *x^vāθravana darəyō.haxəδrayana* (for the latter cf. note 79[2]) are perhaps adjectives that go with the unknown subst. *piθana* in the instr.

'a supernatural realm or system of things'. Correspondingly, it is sometimes convenient to translate *gaēθya-*, the opposite of *mainyava-*, by 'natural' or 'empirical'.

13². *āsnaoiti*, spelled *āsənaoiti* in *Vend*. 19.28 (quoted below, note 143⁶). If, as is likely, Bth. has correctly defined the meaning of this verb as 'to approach', his connection of it with Av. *ăsna-* 'near' must needs be maintained; not, however, his etymology (to base *had-*, see *Wb.* 1755), in proposing which he was not aware that OPers. *ăšnai* means 'near', and is therefore etymologically identical with Av. *asne*, *āsnaē(ča)*. Kent, who duly noted Av. *ăsna-*, but has no reference to *āsnaoiti*, considered three etymologies for *ăšnai*, of which he preferred the third: (1) OInd. *aśnóti*, Av. *ašnaoiti* 'to attain, reach'; (2) OInd. *ájati*, Av. *azaiti* 'to drive'; and (3) OInd. *ámhas*, Av. *ązah-* 'anxiety', comparing Gr. ἄγχι. To me his first connection seems correct, Av. base *ąs-* with preverb *ā*. The *s* of *āsnaoiti* (against *ašnaoiti* with preverbs *ava*, *us*, *paiti*, *frā*) can be ascribed to the analogical influence of *āsna-* 'near', with which the *nav/nu* present stem of *ąs-* + *ā* was semantically so closely connected that it did not follow the pattern of the *nav/nu* present stem of *as-* + *ava*, *us*, *paiti*, or *frā*. In *ăsna- s* is, of course, as justified as in *vasna*, where the suffix is the same.

In intervocalic position the *s* of *ās-* ought to appear as *θ* in OPers., and does so, most likely, in the mysterious *aθiy* of Beh. I, 91, which is usually read *aθi*, and considered to be a doublet of the preverb *ati*.† To OPers. *aθiy bābairu[m yaθā naiy u]pāyam* corresponds Akkad. *ana babili la kašadu* 'at the not reaching (*or* approaching) Babylon'. One is at once struck by the discrepancy between Av. *upāy-*, which takes the acc. without a preposition, and OPers. *upāy-*, which is here said to require a preposition, and an unknown one at that. The discrepancy disappears if we read *aθiy* as *āθiyʰ*, and interpret it as the adverbial neuter comparative of *ăšnai*, built directly on the base *as-* with the preverb *ā*. The meaning will then be: 'Nearer to Babylon when I had not come.' ‡ The relation

† The reference to an alleged Sogd. preverb δ- (= θ-) < *ati* (Meillet–Benv., *Grammaire du vieux perse*, 227, where 99 is a misprint for 59) is out of date, since Sogd. δβr- goes back to *fra-bara-*, see *GMS*, p. 246.621.

‡ I have pleasure in acknowledging the kind assistance of Prof. Sidney Smith, whom I had asked if he could see any confirmation in the Akkadian version of my suspicion that *aθiy* meant 'nearer'. He wrote (8 Jan. 1955): 'In Akkadian *ana* with the infinitive can be used of purpose or temporally, and Weissbach's translation is formally correct; but literally, "at the not reaching (or approaching) Babylon", might well mean that after the first battle to gain access to the river, when Darius was no nearer Babylon, another battle had to be fought at Zazzannu, which would then be a fortified place on the canal system which defended Babylon's north-east in Nebuchadrezzar II's fortifications. Though this is opposed to your view that Darius was nearer Babylon at this point, it would agree with your view of the Old Persian word *aθiy*. The Akkadian *kašadu*, used with the accusative of person means "to overtake, reach", with the accusative of

between Proto-Iran. *āś-na- and *āś-īyah- is comparable to *kambna- (Av. kamna-) : *kamb-(ī)yah- (Sogd. kmbyy, cf. GMS, §1302), or Lat. mag-nus: maius (< *mag-i̯os-). For the derivation of comparatives and superlatives from bases prefixed by preverbs cf. Wn.–Debr. II 2, 446 sq. *āś(ī)yah- was doomed to be displaced by nazdyah- or other synonyms, because of the confusing homophony with *āś(ī)yah- 'swifter'.†

13³. For a suggestion that OIr. *aru̯ant- (Av. aurvant-) 'brave, swift' survives in Oss. uændag 'daring, brave, swift' see BSOAS, XVII, 484. [See Addenda.]

14¹. āfəntō forms the object of a recent note by Benv., Donum Natalicium H. S. Nyberg Oblatum, 17 sqq. From the comparison of the two Av. passages

baγō.baxtəm paiti yaonəm   baγō.baxtəm paiti yaonəm
fraθwarštəm paiti āfəntəm   fraθwarštəm paiti zrvānəm
(Yt 8.35, Yt 13.54, also Vend. 21.5)   (Yt 13.56)

Benv. concludes that āfant- has approximately the same meaning as yaona- (cf. below, note 60¹, p. 207), and translates it 'parcours'. Accordingly in the present stanza āfəntō is for Benv. a substantive, and the translation proposed for lines 3 to 5 runs: '(le pays aryen) où les hautes montagnes, les régions aux vastes pâturages, sont nourricières(?) pour le bétail'.

Before reading M. Benv.'s article, the comparison of the same two passages had led me to a different conclusion, viz. that āfəntəm is a synonym of zrvānəm 'time', and must be connected with Oss. afon (Iron), afonæ (Dig.), 'time'.‡ One may either assume two participles, active and middle, respectively *āfant- and *āfāna-, of a present stem *āf- or *āfa-, or a relationship *ā-fən-ta-: *ā-fān-ah- as obtains between spən-ta- and spān-ah- (on which see literature in note 3²). I should opt for the former alternative, on the strength of the Oss. word for 'year', Iron afædz, Dig.

place "to reach" or "approach", and seems to me, if your view is correct, to render aθiy...upāyam, adopting Weissbach's view; "while I was still not approaching Babylon Nidintu-Bel went to Zazannu and we made battle again (infixed -ta- form of epešu)".'

† I have, of course, considered the alternative of taking OPers. aθiy as belonging to Av. āsu-, etc. But apart from there being no mention or implication of speed in the Akkad. version, one would have to interpret the form as either āθiyaʰ 'quicker', which is an impossible meaning in this context, or as āθi (unthem. loc. of *āθ-) or āθai (them. loc. of *āθa-), both meaning 'quickly'. Either *āθ- or *āθa- would be, however, most improbable stems of the positive, which is generally associated with the determinative -u- (Av. āsu-, OInd, āśú, Gr. ὠκύs).

‡ Prof. Henning tells me that the same comparison had led him to the same conclusion.

afæy, whose endings, if compared with Iron ssædz, Dig. insæy 'twenty' < *u̯insati, suggest that the OIr. form was *āfatī-, fem. of *āfant-.††

Afonæ, as defined by my Digor friend, indicates a definite time, in contrast with ræstæg, which refers to time in general. Examples: ræstæg min yes 'I have time' (not *afonæ min yes); ænafoni ma ærbaco 'do not come at the improper time' (not *ænæræstægi); ci afonæ'y 'what is the time?'; ci ræstægæy 'what time is now (summer or winter)?'; ænglisag særdton afoni nimædzæ 'British Summer Time'; æ afonæ ku ærqærttæy (Памятники, II, 10.21) = Iron yæ afon kŭ 'rcȋdi (Nartı̂ Kadzdzı̂tæ 63.8) 'when its time§ arrived'; Iron ærtæ afon 'three times' (colloquial, = 'for a long time'), Iron Adæmon Sfældı̂stad 46.17; Iron æmæ næ balconæn ærcæun afon ku næ ma ū 'it is not yet time for our traveller to come back', V. Miller, Осетинские этюды, I, 20.17.‖ This definition of afon(æ) supports the proposed etymological connection with afædz/afæy 'year', viz. a definite span of time, and is in harmony with the Av. epithet fraθwarštəm 'appointed, exactly defined', which qualifies āfəntəm in the formula quoted above.¶

It is now clear that āfəntō in the present stanza must be a different word, which moreover, contrary to Benv.'s opinion, cannot be a substantive. The latter point is proved by the wording in the almost identical passage Yt 13.9:

garayasča yōi bərəzantō
pouru.vāstrå̄ŋhō āfəntō;

here the absence of a second ča, and the single relat. pronoun in Iḍāfat function, show that the three last words are all adjectives. Benv. was,

† Similarly Iron kŭdz, Dig. kui 'dog' lead back to OIr. *kuti-. Sogd. kwt- (abl. 'kwt' P3, 188) is then an -i- stem passed into the thematic inflection, like γr- 'mountain'<gari-, cf. GMS, §947.

‡ [In the meantime Bailey, TPS, 1956, 120, has connected Oss. afonæ with an Iranian verbal base of movement attested in Khot. phan-. It is therefore advisable to analyse Av. āfəntəm 'time' as acc. of a thematic neuter ā-fən-ta- (formed like spən-ta-), Oss. afonæ as from *ā-fān-ah- (cf. spān-ah-), and Oss. afædz as from *ā-fa-ti- (with a < IE ṇ), the -ti- abstract of *ā-fan-.]

§ Viz. the time of the stone out of which Soslan was to be born after a period of 'pregnancy' lasting nine months.

‖ On the other hand, 'moment' is ræstæg (as well as usm, Dig. usmæ); cf., in Iron, Batraz udi ucı̂ ræstædźı̂ Kurdalægommæ 'Batradz at that time (=moment) was with Kurdalægon', Miller, op. cit. I, 20.24; ŭcı̂ ræstædźı̂ 'at that moment (while something else was happening), meanwhile', Nartı̂ Kaddzı̂tæ 9.7, 10.20. Hence, in context similar to yæ afon kŭ 'rcîdi quoted above, we find in Iron: ŭcı̂ fætk'uyæn iu ræstæg ŭd (Iron Adæmon Sfældı̂stad 7.11) 'the moment for the apple (=when it was ripe to be picked) had come', fætk'uyæn yæ ræstæg baqŭdı̂ kodta (ibid. 8.4) 'she watched the moment of the apple (when it would be ripe)'. Cf. V. I. Abayev, Русско-осетинский словарь under время and момент.

¶ According to A. A. Freiman a Khwārazmian cognate of Oss. afon 'time' is attested as ufān-, see his Хорезмийский язык (1951), p. 67. ufān-, too, refers to a definite time.

however, right in rejecting Bth.'s explanation of āfəntō as *āp-vant-ah 'rich in waters', and in deriving from *āpyantai the 3rd plur. āfənte 'they are reached', Y 57.29, which Bth. had derived from *āpvantai. The door for the correct understanding of āfəntō is now open: the word stands for *ā-fyantō, nom. pl. of the active participle of the present stem fya- of the base pā(y)- 'to protect'. This present stem occurs, again with the preverb ā, in the infinitive āfyeiδyāi in the sentence ašavanəm tē ašaonat̰ āfyeiδyāi mraomi, Y 71.13, which Bth., with Benv.'s approval (Inf. 76), translated iustum tibi a iusto curandum declaro;† the preverb ā is possibly to be found with another derivative of the base pā(y)- in Gāthic apavant-, see below, note 46². [See Addenda.]

14². θātairō. Bailey, Zor. Prob. 18, n. 3 [and TPS, 1956, 119], follows Scheftelowitz in translating this word by 'grass, fodder'. As the presence of i is unjustifiable by Avestan phonology, the correct reading is most likely θātārō, with the i stroke joined to the a, cf. the variant θrātārō. From θamnahvant- 'solicitous' a base θam- must be deduced, to which θātar- stands as Av. yāta-, Sogd. ny't-, to the base yam-, cf. GMS, §539, n. 3.

14³. Bailey has obtained an attractive meaning for urvāpa- by comparing Oss. fæildun 'to surge' < *pati-vi-lāv-, and ulæn 'wave' (BSOAS, XII, 331).

14⁴. On nāvaya- cf. Henning, BSO(A)S, IX, 91; XII, 309; Zaehner, Zurvan, 214. Differently Benv., Vr̥tra, 60, n. 3.

14⁵. xšaoδaŋha is taken as nom. pl. 'streams, torrents' by Da., Bth., Lo., and Zaehner (loc. cit.). I follow Wi. and Gdn.'s 'mit (wogen)schwall', which Htl. and Hzf., p. 453, adopted also; cf. the meaning of Ved. kṣódas.

14⁶. This is to assume that iškatəm pourutəm, like mourum hārōyum and gaom(ča) suxδəm, consists of a name of a region followed by an ethnical adjective. pouruta-, which only occurs here, may then mean either 'which lies in, or close to, (the region of) the Parutas', or 'inhabited by Parutas'.

Iškata- is also known in the name of the mountain iškata upāiri.saēna‡ Y 10.11 and Yt 19.3. If in this name iškata is adjective, 'Iškatian

---

† Karl Hoffmann's attempt (Corolla Linguistica, Festschrift F. Sommer, p. 81, n. 2) to separate this infinitive from pā(y)- carries no conviction, in view of the unsatisfactory meaning he obtains: 'ich sage dir, daß ein Ašahafter auf Seiten eines Ašahaften erreicht wird'.

‡ Upāiri.saēna (on the long ā cf. Wn., BSOS, VIII, 830), viz. the Av. name of the Hindukuš, was recognized by Marquart, UGE, II, 73 sqq., in the following names: (1) Hüan Tsang's P'o-lo-si-na (according to Marquart that part of the Hindukuš which separates the valleys of the Panjšir and Ghōrband from the Andarāb valley); (2) Ptolemy's Παροπάνισος understood as *para-uparisaina- '(the region) in front (= South) of the Hindukuš', viz. Gāndhāra; and (3) the name of Gāndhāra in the Akkad. version of the Behistun Inscription, §6, pa-ar-ú-pa-ra-e-sa-an-na = *para-uparisaina.

Upāirisaēna' would designate that part of the Hindukuš which lies in the region Iškata. If the adj. is *upāiri.saēna*, then '(mount) Iškata, which forms part of the Hindukuš' would be meant. In the latter case Iškata would be the name of (*a*) a mountain, and (*b*) a region dominated by this mountain.† Assuming with Marquart, *UGE*, II, 74, that the mountain Iškata is the Kōh-i Bābā, the region Iškata which it dominated will be the plain of the Upper Helmand.

The Parutas, if identical with the Παροῦται or Πάραυτοι of Ptolemy VI, 17, 3, are said by that author to have occupied the part of Areia-Haraiva which lay by the Hindukuš, that is, according to Marquart, *UGE*, II, 74, 175, Ghōr.‡ In the remote days to which the Mithra Yašt refers, the Parutas may have held also part, or the whole, of the region Iškata, situated to the east of Ghōr. At any rate Iškata, whether called 'Parutian' because adjacent to the Paruta-inhabited region of Haraiva, or because it was itself Paruta-inhabited, should be located to the south of the Western Hindukuš, between Haraiva and Gāndhāra.

If the 'Απαρύται of Herodotus III, 91 are also Parutas (as assumed by Justi, *GIP*, II, 438, and Benv., *BSOS*, VII, 269), we should put to some use the information that they formed one νομός with the Σατταγύδαι (= OPers. *θatagu*), Γανδάριοι, and Δαδίκαι (who, according to Marquart, *op. cit.* 175, are the ancestors of the Dardic tribes). Since Marquart, *loc. cit.*, thought that the *θatagu* should be sought to the north of Arachosia and west of Gāndhāra, viz. in the area of the Kōh-i Bābā and the upper course of the Helmand, where we saw there is reason for locating the region called Iškata, it is permissible to surmise that Iškata is the Av. name of the country which in Darius' time was inhabited by the Σατταγύδαι = *θatagu*.§

Confirmation that Iškata should be sought by approaching Gāndhāra from Haraiva may be found in the first chapter of the *Vendidad*. Benv., *BSOS*, VII, 265 *sqq.* has ingeniously recognized additional evidence that

† To see in *iškatəm* and *pourutəm* of our stanza two mountain branches, as Jackson, *Cambridge History of India*, I, 327, n. 2, contemplated, is inadvisable, as rivers do not usually 'rush with a swell' *towards* mountains. The same objection applies to Bth.'s translation of *ā iškatəm pourutəmča* by 'hin zu Fels und Berg'. The fragmentary Chr. Sogd. word *šq'*[(Hansen, *AAWL*, 1954, 867.49), corresponding to the first word of the Syriac *wšqyf' wšw''* 'rocks and rocks', has scarcely anything to do with Av. *iškata-*, but may be restored as *šq'[f(t)*..., cf. B. Sogd. *šk"β* 'precipice' P 6.8, NPers. *šikāf* 'cleft, crevice'. For attempts to connect *iškata-* with the name of the fortress Κυρέσχατα (for which Benv., *JAs.* 1943-5, 165, has a more satisfying explanation), see A. Christensen, *Le premier chapitre du Vendidad*, 69.

‡ On the supposed 'Παρουῆται' cf. Marquart, *UGE*, II, 178.

§ The *θatagu* were, according to Marquart, *op. cit.* 177, one of the tribes of the Πακτύες. This is not very likely if the name of the latter is connected with that of the present-day Pātu in North Chitral, as suggested by Bailey, *BSOAS*, XIV, 428 *sqq.*

Aryana Vaējah is Chorasmia, in the reverse order in which the countries mentioned in the present stanza are listed in *Vend*. 1. If we pursue his analysis we note that as *mourum hārōyum* is preceded in *Yt* 10.14 by *iškatəm pourutəm*, so *mōurum–(bāxδīm–nisāim)–harōyūm* are followed in *Vend*. 1.9 by *vaēkərətəm*, which is Gāndhāra, or part thereof (cf. Christensen, *Le premier chapitre du Vendidad*, 28, Henning, *BSOAS*, XII, 52). There is thus a presumption that Iškata and Vaikṛta were situated close to each other.

14⁷. Viz. 'Margu, which forms part of Haraiva', as under Darius it formed part of Bactria, cf. Justi, *GIP*, II, 454.

14⁸. On *gaom suxδəm* cf. *Asia Major* (N.S.), II, 137 *sqq.*, where the reference to Γάβαι (etc.) should be dropped: see Henning's note, *ibid.* 144, who in addition has a more attractive explanation of Elamite *kam-ba-ti-ia*. If the El. word has nothing to do with *gaom* we should revert to Bth.'s assumption that the stem of *gaom* is *gava-*, and analyse *gavašayana-* and *gavašiti-* (see note 15²) as *gava-š°*. The meaning to be postulated for *gava-* is (*a*) 'a settlement' (in particular a village or cattle-station), and (*b*) κατ' ἐξοχήν 'the Settlement', as a name of Sogdiana or part of it. Dig. *kehog*, see *art. cit.* 138, if correctly explained, may be from °*gaµaka-* with early loss of the second *a*. The existence of yet another word **gav(a)-*, meaning 'servant', is conjectured by Benv., *JAs.* 1954, 308 *sq.*

15¹. *Xᵛaniraθa-*, the central clime. *niraθa-* < **ni-rṃtho-* may belong to MPers. *nyr'myšn* 'layer' or 'deposit', cf. Andreas–Henning, *Mir. Man.* I, 177.8, Jackson, *Researches in Manichaeism*, 50, F. W. K. Müller, *Handschriften-Reste*, II, pp. 19, 40, 111; on the base of the MPers. word see Henning, *ZII*, IX, 190, Nyberg, *Hilfsbuch*, II, 5, Bailey, *Oriental Studies...Pavry*, 22 *sq*. Accordingly *xᵛa-niraθa-* can be understood as literally meaning 'self-founded, being its own foundation, not resting on anything else'.

15². On *gava-šayana-*, *gava-šiti-* see Bailey, *TPS*, 1945, 14, who translates both words by 'dwelling in villages', and cf. note 14⁸. *Xᵛaniraθa* is the 'land of settled dwelling, etc.', as opposed to the Northern Steppes where life was nomadic, cf. E. Minns, *The Art of the Northern Nomads*, *Proceedings of the British Academy*, 28 (1942), 5 *sqq*. Gdn., Bth., Lo., and Duch. (*Comp.* 135, 151) take *gavašayanəm* as a bahuvrīhi agreeing with *xᵛaniraθəm*, and *gavašitīm* as a determinative compound: 'diesen Erdteil...den Xᵛaniraθa, den...von Rindern bewohnten und die heiltätige Wohnstätte der Rinder' (Wo.). It would, however, be equally surprising if °*ča* connected a substantive with an adjective, or if *gavašitīm* constituted an item to be added to *xᵛaniraθəm*. It is more likely that *gavašayanəm*, too, is a determinative compound (like *airyō.šayanəm* in st. 13), and both compounds are substantives in apposition to *xᵛaniraθəm*; this was realized by Wi., who translated 'den Sitz der Rinder, den Wohnplatz der Rinder'.

COMMENTARY [16.1–20.3

16¹. On the hendiadys *dahma vīduš.aša* see note 47¹.
18¹. *fraša upa.sčandayeiti*, Lo. 'miraculously smashes'; Gdn. 'stracks zerstört', and (*Lesebuch*) 'vernichtet...immer weiter'.
18². The fivefold division of temporal authority as attested in this stanza and st. 87 is discussed in note 145¹ below.
19¹. All translators take *naēδa* for 'and not', and imply the copula after *miθrō.druxš*; cf. Wo.'s rendering: 'Nach dér Seite wendet sich... Miθra, auf welcher der Seiten (sich) der Miθrabetrüger (befindet); und nicht versieht (der) sich des im Geist.' Against such interpretation, apart from the awkward locative function which the dat. *yahmāi* is here supposed to have, stands the fact that *naēδa* means 'and not' only after a previous negation. Evidently *naēδa* in the present passage, and the first *naēδa* in st. 71, mean 'not, not at all', as Lat. *nec* sometimes does, cf. Stolz–Schmalz, *Lat. Gramm.*, 5th ed. (1928), pp. 640 *sq.*, 856. Another Av. example is *Vend.* 18.11: *aēvahe zī aš*[*əm*]*aoyahe ayahe anašaonō zānu.- drājå̆ asti āfritiš, dvayå̆ hizu.drājå̆, θrayąm naēδa.čiš, čaθwārō x*ᵛ*atō zavante*, 'the greeting (thus Gdn., in his transl. of this chapter, *Avesta... Studies...Sanjana* (1904), 201 *sqq.*) of one evil heretic teacher who does not own Truth is a protrusion of the chin, of two a protrusion of the tongue; of three (there is) none at all (*lit.* '(there is) not any (greeting)', cf. note 71⁵ below); four curse themselves'.
19². For the meaning of *paiti.pāite* OPers. *hačā draugā dršam patipaya*ʰ*uva*, Beh. IV, 37, should be compared, 'guard yourself strongly from Falsehood'.
20¹. *frastanvanti* seems to belong to the base *stan-* of MPers. '*st*'*n*-, '*std*, NPers. *sitǎn-*, *sitadan*, 'to carry off', cf. Henning, *ZII*, IX, 189. Wo.: 'laufend kommen sie (die Betrüger) nicht vom Fleck, reitend machen sie keinen Fortschritt, fahrend gewinnen sie keinen Vorsprung'; Lo.: '...die Rosse...entkommen nicht im Lauf, kommen im Reiten nicht voran, bleiben im Fahren nicht voraus'; Htl.: 'laufend kommen sie nicht von der Stelle, tragend kommen sie nicht vorwärts, ziehend harren sie nicht aus'. Since in clear cases *bar-* 'to ride' is used in the middle, cf. *Wb.* 936; 942, n. 14, there is a strong presumption that the active *barəntō* refers to the horses as bearing their riders. Accordingly we should side with Htl. as far as his 'tragend' is concerned.
20². There is no reason to postulate with Bth. for *framanyente* a base ³*man-*, derivatives of which do, however, occur in Av., see below, note 79¹, p. 224. *framanyente* should not be separated from *Y* 68.13 *yōi vaŋhaθra framanyeinte* 'who persevere at their post', as Htl. evidently realized, see preceding note. Duch., *Comp.* 75, has a different view; yet another interpretation was put forward in *Ap.* I. 148, 363, by Hzf., who later abandoned it (*Zor.* II, 435).
20³. *frəna ayaną̇m mąθraną̇m*, according to Bth. 'because of the abundance of evil spells'. The translation here proposed is a revival of

the one which Wi. adopted as a guess. We must, I think, separate *frəna* in the present passage from its two homonyms in *Vend.* 5.4, 8.34, and *Yt* 5.129 respectively. One does not expect the *number* of evil spells to turn back the spear, but the fact that they are cast. *frəna* may therefore be the adverbial instr. of *frăna-, a derivative of *fră* that corresponds in form to Lat. *prōnus*, in meaning to Lat. *prō* and Parth. *frh'ḫ* 'because of' (on the latter see Mary Boyce, *MHCP*, 187). [See Addenda.]

In the *Vend.* passages the meaning 'abundance, quantity' is assured for *frəna* by the Pahl. transl. *hač frāyistīh*, glossed by *hač vasīh*. However, Bth.'s interpretation of the word as loc. of a stem *fr-ani-* is unsatisfactory. With this suffix the base should have appeared in its full form **par(ani)*-, cf. Wn.–Debr., II, 2, 207 sq. What made Bth. resort to this suffix was his notion that *ə* before *n* must represent a brief *a*, cf. *GIP*, I, 175 top; he was therefore unable to equate *frəna* with *frāna-* 'fullness' (in *zastō.-frānō.masah-*), cf. his *Studien*, II, 102 sq. By now it is known that *frāna-* was almost bound to acquire in Avestan a side-form **frana-* (cf. *upamana-*, etc., note 9[4], p. 167), of which *frəna* would be a possible variant.

Different again is *frəna* in *Yt* 5.129, where one expects the word to agree with *čarəmå*; it is perhaps the middle participle of *par-* 'to fill' (*frăna-*), construed with the acc. of relation:

čarəmå vaēnantō brāzəṇta
frəna ərəzatəm zaranim

'the furs shine upon the viewers, being replete with (regard to) silver and gold'.

22[1]. Benv., *Donum Natalicium H. S. Nyberg Oblatum*, 20 sqq., proposes to replace Bth.'s translation of *iθyajah-* as 'danger' by '*abandon*', which to my mind imposes too subtle an interpretation on most passages concerned. The evidence collected by Oldenberg, *ZDMG*, 55 (1901), 281, on Ved. *tyájas*, which Benv. uses, had been considered by Bth., who quoted it. Unless fresh evidence comes to light, there is no point in departing from the generally suitable meaning which the comparison of all Av. passages had led Bth. to postulate.

23[1]. *ana* = 'in this (viz. the following) way, hereby'.

23[2]. *xšayamnō barahi*. It seems that the middle partc. *xšayamna-* with a finite verb always indicates that the subject can, or may, perform the action expressed by the finite verb. At any rate I have attempted to translate accordingly here and in sts. 37, 101, 108, and 110. In st. 37 also Gdn., *Lesebuch*, 20, has 'über sie vermag er Not und Schrecken zu bringen'.

24[1]. *nōiṯ...ava.ašnaoiti* 'one does not hit' (thus rightly Wo.), cf. note 48[3].

24[2]. *šanman-* < **šam-man-* (base **šam-* 'mittere'), see Henning, *Sogdica*, 24, and cf. *sravašəmna-*, below, note 30[1]. For *-nm-* < *-mm-* cf. *xšąnmānē*, Benv., *Inf.* 70. Da. and Htl. unconvincingly connected

COMMENTARY [24.2–25.1

šanman- with OInd. kṣanóti 'to injure, wound', translating it by 'blow';
cf. note 39¹.
24³. huxšnuta-. Attention was drawn in *BSOAS*, XVII, 481, to the
existence in Iranian of forms belonging to the base *xšu- (Gr. ξύω)
without the nasal infix. [See Addenda.]
24⁴. Note Morg.'s connection of Sogd. *p'δδ*, Oss. *fat*, etc., 'arrow'
(cf. Bailey, *TPS*, 1945, 11), with Av. (*para.*)*paθwant-*, *IIFL*, II, 242*b*.
24⁵. We find *baēvarə.spasanō* here, and °*spasānō* in st. 60, both times
in the nom. sg. Bth., followed by Duch., *Comp.* 177, justifiably
assumed thematic treatment of a stem °*spasan-*, since °*spasānō* has a
parallel in the nom. sg. *baēvarə.čašmanō* (stem °*čašman-*) in st. 141. The
history of these forms can be pursued more closely: the *ā* of *baēvarə.-
spasānō* is derived from the unthematic acc. sg. °*spasānəm*, which
evidently was the cause of the transfer to the them. inflection; similarly
the them. gen. *činmānahe* in *A.* 3.7 (*Wb.* 596) is built on a stem **činmāna-*,
which was extracted from the unthem. acc. *činmānəm; there is thus
a chance that the nom. °*spasanō* resulted from a shortening of *ā* in the
nom. °*spasānō*, cf. *upamana-*, etc., note 9⁴, p. 167. The them. inflection
°*spasānō*,* °*spasānāi*, etc., may explain the 'datif à vocalisme prédésinen-
tiel long' (Benv., *Inf.* 50) *puθrāne* in *Vend.* 4.47: it is the unthem.
*puθrane which has borrowed the long *ā* of the them. stem **puθrāna-*,
itself an offshoot of the unthem. acc. *puθrānəm. The same explanation
may apply to *čimāne*, discussed below, note 32¹. [See Addenda.]
25¹. *vahmō.səndah-*, 'das Gebet erfüllend' (Bth., to ²*sand-*), 'qui se
plaît aux prières' (Da., to NPers. *pasand*, see *ZA*, I, 463, n. 2), 'an
Dankgebeten sich erfreuenden' (Gdn. to OInd. *chandas*). Of all the
forms quoted by Bth., *s.v.* ²*sand-* 'efficere, perficere', Lo. was prepared to
leave only *səndā* in *Y* 51.14, assigning *sąs*(*tā*) and *asąsaṱ* to *sąh-*, see
NGGW, 1934, 72; 1935, 159. Duch., *Zor.* 278, removes even *səndā*, by
taking it as causative of ¹*sand-* 'videri'. Bth.'s ²*sand-* is thus abolished.
As regards Bth.'s entry ¹*sand-*, after eliminating from it *sanaṱ* (on which
cf. note 104⁷), we must distinguish two meanings, (*a*) 'videri' and (*b*) 'to
please'; cf. Wn., *IF*, 45, 322 *sq.*, who compared the use of OInd. *chand-*
for both 'to seem' and 'to seem good, please'. Outside Avestan, meaning
(*b*) is found in NPers. *pasandīdan* 'to approve, be content', Parth. *psynd-*
'apprécier, goûter' (see Ghilain, *Essai*, 55), Sogd. *ptsynd-* 'to be pleased,
agree, approve', Pahl. Ps. *wysndkyh* 'dislike', MPers. *hwnsnd* 'content'
(*hu-* + **ni-sand-*, see Salemann, *Manichaeische Studien*, I, 83), Sogd.
xwsndy' 'contentment', see *GMS*, §338.
In Avestan the following forms have the 'pleasing' connotation, apart
from *vahmō.səndah-*:
(1) The finite forms *səndayaṇha*, *Visp.* 8.1 (correctly translated by
Bth., *s.v.* ¹*sand-*; see Da., *loc. cit.*, and Salemann, *loc. cit.*), *asąsaṱ Vend.*
19.15 ('Zarathuštra agreed to (=obeyed) my order'), possibly also

sạndā Y 51.14 ('do approve the sentence'), and sąs Y 46.19 ('to these things you will agree for my sake') [see Addenda];
(2) the infinitive sazdyāi Y 30.2, translated as 'gefallen' by Andreas and Wn., *NGGW*, 1909, 48 (sazdyāi in Y 51.16 belongs to sah- according to Lo., *NGGW*, 1935, 160);
(3) the adj. *paitī.sạnda-* in Y 38.5: *apasčā vå...avaočāmā vahištå sraēštå avā vạ vanuhīš rātōiš darạyō.bāzāuš nāšū paitī.vyādå paitī.sạndå* 'Und euch, Wasser, ...die...besten (und) schönsten, herab rufen wir, ihr guten, mittelst der langarmigen Opferspende, (die ihr) im Unglück (nāšū) Vergeltung übend (paitī.vyādå) Abhilfe schafft (paitī.sạndå)' (Wo.). The 'calamities' (nāšū, loc. pl.) are uncalled for at this juncture, the gen. *rātōiš darạyō.bāzāuš* is left hanging in the air, and *vyādā-* elsewhere means 'reward, thanks'. The difficulties disappear if we interpret *paitī.vyādå* as 'grateful', and nāšū as belonging not to ¹nas- 'to perish' but to ²nas- 'to attain', hence 'what is attained, obtained, = share', cf. Ved. *dūṇḍśa* 'unattainable'; *rātōiš* will then depend on *nāšū*. Translate: 'we call you down...(to be = show yourself) grateful for, and pleased with, the (= your) shares † of the long-armed sacrificial offering'.

25². *tanumạθra-* according to Bth. (and Duch., *Comp.* 155) has *tanu-* virtually in the loc., 'having the *maθra-* in his body (person), with whose body (person) the *m.* is connected, who has absorbed it, has become one with it'. If this lengthy explanation were correct, one might reasonably expect to find the epithet applied to Ahura Mazdāh, since the *maθra-* is Mazdāh's 'soul' according to *Yt* 13.81, *Vend.* 19.14. Yet *tanumạθra-* is only found as epithet of Mithra and Sraoša, and of men. Moreover, its counterpart *tanu.drug-* (*drvantō tanu.drujō, Vend.* 16.18) is equally unlikely to contain a virtual loc.: the 'Drug in the body' would suggest the demon of putrefaction (cf. *Wb.* 780.2–11), rather than the demon Falsehood who bedevils the 'owners of Falsehood'.

Htl.'s interpretation, 'dessen Leib aus den Liedern besteht' was rejected by Duch., *Comp.* 155, on the ground that with such a meaning one should have expected to find **maθra-tanū-*. This objection, however, is not valid if our compound belongs to a comparatively rare type of bahuvrīhi which involves a metaphoric identification, as is the case with Ved. *vṛkṣá-keśa* 'whose trees are (like) hair' = 'wooded (mountain)', cf. Macdonell, *Vedic Grammar*, 172.‡ Accordingly the *tanu.maθra-* and the

† Prof. Bailey aptly recalls Khot. *nasa* 'portion, share' < ²*nas-*, cf. S. Konow, *Saka Studies*, 159.

‡ Wn., II, 1, 276 top, translates 'whose hair are trees', but defines like Macdonell: it is the second member of the compound which provides the figurative term of the identification. Such bahuvrīhis are for Wn. adjectivized tatpuruṣas of the type *mukha-candra* 'a mouth which is like the moon, a moon-mouth'. *tanu.drug-* and *tanu.maθra-* are similarly understandable as displaying attributive usage of determinative compounds meaning 'a *drug-* person (body)', 'a *maθra-* person (body)', respectively.

COMMENTARY   [25.2–27.1

tanu.drug can be taken as individuals whose bodies or persons are, as it were, the *mąθra-* or the *drug-*. For the idea of personification or incarnation here involved, one may compare the name of the Saošyant *Astvaṱ.-ərəta-*, which literally means 'who is bodily, embodied, Truth'.

Into this pattern fits also *tanu.kəhrpa*, *Vyt.* 3, '(sons) whose bodies are (like your) shape', an epithet which is sufficiently ambiguous for *Vyt.* 1 to carry the late gloss *kəhrpa xᵛōuš* '(your = Vištāspa's) own shape', *xᵛōuš* (thus L 5; Pahl. transl. 'xᵛēš, cf. *Wb.* s.v. *xayōuš*) being the Pazand form which corresponds to Man. MPers. *xwybš* ($x^v\bar{e}β(a)š$).

25³. *bāzuš.aojah-*. Of the four possible explanations mentioned by Duch., *Comp.* 15 *sq.*, I prefer the last, 'graphie d'un *bāzauš°*, Gen. Sg.', without, however, accepting the 'graphie': *bāzuš°*, instead of the expected gen. sg. *bāzāuš°* (*sic*), is a mistake which arose from the co-existence of two forms of the nom. sg., *bāzuš* and (*uz*)*bāzāuš*.

26¹. *akatarəm sraošyanąm*, Wi.: 'den Vernichter (*kan-*) der Strafbaren'; Gdn.: 'den härtesten wider die strafbaren'; Bth.: 'der denen überaus böse ist, die sich strafbar machen'; Da.: 'cruel dans le châtiment'; Lo.: 'die schlimmste unter den Strafen (vielleicht auf das Abschlagen des Kopfes zu beziehen)'. One would certainly agree with Lo. that it is sufficient to recognize in *sraošyanąm* a neuter noun *sraošya-*, parallel in meaning and formation to the fem. *sraošyā-* 'disciplinary measure, punishment'; cf. also Benv., *Rev. hist. relig.* 130 (1945), p. 14, n. 2. But *akatarəm* is troublesome. I take it as standing for *akatara*, instr. sg. neuter, changed to °*tarəm* under the influence of the surrounding accusatives, especially (*ačaē*)*tārəm* and (*hamaēs*)*tārəm*; cf. the similar case of *vaṇhānəm*, below, note 124². Hence literally 'with an (even) worse one of punishments', worse, that is, than the smashing of heads. For possible 'worse' punishments cf. sts. 110 *sq.* or 48; to come under the treatment of Mithra's assistant Vərəθraγna (see below, p. 194) may also be less agreeable than having one's head smashed, cf. sts. 71 *sq.* [See Addenda.]

27¹. *rąxšyant-*. I regard *rąxšya-* as a denominative of **rąxša-* 'defiance', which in its turn would be a verbal noun formed from the -*s*-extension of a base **rąx-* or **rąk-* (cf. Av. *daxša-* 'fire, Brand'† to *dag-* 'burn', Sogd. *pwyš* 'cauldron' to *pak-* 'cook', and Av. *fra-daxš-*, below, note 39⁴).‡ The same base is attested in Sogd. *rxn-* 'to dare' (cf. *JRAS*,

† As a present stem *daxša-* is attested in Sogd. δxš-, cf. Henning, *Sogdica*, 48.2, and, with the meaning 'to give pain, hurt, smart' in Yaγn. *daxš-* (*darim adaxš* 'my belly (cf. Morg., *IIFL*, II, 390, s.v. *dēr*) hurt', *pōdam* (*sarš*) *daxci* 'my foot (his head) hurts', Junker, *Yaghnōbī Studien*, 19.41, 42, 20.48), and Galinqaya *dašn-* (*dösdöm dašna* 'my hand hurts', Henning, *TPS*, 1954, 176, n. 4; cf. *vašn-* to *uaxš-*, ibid. 171); for the semantics cf. Sogd. *sūdzīn* 'to burn' and 'to feel pain'. [Cf. also Yaγn. *daxša* 'illness, pain', Benv., *JAs.* 1955, 154.]

‡ A different interpretation of *rąxšyant-* was proposed by Nyberg, *Rel.* 466.273, who took the word as a fut. partc. of **ranj-*, 'due to cause pain'. Such

181

1946, 181.134; Benv., *VJ*, 91 *sq.*, *JAs.* 1943-5, 95; Henning, *BSOAS*, XI, 734.621). For the Sogd. present stem two explanations seem possible: either *rxn*- (*raxn*-) is a metathesis of OIr. **ranx(a)*- (= Av. **raxn(a)*-) < IE **l/renkᵘh*-; or *rxn*- has preserved the *nā/n* formative and represents OIr. **raxn(ā)*-, with *rax- from either IE **l/rn̥kᵘ*- or **l/rn̥kᵘh*-.† [Add.]

27². *paiti...vārayeiti*, Wi. 'prevents'; Gdn. 'obscures'; Lo. 'conceals'; Bth. 'destroys'; Da. 'removes'; Bailey 'keeps safe' (*Zor. Prob.* 13); Hzf. 'makes inefficient' (p. 480).

28¹. *vīdārayeiti*, connected by Henning with Sogd. *wδyr*-, see *Asiatica* (*Festschrift Weller*), 289, n. 3. The primary meaning is probably 'to disjoin, hold apart' (cf. st. 48), whence 'to sort out, arrange, etc.'

28². *qiθyā*- 'gate-post, pillar' survives in Dig. *uælindzæ* 'roof', lit. 'what surmounts the pillars', corresponding to Iron *uælxædzar* 'roof', lit. 'what surmounts the house'. For the *i* of °*indzæ* cf. *innæ* 'other' < *anya*-, and *Ir* 'Ossete' < *arya*- (see *BSOAS*, XVII, 486, n. 1); *dz* < *c* = *ts* (< *θy, cf. *æcæg* 'true' < *haθya*-) after *n*, as in *fondz* 'five'. Dig. *cægindzæ*, Iron *cædẓîndz* 'column, post' also belong here if the literal meaning is 'ring(shaped = round) pillar', cf. *cæg* 'link, ring', *cægkag* 'suitable for a ring', and *cæfxad* 'horseshoe' < *cæg-fad 'foot-ring' (see Abayev, Русско-осетинский словарь, 613). Closer to Lat. *antae* is *pati-antā-, which Henning has recognized in Sogd. *pδynd* 'threshold', NPers. *palindīn* 'door-frame, lintel', see *BSOS*, X, 99 *sq.*

28³. *vīra*- when contrasted with cattle means 'slaves', see Lüders, *SPAW*, 1917, 368.

28⁴. *yahva*; the plural is due to anticipation of *anyǎ...yāhva* in the last two lines; the scribe who introduced it wrote the correct *yahmi* as far as *yah*, then changed his mind, but forgot to replace -*a*- by -*ā*-.

30¹. On *srao*° (cf. also *sraotanū*- and *sravašəmna*-) we have a variety of suggestions: 'berühmt' (Wi.); 'flink' (Gdn.); 'beau' (Da.); 'tüchtig' (Bth.); 'trefflich' (Lo.); 'riche' (Benv., *Vr̥tra*, 52, n. 2); 'fair(?)' (Hzf., 433); cf. also Duch., *Comp.* 166. The explanation of *srava*-, *srao*- (< *sravō-) may lie in a closer semantic connection than has hitherto been a meaning hardly suits the context. All the forms belonging to *ranj- that have meanings which might fit *rəxšyant*- (cf. Henning, *ZII*, IX, 199, but note *BSOAS*, XI, 484, n. 5) have the preverb *adi*-.

† The opportunity may be taken to draw attention to the probable etymological connection of NPers. *lagad* 'kick' with Gr. λάξ, λάγδην 'kicking with the heel'. *lagad* should represent OIr. **l/rakata*-. For the suffix cf. Av. *nəmata*- 'felt' (see Lüders, *APAW*, 1936, No. 3, 16 *sqq.*), and Sogd. *šykth*, OInd. *sikatā* 'sand', beside *sikitā- (in Yd. *sēyio* and Oss. *sîdẓît*) and *sikā- (in OPers. *θikā*- and Bal. *six*), see Morg., *IIFL*, II, 245, *NTS*, V, 51. If the somewhat artificial connection of λάξ with Skt. *lakuṭa* 'club' (supposedly < **lakr̥ta*) and Lat. *lacertus* (cf. Walde–Hofmann, *LEW*, s.v.) should happen to be correct, one may assume an *r/n* alternation **lekr̥-to*- (> Ind. **lakr̥ta*-): **lekn̥-to*- (> Iran. *lakata-). The Iranian adj. suffix -*ata*- mentioned by Bailey, *BSOAS*, XIII, 933, does not belong here. [See Addenda.]

attempted, of Av. *sravant-*, *upa.srvant-*, with Pahl. *ōštāp, apar ōštāp kar*, which respectively translate the Av. words. For *ōštāp*, Arm. *štap*, 'haste, hurry' (cf. Bth., *ZWb*. 236, n. 3, Henning, *ZII*, IX, 190; 224) stand to *ōštāp kar* (cf. Pahl. Ps. *'wšt'pykly* 'oppressor') as 'pressing' (cf. French *presse*, Ital. *premura*, 'haste, urgency') to 'oppressing'. If 'to press' is the basic meaning of both **stap-* and Av. **sru-*, we can understand *sravatō* in *F*. 12 as 'hurrying', and *zəmō*...*upa.srvatō* in *Vend*. 7.27 as 'oppressive winter'; *vəhrkā̊ŋhō sravaṇhavō*, *Vend*. 18.65, of which the Pahl. transl. is still obscure, can then be taken to mean 'importunate, insistent, harassing, wolves', cf. French *pressant*. For suggestions on the possible meaning of **sravō-*, *srava-*, we must again turn to French *pressant* (this time in the sense of 'urgent'), as well as to Ital. *premuroso* 'solicitous' and Arm. *štap* 'diligence, haste, promptitude, urgency': *sraoyəna-* 'having bustling, fussing, solicitous, women'; *sraoraθa-* 'where cars are hurrying (=fast), *or* prompt (at the owner's command)'; *sraotanū-* 'whose body is impetuous, prompt, of quick motion'; *sravašəmna-* 'whose thrusts are pressing, impetuous, vehement', °*šəmna-* having arisen from compound-thematization (cf. Duch., *Comp*. 37) of °*šamnah-* 'thrust' (suffix as in *rafnah-*, *tafnah-*), cf. *šanman-* 'thrust', above, note 24². Bth.'s etymological connection of *sr(a)vant-* with OInd. *tsárati* (*Wb*. 1649 *sq*.) can be abandoned without regret.

30². *ništarətō.spaya-*. Since Av. ²*fraspāt-* very likely means 'rug' (see Henning, *BSOAS*, XII, 315), the same can be assumed for the etymologically related *spaya-*. Both words may belong to OIr. **span-* (cf. Av. *zaya-*, *zāta-*, to *zan-*) < IE **(s)pen-* (OHG *spinnan*, *spannan*, Arm. *henum*, etc., cf. Walde–Hofmann, *LEW*, II, 279 *sq*.). Bth.'s ²*span-* 'spider' (postulated for the sake of the gen. sg. and nom. pl. *sūnō*), which would have provided a suitable alternative root for *spaya-* and °*spāt-* (IE **ku̯en-*), was eliminated by S. H. Taqizadeh, *BSOS*, IX, 321 *sqq*. A different base is seen in *upairi.spāt-*, see below, note 125⁵.

30³. *niδātō.barəzišta-*, lit. 'where cushion-heaps have been deposited (=piled up, cf. "throwing": "erecting" in OPers., *Asia Major* (N.S.), II, 135 *sq*.)'. I take *barəzišta-* as a compound of *barəziš-* 'cushion' and *stā(y)-* 'heap', whose only attested form *stāiš(ča)* also shows thematic inflection.

30⁴. Lit. '...(of him [see above, p. 161, n.†]) who (being a) Truth-owning (man) worships you with prayer in which (your) name is mentioned with regular utterance, offering libations'. This is the full definition of what later on is concisely referred to as *miθrahe vāxš* 'the utterance of Mithra', see note 137³.

32¹. *hąm hīš čimāne barąnuha* 'trage sie hin zum Sammel-Orte (*Činvaṱ*)' (Wi.); 'zur Freude' (Spi. doubtfully); 'schreibe sie uns auf rechnung' (Gdn.); 'ramasse-les ensemble' (Da.); 'sammle sie auf zur Bußzahlung' (Bth., followed by Htl.); 'bring zie zusammen zum

Vorrat' (Lo.). The assumed base of *čimāne* is thus either *kay-* 'to heap up', or *kāy-* 'to repay'. Benv., *Inf.* 50, without himself pronouncing on the base, rejects Bth.'s interpretation of the form as an infinitive, and posits a nominal stem *čiman-*, with long *ā* as in *puθrāne* (on which see above, note 24[5]). This stem I would analyse as an *-an-* extension (cf. *uxšne*, Benv., *op. cit.* 51) of the OIr. base **čam-* 'to sip, drink', cf. Skt. *camati*, NPers. *čam, čamīdan*: Mithra collects the libations in his chariot (see st. 136) and takes them to his abode in Paradise (see note 50[1]), where he consumes them at leisure. [See Addenda.]

33[1]. *urvaiti dātanąm sravanhąm* 'im Festhalten an den gegebenen Sprüchen' (Wo.); 'um des Bundes der gegebenen Verheißungen willen' (Lo.); 'treu dem gegebenen worte' (Gdn.). In *A* 3.8 a layman who has failed in his religious duty is declared *a-vačō.urvaiti-* 'disqualified from entering a stipulation by mere verbal promise', cf. *vačō.urvati-*, *Wb.* 1343. In our passage the speakers are not so disqualified and consider, accordingly, that the god of Contract is bound to assist them by virtue of the vow *they* have made to honour him.

urvaiti- 'fœdus, fides', which is etymologically related to Av. *urvata-*, Ved. *vratá* 'vow', may be of help in the understanding of the OPers. expression *aurmzdahd^{u}gm ⌜v⌝rtiyiy* in *Beh.* IV, 44 (as briefly indicated in *OIr. Lit.* §11). Benv. has recognized that these two words constitute a form of oath: 'I swear by Ahuramazdāh', *BSL*, 1951, 35 *sqq.*; 50. That he was right in treating the long word as a compound 'Ahuramazdāhian oath' (better 'proclamation'), is shown by the El. version, which, as Weissbach pointed out (*Die Keilinschriften der Achämeniden*, 63, n.), has 'Ahuramazdāhian', not 'Ahuramazdāh'. However, the spelling ⌜v⌝rtiyiy does not favour the reading *vartaiyaiy* (which should have been written *⌜v⌝rtyiy*) adopted by Benv., and the meaning of *vart-* 'to turn' does not suit the context. Until proof to the contrary we may assume that the Indo-Iranian initial group *u̯ra-* survived unchanged in OPers.; by reading *vratiyaiy* we obtain a denominative present stem *vrati-ya-*, in which *vrati-* corresponds to Av. *urvaiti.*† The meaning will then be: 'I vow, state my (good) faith by, a proclamation (made in the name) of Ahuramazdāh.'

33[2]. *paiti.paršti-* 'interpretation'; cf. OPers. *patiprsa-* 'ro read', Man. MPers. *phypwrs-*, Parth. *pdbwrs-*, Sogd. *ptβs-*, Khot. *pūś-* (Bailey, *BSOAS*, XIII, 128), and Av. *paiti.fraxštar-* 'interpreter' (*Yt* 13.92, quoted below, note 92[8]).‡ Other suggestions for *paiti.paršti-* are 'Erfragung' (Bth.), 'Erforschen' (Lo.), 'Révélation' (Da.), 'Gegenfrage' (Htl.), 'Unterricht' (Wi.).

† *vratiyaiy* could alternatively be a denominative of *vrata-*, cf. Bth., *GIP*, I, p. 85, §152, a, β.

‡ I do not see why, to quote Benv., *Noms d'agent*, 22, *paiti.fraxštar-* 'doit signifier "qui informe publiquement" (parallèle à *staotar-*)'.

COMMENTARY   [34.1–34.5

34¹. *framanah-*, cf. OInd. *pramánas* 'cheerful', MPers. *prmyn-* 'to be happy, cheerful', *prmyn* adj., Parth. *frmnywg* 'hope', see Mary Boyce, *MHCP*, 187.

34². *harəθā*, corrected by Caland to *haməraθā*.

34³. Lines 9–10 must either be read

   vanāma ⟨vīspå ṯbaēšå⟩
   vīspå ṯbaēšå taurvayama,

or else *vanāma* should be deleted as a thoughtless insertion before *vīspå*, due to analogy with the two preceding clauses.

34⁴. *sātar-* only occurs in the present formula, where Bth. assigned to it the derogatory meaning 'tyrant', and in sts. 109 and 111, where he took it in the neutral sense of 'ruler'. We shall see, in note 109⁶, that there is every reason to translate it as 'tyrant' throughout. Etymologically *sātar-* perhaps belongs to NPers. *sān* 'whetstone', OInd. *śiśāti* 'to sharpen', and means 'enjoiner', cf. Germ. *Einschärfer*. Wi. (p. 47) connected the word with Ved. *śátru* 'enemy, rival, hostile king', F. B. J. Kuiper, *Acta Orientalia*, XII, 196, with OPers. *θāti*.

34⁵. It is generally held that the *kavi*-s of the present formula and the *kavi*-s collectively branded by Zarathuštra in his Gāthās were princes or rulers (see A. Christensen, *Les Kayanides*, 9 *sq.*).† This assumption implies three distinct meanings of the word *kavi-*, or an unproven development in three stages: (1) OInd. *kaví*, a wise composer of hymns;‡ (2) Av. *kavi-*, members of a class of rulers badly spoken of, because they protected religions, or a religion, opposed to Zarathuštra's preaching; and (3) Av. *kavi-*, title of a dynasty well spoken of, because Vištāspa, the last holder of the title, adopted Zarathuštra's religion. The situation gains in plausibility if we eliminate the second link in this chain. In the Gāthic *kavi*-s, mentioned on a par with *karapan*-s and *usig*-s, who are agreed to have been respective members of two priestly classes, we may as well recognize the Iranian counterparts of the Vedic *kavi*-s: composers of hymns to various gods, who in addition had perhaps assumed certain priestly functions; they would incur Zarathuštra's disapproval because of their insistence on the traditional Indo-Iranian ritual. One particular family of *kavi*-s, whose home was in Sīstān, rose to temporal power and came to rule over the Chorasmian state or part of it; this family used the professional description *kavi* as personal dynastic surname. Zarathuštra, in addressing his protector as Kavi Vištāspa, would not associate the king's dynastic name with the class-name of the priestly hymn-writers

---

† For K. Barr, *Avesta*, 24, the Gāthic *kavi*-s are princelets who combined in their person religious and worldly authority, *sakrale Stammefyrster* (p. 206).

‡ 'Poète qui comprend les énigmes', L. Renou, *JAs.*, 1953, 180 *sqq.*; cf. Lydian *kaveś* 'prêtre, devin, mage', Olivier Masson, *Jahrbuch für Kleinasiatische Forschung*, I, 182 *sqq.* (reference kindly supplied by H. W. Bailey).

('hymn-mongers' to him) who were bent on frustrating his religious reform, any more than we are apt to think of parsons when addressing Mr John Parson. One is reminded of the Barmecide family, which, as H. W. Bailey has shown (*BSOAS*, XI, 2), was named after the head of a vihāra, *pramukha*.† In later times, when the Kavis had taken their place in the remote legendary history of Iran, their title assumed a wider meaning, cf. Christensen, *op. cit.* 61. In Zoroastrian Middle Persian literature kavis are associated with heroes, cf. *kayān u yalān*, Gt. Bd. 177.10, cf. Da., *ZA*, II, 319, n. 132, and *Gt. Bd.* 75.15 (reference kindly supplied by H. W. Bailey, who suggests *vīrān* as a possible alternative to the reading *yalān*); in Manicheism the kavis became 'giants', see Henning, *BSOAS*, XI, 53 *sq.*; Chr. Sogd. *pr qwy'q*, Hansen, *AAWL*, 1954, 843.37, 875.16, translates Syriac *ganbārā'īθ* '*fortiter*', lit. 'in a manly way, heroically', thus indicating that the abstract *qwy'q* 'Kavi-ism' means 'heroism'. Cf. also the *drafš-i kāvyān*, Christensen, *L'Iran sous les Sassanides*, 502, n. 5.

34⁶. *karapan-*, derogatory term for the members of a class of priests of whom Zarathuštra disapproved. W. B. Henning, *Zor.* 45, has recognized its connection with Khwarezmian *karb-* 'to moan' or 'mumble'. K. Barr, *Avesta*, 24, has adopted the translation 'Mumblers'. [Differently V. I. Abayev, *Archiv Orientální*, XXIV (1956), 53.]

35¹. *arənaṱ.čaēša-* 'punisher of guilt', thus Wi.; cf. also B. Geiger, *Die Aməša Spəntas*, 217, n. 1, Bailey, *BSOS*, VI, 71 *sq.*

Geiger rightly pointed out that the present stem from which °(*ṱ*)*čaēša*- was derived (*s*-extension of *kāy-*) is attested in *čaēšəmna-*, *Yt* 19.93: (*kava vīštāspō*) *ašahe haēnayā̊ čaēšəmnō*. However, the clause has been misunderstood by most translators: 'quand il convertissait les hordes au Bien (*litt.* instructeur de l'Asha à la horde)' (Da.); 'um das (beleidigte) Aša zu rächen an dem Feindesheer' (Bth., *Wb.* 464); 'der Lehrer der Wahrheit für sein Heer' (Lo.). Only Gdn., *Drei Yasht*, 58, approximated the obviously correct meaning: 'als er die Feinde des Gerechten

† It can hardly be objected to the view here put forward, that the 'royal Fortune', *kavaēm xᵛarənō* (see Bailey, *Zor. Prob.* 13; 23, n. 1; 30), having belonged already to Haošyaŋha, Yima, Θraētaona, and Kərəsāspa (*Yt* 19.26, 31, 36, 38), all of whom chronologically precede the Kavi dynasty, the 'royal' connotation of *kavi-* must have preceded the foundation of the dynasty. With these early princes the use of *kavaēm* is most likely an anachronism, as it is when Ahura Mazdāh and the Incremental Immortals are said to own *kavaēm xᵛarənō* (*Yt* 19.10, 15). Since the arch-enemy of the Kavis, Fraŋrasyan, who surely coveted the *kavaēm xᵛarənō*, is said instead to have desired the *xᵛarənō yaṱ asti airyanąm dahyunąm*, *Yt* 19.56, *Yt* 5.42, it seems that the 'Kavian Fortune' is simply the 'Fortune of the Iranian Lands' after it had passed into the hands of the Kavi dynasty. Once the *kavaēm xᵛarənō* had become the prerogative of a line of 'Caesars', its meaning could easily be interpreted as 'royal, Caesarean, Fortune', and poets were free to confer it on any human or divine leader of the past who in their opinion deserved it.

zu besiegen im Begriff war'. *haēnā-*, usually '(enemy) army', here stands for 'enemy (army), enemies constituting an invading army, enemies organized for raids'. Hence we may translate: '(Kavi Vištāspa,) avenger of the marauding enemies of Truth.' [See Addenda.]

35². There is no doubt that Gdn.'s definition of *yaoxšti-* as 'sense, perception' (*Studien zum Avesta*, 61 *sqq.*), with which Da.'s agrees, see *ZA*, I, 87, n. 22, fits the few contexts where precision matters (cf. sts. 82, 107) far better than Bth.'s 'skill'.† The etymological connection with the Arm. LW *yuzem* 'to search, ferret out, hunt after, examine' (which scarcely belongs to Av. *yaoz-* 'to be in commotion', as Hübschmann claimed, *Arm. Gr.* 199), is defensible on the assumption that the abstract *yaoxšti-* primarily meant 'faculty of investigating' (cf. OInd. *dŕṣṭi* 'faculty of seeing'), hence 'perception, perceptive quality'. *yaoxšti-* will then have intrusive *-x-*, like *frapixšta-*, etc. [See Addenda.]

With regard to *Yt* 8.8, however, one wonders if Bth. was right in emending (*vāta*)...*yaoxtivantō* to *yaoxštivantō*; this may well be the one case where Bth.'s translation by 'skill(ed)', and etymological derivation from the base *yaog-*, should be upheld as referring to the straightforward *-ti-* abstract of this base.

36¹. The image is taken from the sea in commotion; *Y* 65.4 has almost the same wording: *yaozənti vīspe karanō zrayā vouru.kašaya, ā vīspō maiδyō yaozaiti* 'all the sides of (*lit.* in) the sea Vourukaša are surging, the whole centre is surging'. Cf. also *Yt* 5.38.

36². *arəzō.šūta-* 'battle-tossed', see Benv., *Donum Natalicium H. S. Nyberg Oblatum*, 25, who rightly assigns °*šūta-* to *š(y)av-*, separating it from Bth.'s ²*hav-*. Benv., moreover, thinks (*Asiatica, Festschrift Weller*, 32) that the present stem *apa.xᵛanva-* in *Yt* 14.46 which Bth. assigned to ²*hav-*, means '*repello*', and provides the etymology of Man. Sogd. *x'w-* (B. Sogd. *γw'w-*). However, *apa.xᵛanvainti* has as subject *vāčō* 'words', and is therefore just as likely to belong to the base *xᵛan-* and mean 'to recall'. For Henning's interpretation of B. *γw'w-* see *GMS*, §233, n.

† B. Geiger's theory (*Die Aməša Spəntas*, 216 *sqq.*) that Av. *yaoxšti-* has the meaning of Ved. *māyá* 'magic skill (etc.)' (cf. above, Introduction, p. 5 *bis*), strikes me as unlikely. Why should such a characteristic quality as *māyá*, if it had survived as an Indo-Iranian inheritance in the Avestan religious repertoire, not have been called by its name, which in Av. would be **māyā-*? At any rate Old Indian *yukti* 'union', to which Geiger looks for etymological support, is in no way a synonym of *māyá*, the meaning 'cunning device, magic' being only attested in classical Sanskrit. Geiger's argument that the Ādityas use *māyá* for the purpose, *among others*, of watching and spying, proves nothing for the Avestan Mithra's *yaoxšti-*, which as far as a purpose is discernible *only* serves to perceive, and was therefore more appropriately defined by Gdn. and Da. as 'perception'. Significantly, when Av. *yaoxšti-* is a quality of divinities other than Mithra, Geiger wants it to have a different connotation from Mithra's *yaoxšti-* (*op. cit.* 223, n.), though he does not say which.

36.3–38.1]   COMMENTARY

36³. The *hapax legomenon frā...xrå̄nhayete* was taken by Bth. (after Trautmann) as an iterative of a base **xrāh-* related to OE *hrōrian* 'to stir', see *ZWb*. 164.

37¹. The meaning of *āiθi-* is given by Arm. *ah* 'fear', as pointed out by Benv., *TPS*, 1945, 68 *sq*. On *xšayamnō* see note 23² above.

38¹. *xrūmā̊ šaitayō frazinte* 'die grausigen Wohnstätten werden verwüstet' (Bth.); 'grausig werden ihre Wohnstätten vernichtet' (Lo.); 'sinistres sont les demeures, désolées' (Da.); 'to earth the dwellings are destroyed' (Hzf., p. 425). This sentence closely resembles the end of *Yt* 13.38:

   yūžəm taδa taurvayata
   vərəθrəm dānunąm tūranąm
   . . . . . .
   yūšmaoyō parō karšnazō
   hvīra baon səvišta
   . . . . . .
   xrūmā̊ asəbiš frazinta
   dānunąm baēvarə.paitinąm

'die grausigen Stätten der über zehntausend gebietenden Dānu's werden verwüstet' (Wo.); 'die grausigen (?fest? vgl. *paxruma-*) Städte der Dānus, die zehntausende beherrschen, werden verwüstet (?verlassen?)' (Lo.); 'verwüstet stehen die Flecken der zu Myriaden gefallenen Dānu leer' (Gdn., *Drei Yasht*, 136): 'sinistres sont les demeures ravagées des chefs des myriades des Dānus' (Da.).

Bth. saw in *frazinte/a* 3rd plurals pres./injunctive of a passive present stem *zya-* from the base *zyā-*. In fact the only present stem attested both clearly and widely is Proto-Iranian **źin(ā)-*: Av. *zinā-*, OPers. *dinā-* (and *dina-*, thematic), Sogd. *zyn-*, Khot. *ysin-*. The suspicion thus aroused that the *n* of *frazinte/a* belongs to the stem, rather than to the ending, is strengthened by *Yt* 13.38, where *asəbiš*, if subject, would be the only substantival instr. plur. in *-biš* of the Younger Avesta that *must* be taken as fulfilling the function of a nominative.† If one takes into account that the two preceding sentences in *Yt* 13.38 have *yūžəm* as logical, the first also as grammatical subject, *frazinta* will be most easily understood as 2nd pl. pret. act. of *fra-zinā-*. Such a form has not yet been recognized in Avestan, but it has long been realized that where the present-formative of the (Indian) ninth class appears as *-nī-* before a consonant, Avestan has

---

† See E. Schwyzer, *IF*, 47, 251. The supposed nominative *dāmə̄bīš* (see Schwyzer, p. 249 bottom), occurring in *Y* 19.19, was translated as an instr. by Wo., a loc. by Da. Presumably it does not at all mean 'creatures', but 'intentions', cf. *huδāman-* (etc.). Accordingly *kaṱ hvarštəm? staotāiš aša.paoiryāišča dāmə̄bīš* means: 'What is done well? (What is done) through prayers and intentions headed by (see Duch., *Comp.* 148) Truth.'

merely -n-, cf. Jackson, *Avesta Grammar*, 166, Bth., *GIP*, I, p. 74; hence °zinta corresponds to Skt. **(a)jinīta*. For *frazinte* as 3 sg. pres. middle, corresponding to Skt. **jinīte*, there is already an Av. parallel in *vərənte* 'he chooses'.

As regards *xrūma-* already Lo. queryingly compared Av. *paxruma-* (*Die Yäšt's*, 116, n. 4), which I suggested in *JRAS*, 1942, 101, should be connected with Sogd. *xrwm, xwrm* 'earth, soil', cf. also Bailey, *TPS*, 1945, 34. This suggestion was taken up by Hzf., p. 429, and extended to *xrūmą̇* and *xrūmīm* in the present stanza. *xrūmą̇* was for Hzf. 'loc. or instr. of the noun', and *xrūmiya-* 'earthy in the sense of sandy or stony'; hence he translated lines 5 sq. 'the cattle...is driven the stony way of captivity'. One may hazard the opinion that *xrūma-*,† as a derivative of the base **kreu-* 'to collapse, break up' (Pokorny, *IEW*, 622 sq.), means 'crumbling'; substantivized as a neuter, 'the crumbling (substance)' κατ' ἐξοχήν came to mean 'soil, earth, dust'; from this noun a further adj. *xrūmya-* was derived, see note 38[4].

We may accordingly translate *Yt* 13.38 as follows: 'You then overcame the valour of the Turanian Dānus...;‡ through you the Karšnazids became the strongest good warriors...; you swept away the crumbling sites§ of the chiefs of the myriads of Dānus'.||

38[2]. *šaitayō*. To the examples quoted by Bth. of Av. *-ai-* standing for original *-i-* (*Wb.*, s.v. *bawrini-*), *saite* should be added in *Yt* 5.102: *kəm kəmčiṯ aipi nmāne gātu saite x^vaēui.starətəm*. Here according to Bth., *Wb.* 1571, the subject of the sentence is missing. Hence Wo. translates 'bei jedwedem im Haus liegt auf dem Divan mit schöner Decke... (wer?)'. Lo. has 'in jedem Haus steht (*but should be*: liegt) ein schöngebreitetes Lager'. It is clear that the emendation to *saēte*, proposed by Gdn., yields no satisfactory meaning. We must imply the copula, recognize in *saite* an infinitive standing for expected **site* (cf. *ite, mruite, daste*), and translate: 'in everybody's house, spread astride,¶ (is) a couch to lie on.'

† The length of *ū* is vouchsafed by Sogd. *xrwm*'s being a heavy stem, cf. *GMS*, §484 *sqq*.

‡ [According to V. I. Abayev's attractive suggestion, *Archiv Orientální*, xxiv (1956), 44 sq., the Dānus are the inhabitants of the banks of the Yaxartes = *Τάναϊς* = Bth.'s ¹*dānu-*.]

§ Proleptically, = the sites which crumbled as a result of your onslaught.

|| Bth. rightly doubted W. Geiger's and dismissed Gdn.'s translation of *dānunąm baēvarə.paitinąm*. However, by ignoring, according to his usual practice, Da.'s opinion, he missed what is almost certainly the correct interpretation. For genitives governed by first members of compounds see above, p. 160, n. †.

¶ Thus I tentatively translate *x^vaēui.starətəm*, instead of emending it to *x^vaini.starətəm* with Bth. Cf. Oss. *xiŭ*, Dig. *xeŭ*, 'with straddled legs', which phonologically could be derived from *x^vaēui-*. *starəta-* is said of a *gātu-* also in *Yt* 17.7 (quoted below, p. 208, n. †).

38³. We find *haiθīm . ašavajanas°* here and in st. 45, and *haiθīm . ašavan-* in Y 11.1:

  θrāyō haiθīm . ašavanō
  āfrivačaŋhō zavainti
  gāušča aspasča haomasča

'Drei wirklich der Wahrheit Ergebene fluchen Verwünschungsworte: das Rind, das Pferd und der Hauma' (Lo.). Accordingly *haiθīm . - ašavajanas°* is by everybody taken to mean 'who slay those who are really (*or* truly) owners of Truth'. One immediately asks: who are the 'really' *ašavan*, and in what respect do they differ from ordinary *ašavan*-s? Are there beings more 'really *ašavan*' than Ahura Mazdāh, who yet is simply called *ašavan*? Y 11.1, the only passage from which we learn what sort of beings are considered *haiθīm . ašavan-*, suggests that 'really' or 'truly' is not in this case an exact rendering of *haiθīm*. It should be noted that *haoma-* in Y 11.1 is not the god, or even the plant, but as appears from Y 11.3, the drink prepared of the plant; this drink, naturally enough, is identified with the *ašavan* god Haoma in Y 11.3, but it nevertheless can be called *ašavan* only by extension of the proper original use of the word.† The same extension has to be assumed in the case of ordinary cows and horses.

  Paradoxically enough the 'really' *ašavan* thus appear to be less qualified to bear this title than ordinary *ašavan*-s, but the paradox is readily explained. We, too, are apt to say of somebody that 'he is a real saint (or devil)', meaning, actually, that although he is not quite a saint (or devil) he may be considered one for practical purposes. *haiθya-* 'what is, what is real', could easily develop the connotation of 'what is virtual, *de facto*'. *haiθīm . ašavan-*, being an adjective, may well act as a neuter in the compound *haiθīm . ašava-gan-*: not 'he who slays persons that are *haiθīm . ašavan-*', but 'he who strikes at what is *haiθīm . ašavan-*'.

  With this interpretation, the episode of the suffering cow in the second half of the stanza arises without strain from what precedes, and need not be considered an interpolation, as is usually done. In the eyes of a Zoroastrian, we may gather from Y 11.1, the chief item that is 'virtually Truth-owning' is the cow, and it would not be surprising if the compound *haiθīm . ašavan-* had been invented primarily with the cow in view. We may accordingly thus paraphrase the first four lines of the stanza: Mithra makes havoc of the contract-breakers' dwellings, because they strike at the most-nearly-*ašavan* piece of property of the *ašavan*-s, that is, the cow, whose sufferings are as follows.

  In st. 45 the cow is not mentioned, but there is no objection to maintaining also in that passage the proposed interpretation: 'those whom seek out the owners of Falsehood' are, of course, the owners of Truth them-

† The drink *haoma* can scarcely be said to 'own Truth'; but its function, according to Y 10.8, is to produce an intoxication that is 'accompanied by gladdening Truth'; cf. also Haoma's epithet *aša-vāzah-* 'inducing Truth'.

COMMENTARY   [38.3–38.6

selves, but the owners of Falsehood, whose 'seeking out' the *ašavan*-s includes ill-treating their cattle, can still be correctly described as 'striking at what virtually owns Truth', i.e. cows.†
   The above version of line 4 also differs from all others in that I take °*ča* here and in st. 45 as co-ordinating two adjectives, *miθrō.drujō* and *haiθīm.ašavajanas*°, and not two substantives, *miθrō.drujō* and *drvantō*.
   38^4. The *hapax legomenon xrūmīm*, rather than an adverb in original -*īm*, as Bth. took it to be, is acc. sg. masc. of *xrūmya*-, parallel to *varaiθīm* from *varaiθya*-. As a -*ya*- adj. derived from *xrūma*- 'earth, soil, dust', it may have the meaning 'dusty' as suggested by Hzf., see note 38^1, p. 189.
   38^5. None of the translations proposed for *yā darənāhu miθrō.-drujạm mašyānạm frazaršta aēšạm raiθya* seems to me satisfactory: 'welche an stricken von den treubrüchigen menschen mit dem heereszug (instr. of *raiθya*- "wagenzug") fortgeschleppt wird' (Gdn.); '(wenn) es zu den Schlupfwinkeln der miθrabetrügenden Menschen an ihrem Wagen fortgezerrt (wird)' (Wo.); 'die in die Wohnungen der m. Menschen fortgeschleppt wird an deren Wagen' (Lo.); 'qui paît dans les vallons des hommes Mithrô-druj: traîné sur la route...' (Da.); for *darənāhu* and *raiθya* Htl. and Hzf. (425 *sq*.) have respectively 'Befestigungen', 'Straße', and 'pens (hovels?)', 'chariot'; Spi. connected *darənāhu* (as *darna-) with NPers. *darra* 'valley', cf. Horn, *Npers. Et.* 124.
   Da.'s 'on the road' is not supported by the firmly attested long *ā* of the word for 'road', cf. NPers. *rāh*, Parth. *r'h̠*, Sogd. *r'δδ*, etc. On the other hand, as a loc. of *raθa*- 'chariot' *raiθya* should not mean 'at, *or* by means of, the chariot', which in any case is an odd idea; as to having the cow 'on, *or* in, the chariot', this would be absurd. I therefore prefer to take *raiθya* as nom. sg. fem. of an adj. *raiθya*- meaning 'attached to a chariot = draught-animal'. As to *darəna*-, if we follow Bth. and identify it with Skt. *dharaṇa* 'act of holding', the obvious guess at its meaning will be 'clutch'.
   38^6. From O. Friš's discussion of the last two lines of this stanza, *Archiv Orientální*, xx (1–2), 1952, 598 *sqq*., I accept the interpretation of °*azānō* (which forms a compound with *asrū*°) as belonging to the base *az*-; cf. *nyāza*- 'to tighten', *urvizō.maiδya*-, 'having a narrowly laced waist', < *urvyaza- (Henning, *Sogdica*, p. 34, n. 4), Sogd. *'zγw(h)* 'anxiety' (see *GMS*, §403, n.), and Yaγn. *azōn*- 'to torment'‡ (= *az*- + causative formative -*ōn*-).

† It is, of course, no objection to our argument that the mythical three-legged ass that stands in the middle of the ocean is referred to in *Y* 42.4 as *xarəm yim ašavanəm*, and not as **haiθīm.ašavanəm*. This particular *xara*- is a divine being, whereas the cow and the horse of *Y* 11.1 are ordinary specimens of cattle.
‡ 61st sentence of *Yagnobskaya Skazka*, published by S. I. Klimchitski in *Akademia Nauk SSSR, Trudy Tadzhikistanskoi Bazy*, ix (1938), pp. 94 *sqq*.: *déhčę azónčę badán-š siyóh kúnčę* 'он его бьет, мучит и чернит его тело'; the Glossary has *azōn*-, past partc. *azōnta*, 'мучить'; perhaps 'to torment by strangling, causing to choke'.

38⁷. *hištənte* is here scarcely, as Friš thinks (*art. cit.*), an auxiliary verb ('sind ständig von Tränen bedrängt'), since the act of obstinately *standing* fits in well with the cows being forcibly dragged forward (*frazaršta*) by the *miθrō.druǰ*-s.

39¹. *asəmnō.vid-* and (st. 40) *asəmnō.gan-*, according to Bth. 'not finding (hitting) = missing, the target', in support of which explanation he later produced a very uncertain etymological connection of *šəmna- with Arm. *nšavak* and MPers. *nišān*, see *ZWb.* 97. One may, however, doubt whether *šəmna- means 'target'; for in st. 40 the loc. *sarahu*, rather than indicating the direction in which knives and maces are thrown or swung, appears to refer to the final point on which these weapons come to rest, in which case the weapons cannot be said to have 'missed the target'. It seems that Da. had the right intuition when although translating 'manquent le but' he commented: 'peut-être...n'obtenant pas le *shanman*, le coup', and connected *šanman- with OInd. *kṣaṇóti* 'to hurt, injure, wound'. We need not follow Da. in assigning to the same base *šanman-* 'thrust', on which see above, note 24². But we may say that as *šam-man 'thrust' (beside **šam-nah-*, note 30¹) became *šanman-*, so **šan-man-* 'wound, injury' through nasal metathesis became **šamnan-*, which in compound junction was thematized to *šəmnō°* (cf. similar cases quoted by Duch., *Comp.* 9). That OInd. *kṣan-* goes back to IE **k̂sen-*, which in Iranian was bound to become *šan-*, is likely in view of Sogd. *pšyn-* 'to trap', *ptšng* 'cross, torture-instrument', which according to Henning belong to this base, see *GMS*, §616, n. The idiom implied in **šəmno.jan-* 'striking a wound' can be compared with *pištrəm*† *jainti* 'he strikes an injury' in *Vend.* 13.10 sq. *°vid-* in *asəmnō.vid-* is then likely to mean much the same as *°jan-*, and should therefore not belong to *vid-* 'to find', but to OInd. *vídhyati* 'to pierce, hit, strike', cf. Khot. *bid-/bist-* 'to pierce'.

39². *yaθa grantō...miθnāiti miθrō* 'wenn M....(ungnädig) entgegentritt' (Gdn.), 'wenn M....beleidigt bleibt' (Wo.), 'wenn grimmig... M....es vereitelt' (Lo.), '...unter ihnen weilt' (Htl.). As Gdn. saw, *miθnāiti*, which seems to mean 'is opposed, antagonistic, hostile', most likely belongs to Ved. *mith-* 'to oppose, combat', to which Hzf., *Ap. I.* 185 sq., assigned Av. *hamaēstar-* 'opponent, combatant', *hamista-* 'combated', *hamisti-* 'engagement'. Whether OPers. *hamiθriya-* and *miθah-* also belong here, as Hzf. claimed, is an open question. For other views on *hamiθriya-*, with which Bailey connects Arm. *amehi* 'unruly' (*JRAS*, 1951, 194), see Henning, *BBB*, p. 111a, and Kent, *Old Persian*, *s.v.*

39³. *apaiti.zanta-*. On the meaning of *pati-zan-* see above, p. 160.

39⁴. *fradaxšanya-*, adj. to *fradaxšanā-* 'sling'. This, according to Da., *ZA*, II, 215, is glossed by NPers. *falāxan*. The latter, if it is a Sogdian

† On the ultimate base of *pištrəm* cf. *BSOAS*, XVII, 480, n. 1.

COMMENTARY   [40.1–41.1

LW, belongs perhaps to the same base as Av. *fra-daxš-*, without the *s*-extension (cf. Av. *rąxš-*, *daxš-*, Sogd. *puxš-*, above, note 27[1]), as it may represent Sogd. **fδ'xn* < **fra-dāx(a)na-*. [See Addenda.]

40[1]. [Instead of 'thrown at' read 'brought down upon', in accordance with H. W. Bailey's analysis (*TPS*, 1956, 97) of *niγrāire* as belonging to the base *gar-* 'to take', which Benveniste had recognized in *abi-gar-*, see below, p. 226, n. †. But cf. Burrow, *BSOAS*, xx (1957), 137.]

41[1]. *sraoša-* etymologically means 'obedience', cf. *sraošin-* 'obedient, willing', Gāthic *sǝraošan-* 'obedient', Parth. *srwšyft* (see Henning, *BBB*, p. 82). Since religious obedience demands on the part of a Zarathuštrian constant struggle against all manifestations of evil, as well as suppression of laziness and sleepiness by means of vigilance, the word *sraoša-* acquired the connotation of 'discipline', cf. its derivative(s) *sraošyǎ-* 'disciplinary measure, punishment', above, note 26[1]. The god Sraoša is accordingly the hypostasis and genius of Discipline, see Benv., *Rev. hist. relig.* 130 (1945), II, 13 *sq.*, *JAs.* 1954, 304. In this capacity he is a natural ally of Mithra the guardian of the Contract, and of Rašnu the judge (cf. below, p. 223) whose duty it is to pass sentence on contract-breakers. The divine triad Mithra, Rašnu, and Sraoša, remains throughout the development of Zoroastrianism in charge of prosecuting the wicked (cf. Zaehner, *Zurvan*, 103, n. 2, 363, Pavry, *The Zoroastrian Doctrine of a Future Life*, 82 *sqq.*), of which charge the present passage is the earliest evidence. In addition, we shall find indication later on (p. 223, (1)) that the members of this triad may have acted as tutelary genii of the house. As a result of Sraoša's association with Mithra, the god of Discipline came to share with the god of Contract certain attributes and items of equipment (see B. Geiger, *Die Aməša Spəntas*, 110 *sq.*). Nevertheless, as Lo. observed in *Die Yäšt's*, 86 *sq.*, the 'Zarathuštrian' character of Sraoša is on the whole well preserved. Cf. Introduction, p. 49.

Sraoša's specific function within the triad Mithra–Rašnu–Sraoša must have been that of a punisher. This can be argued not only on etymological grounds, but also from his being replaced in *Yt* 14.47 by Vərəθraγna:

> vərəθraγnəm ahuraδātəm yazamaide
> yō vīrāzaiti antarə rāšta rasmana
> āča parača pərəsaite
> haδa miθra haδa rašnvō:
> kō miθrəm aiwi.družaiti
> kō rašnūm paiti.irinaxti
> kahmāi yaskəmča mahrkəmča
> azəm baxšāni xšayamnō?

'Ahura-created Vərəθraγna we worship, who goes up and down between the lined-up regiments. Together with Mithra, together with Rašnu he inquires right and left: who is false to Mithra, who deserts Rašnu? On whom may I bestow illness and death?' (cf. *Yt* 10.110). The purpose for which Vərəθraɣna makes these inquiries is obvious from *Yt* 10.70–2: he is going to exact fearful retribution from the offenders of Mithra and Rašnu.† Hence, where in his stead Sraoša appears we may expect that he, too, is fulfilling the role of persecutor of *miθrō.drug*-s.

In st. 52 the third member of the triad is not Rašnu but Nairyō.saŋha. The replacement is understandable as arising from a contamination of the two distinct triads to which Sraoša belongs: (1) Miθra–Sraoša–Rašnu, and (2) Sraoša–Aši–Nairyō.saŋha (attested in *Y* 57.3 = *Yt* 11.8). Sraoša and Aši alone are found together in *Y* 10.1, and above all in Zarathuštra's verse *Y* 43.12, where Sraoša comes with Aši to apportion the awards, the word used for 'award' being *aši-*. There is thus a presumption that Sraoša's standing epithet *ašya-*, which only occurs as epithet of Sraoša, refers not as Bth. thought to Aša, but to Aši.‡ It is true that Sraoša has also strong ties with Aša: he is *hišārō ašahe gaēθā̊* 'watching over the (living) world of Truth' (*Y* 57.17), and is described in the introductory formula of each *karde* of his Hymn (*Y* 57) as 'the Truth-owning *ratu* of Aša' (*ašavanəm ašahe ratūm*). This very formula, however, of which Sraoša's epithet *ašya-* is also a standing feature, provides further reason for deriving *ašya-* from *Aši-*; for if we connect *ašya-* with *Aša-*, the formula will relate Sraoša to Aša by three different words (*ašīm, ašavanəm, ašahe*), and lack all reference to Sraoša's undoubtedly close connection with Aši. In addition, Sraoša's (and again exclusively Sraoša's) epithet *ašivant-* must be borne in mind. This, according to Bth., is derived from the common noun *aši-* 'award, portion', and therefore means 'the apportioner, awarder'. Bearing in mind *Y* 43.12 referred to above, where Sraoša is mentioned in one breath with Aši the goddess and *ašīš* 'the awards', it seems likely that *ašya-* refers to the goddess, as *ašivant-* refers to the award of which the goddess is the hypostasis. Accordingly I have chosen to translate *ašya-* by 'the friend of Aši'.§ [See Addenda.]

† In *Yt* 14.63 the form this retribution takes is described in terms identical to those used of Mithra in *Yt* 10.48: *āat̰ yat̰ vərəθraɣnō ahuraδātō | θraxtanąm rasmanąm | yūxtanąm šōiθranąm | miθrō.drująm mašyānąm | apąš gavō darəzayeiti | pairi daēma vārayeiti | apa gaoša taošayeiti | nōit̰ pāδa viδārayeiti | nōit̰ paiti.- tavā bavaiti* 'When Ahura-created Vərəθraɣna, once the regiments have closed in and the homesteads are up in arms, fetters behind the evil hands of men false to the treaty, switches off their eyesight, deafens their ears, (then) one no longer disjoins the feet, one has no strength to counter'. (This is, with minor variations, the generally agreed interpretation of this passage, except that in my opinion the situation, and the parallelism with *Yt* 10.48, indicate that *θraxtanąm rasmanąm* and *yūxtanąm šōiθranąm* are two genitive absolute clauses. The passage is happily unequivocal as regards the extent of protasis and apodosis, see notes 48[1] and 48[3].)

‡ This is taken for granted by Nyberg, *Rel.* 66, who on the ground that Sraoša is Aši's brother in *Yt* 17.16 (see *infra*) translates *ašya-* by 'mit Aši verbunden'.

§ *Sraoša- ašya-* became in Middle Persian *Srōš-Ahrai*, the name which the Persian Manichees gave to the 'Column of Glory'; in Sogdian Manichean texts

COMMENTARY [41.1–42.1

In *Yt* 17.16 Aši, described as Ahura Mazdāh's and Ārmaiti's daughter, is said to have as brothers Mithra (who, having been created by Ahura Mazdāh, cf. sts. 1 and 61, was evidently considered to be his son), Sraoša, and Rašnu, as sister the Mazdayasnian Religion. In the light of this statement we can understand Mithra's close connection with both Aši and the Religion (cf. below, note 68[2]); the latter even replaced her brother Sraoša in *Vyt*. 52, where instead of the usual triad Mithra–Rašnu–Sraoša we meet Mithra, Rašnu, and the Mazdayasnian Religion acting in concert.†

Finally we note with Benv. (*The Persian Religion*, 92) and Zaehner (*Zurvan*, 103), that Sraoša, by 'watching over the truces and treaties of Falsehood (and) the most Incremental (Spirit)' (*āxštišča urvaitišča drujō spasyō spəništahe*, *Yt* 11.14), anticipates the function of μεσίτης 'Mediator' which Plutarch attributes to Mithra.

41[2]. *paiti θrātāra yazata tē rasmanō raēčayeinti* '(der ašafromme Sraoša jagt (sie) von allen Seiten zusammen) den schirmenden Yazata's entgegen. Diese Schlachtreihen gibt er (dem Verderben) preis,...' (Bth.-Wo., reading with most MSS. *raēčayeiti*); '...den beiden schützenden Göttern entgegen; sie lassen die Schlachtreihen im Stich...' (Lo.). But what sort of punishment is it to be driven towards one's protective gods? Da., with his usual common sense, noticed the incongruence and proposed to emend *θrātāra* to **θrāstāra* 'terrifying'. A simpler solution is to take Gdn.'s text as it stands and see in *θrātāra yazata* the subject‡ of *paiti...raēčayeinti*; cf. *paiti.irinaxti* 'he deserts' in *Yt* 14.47, quoted above, p. 193. *tē rasmanō* is thus the object not only of *paiti...raēčayeinti*, but of all the preceding verbs as well.

42[1]. *ime nō aurvantō aspa para miθrāδa nayente* 'die da führen, o Miθra, unsere schnellen Rosse hier fort' (Bth.-Wo., emending *miθrāδa* to *miθra iδa*); 'diese schnellen Pferde werden uns, o Miθra (?)...(?) fortgeführt' (Lo.); 'voilà nos chevaux rapides emportés loin de Mithra' (Da.); 'diese unsere Kampfrosse werden von Mithra weggeführt' (Wi.). Of these translations Wi.'s is certainly the most sensible, especially in conjunction with his understanding of the next two lines, see note 42[2].

---

the name turns up as *srwšrt*, in which form it passed into Uighur (see Le Coq, *Türkische Manichaica*, III, p. 6, b R, 3); cf. Henning, *BSOAS*, XI, 216, n. 8, Waldschmidt and Lentz, *SPAW*, 1933, 504.

† A further variation in the membership of the triad is met with in *Yt* 13.3 (quoted above, Introduction, p. 37), where Ahura Mazdāh is said to be escorted by Mithra, Rašnu, and Ārmaiti. Ārmaiti is not only Aši's mother, but also, being Ahura Mazdāh's daughter (cf. above, p. 168, n. ‡), her sister, as well as the sister (and mother?) of Mithra, Rašnu, Sraoša, and the Religion. As such she may be replacing either of the last two divinities in *Yt* 13.3.

‡ *θrātāra* as a thematic nom. pl. falls in with the examples of them. nom., gen., voc. sg., and acc. and gen. pl. of *-ar-* stems which Bth. quoted in *GIP*, I, §400 end.

Nevertheless, one would expect that not the horses (or the arms) of the harassed soldiers are at stake, but their lives. Chased about by the avenging gods they would spur the horses to flee from Mithra, and be horrified at finding that the animals instead of obeying take them straight *to* Mithra. Such disobedience would be in keeping with the perverse behaviour of the contract-breakers' horses and spears in st. 20. In tentatively taking this view of the situation, we shall replace *nayente* by the variant *nayeinti* of K_{15} (cf. also L_{18} and J_{10}). As to *para* with the abl., meaning 'before, in front of', I am assuming an extension to local use of the temporal preposition *para* 'before' + abl.; admittedly elsewhere *para* (OInd. *purā́*) has local value only in the OPers. adverb *avaparā* 'dort davor' (Bth.); but the well-attested local, beside temporal, value of *parō* (OInd. *purás*) may have encouraged the occasional employment of *para* in a local sense. As to a locative expression 'in front of M.' being used with a verb of movement in the sense of 'towards (the presence of) M.', the comparable use of the simple locative case (see Reichelt, *Aw. Eb.* § 512) offers a parallel.

42². In the last two lines I have followed Lo.'s interpretation. Wo.: 'die da machen, o Miθra, unsere starken Arme mit dem Messer zunichte'; similarly Da. and Hfz. 425; Wi: 'diese unsere starken Arme (und) Schwerter werden von Mithra vernichtet'; Htl.: 'jene mit den gewaltigen Armen, Miθra, zerschlagen unsere Schwerter'; Gdn.: 'und metzeln uns mit starkem arm und messer nieder, o Mithra'. There are two possibilities of accounting for *uγra.bāzava* as translated by us: either as nom. pl. (with them. ending) of the bahuvrīhi *uγra.bāzu-*, which is attested in st. 75; or as two words, 'strong arms', metonymically meaning 'strong-armed individuals' (cf. Wn. II, 1, 288 *sq.*). It is in accordance with the latter usage that Gdn. and Lo. translated *Yt* 13.136 *kərəsāspahe... fravašīm yazamaide paitištātāe uγrahe bāzāuš haēnayåsča pərəθu.ainikayå̊* 'we worship Krsāspa's Fravaši, for resisting the strong-armed, and (for resisting) the wide-fronted hostile army...'.

43¹. Lo.'s interpretation of this formula as a *climax ascendens*, see *Die Yäšt's*, p. 37, n. 1, is preferable to Bth.'s *climax descendens*. [Addenda.]

44¹. On this 'abode' of Mithra's, as against the one he has on Mount Harā, cf. note 50¹.

44². *vīδātəm* is also used in connection with Sraoša's house in *Y* 57.21, where the Pahl. transl. has *wyn'rt* 'durably fixed' (cf. on this MPers. verb Henning, *TPS*, 1954, 175, n. 1); similarly *vīδāti-* in *Vend.* 13.49, *nōiṯ mē nmānəm vīδātō hištənti* 'my house(s) would not stand in a state (of being solidly) set'. Here the base is IIr. *dhā-*, whereas in *vīδātəm* in st. 64 it is *dā-*.

44³. It is not usually realized (cf., e.g., Wn.-Debr., II, 2, 166; III, 255) that the neuter nom.-acc. sg. of *mazant-*, corresponding to Ved. *mahát*, does not occur in Avestan, where in its stead *maza* is used, see below, note 64³. *mazaṯ* in the present passage is nothing but an emendation of Westergaard's, which on closer inspection turns out to be concealing a hitherto unrecognized Av. derivative of the same base. We must restore to the text the reading of F_1 (cf., on this MS., Gdn. in the *Prolegomena* to his edition, p. xliv), which is *mazāδa*, and postulate an abstract noun *mazatāt- 'greatness, size', comparable to OInd. *mahatā*. On the analogy of *amərətāt*- from *amərətatāt*-, *haurvāt*- from *haurvatāt*-, and *arštāt*- from *arštatāt- (cf. note 139²), *mazatāt- was shortened to *mazāt-. On the other hand, as beside *fratəmatāt*- in *Yt* 13.95 (quoted above, p. 27) we have *fratəmaδāt*- in st. 18, it is safe to allow two side-forms respectively, *mazaδāt- beside *mazatāt-, and *mazāδ- beside *mazāt-. *mazāδa* then reveals itself as the instr. sg. of *mazāδ*-, governed by *anazō*, and the convincing sense we obtain supports the validity of this reconstruction. Moreover, with the above history of *mazāδa* in mind, all the variants quoted by Gdn. become understandable: *mazdaδa*, *mazdāδa*, *mazdāda*, *mazdāṯa*, are corruptions of *mazaδāδa (itself a contamination of *mazaδāta and *mazāδa*) in which the second *a* was dropped under the influence of *mazdaδāta*- which some scribes adopted altogether (*mazda δāta*, *mazdaδātanąm*); *mazāṯ* is a variant of *mazāδa* as *hamərəθāṯ* of °*θāδa* (st. 71), *zəmāṯ* of °*māδa* (st. 72), or *yānāṯ* of °*nāδa* (st. 137); finally *mazāδadāt* is a contamination of *mazaδāδa, *mazāδa*, and *mazāṯ*.

44¹. *pərəθu aipi vouru.aštəm* 'gar weithin weites Unterkommen gewährend' (Wo.); 'mit breiter umfriedigung umschlossen' (Gdn.); 'breit und von lieben Gehilfen bewohnt' (Gdn., *Lesebuch*); 'das gar breit ist und weite Unterkunft(?) bietet' (Lo.); 'to-far-distances with-wide-shelter' (Hzf., 452, taking *pərəθu.aipi* as a compound). This particular abode seems altogether too vast to offer 'Unterkunft'. Benv., *Inf.* 39, translates *vouru.aštəm* by 'aux larges haltes', on the strength of his interpretation of Gāthic *aštō* as 'halte'; we shall see in note 138¹ that this interpretation is anything but assured. Accepting Bth.'s and Benv.'s connection of °*aštəm* with *as*- 'to reach', we should rather see in *vouru.- ašta*- a determinative compound with a deponential -*ta*- adj. (cf. Wn.–Debr., II, 2, 576 *sqq*.) as second member, hence meaning 'which reaches widely'.

It would be legitimate to imagine the instr. sg. **vouru.ašta* being used adverbially (cf. Reichelt, *Aw. Eb.* §455) in the sense of German 'weithin', hence lit. 'wide-reachingly'. Such an assumption may account for OPers. *avθašta*, to be read *avaθāštā*, interpreted as *avaθā* 'then' + *aštā* (instr.), and translated 'up to then, until then', lit. 'then-reachingly, reaching (the period called) then'. The passage in which the word occurs (*Beh.* IV, 70 *sqq*.) reads as follows:

(*a*) *tuvam kā hyaʰ aparam imām dipim vaināhi tyām adam niyapaišam imaivā patikarā mātya vikanāhi yāvā da-* (line 72) *θa-sa āhi avaθāštā paribarā* 'you who hereafter will see this inscription which I had engraved,

or these sculptures, do not destroy (them); as long as you are *da-θa-sa*, until then do look after (them).'
This sentence must be compared with lines 77 *sq.*:
(*b*) *yadi imām dipim imaivā patikarā vaināhi vikanāhi-diš utā-tai yāvā taumā ahati nai-diš paribarāhi* 'if you see this inscription or these sculptures, (if) you destroy them and do not look after them as long as to you is strength....'†
It may be objected to our interpretation of *avθašta*, that the parallelism with -*diš paribarāhi* in (*b*) demands that *paribarā* in (*a*) be also preceded by a pronoun that refers back to *dipim* and *patikarā*. The answer is that such a pronoun, corresponding to -*diš* after *vikanāhi* in (*b*), would be equally desirable after, or before, *vikanāhi* in (*a*), yet is missing. We can therefore maintain with good cause, that on both occasions when (*b*) has -*diš*, in (*a*) the enclitic pronoun is to be understood from the context.

Before leaving this question, a note must be added on *da-θa-sa*, which in Kent, *Old Persian*, appears as *utava*. In *Journal of Cuneiform Studies*, 5 (1951), 52, Cameron confirmed King and Thompson's reading of the first and last signs as *da* and *sa* respectively. For the middle sign Cameron hesitatingly mentioned the reading *di* as a faint possibility. Asked for further details Prof. Cameron obligingly sent me (28 March 1955) a drawing of the crucial signs, which are given below with his kind consent. By way of comment Prof. Cameron pointed out that 'what it (= the first sign of line 72) actually looks like is ⟨⟨ ⟨ = nothing'. On considering this verdict I came to the conclusion that the sign could only be ⟨⟨⟨ = *θa*. Consulted again, Prof. Cameron kindly wrote back (21 September 1955): 'I quite agree that the first character of line 72 could more easily be *θa* than anything else.'

What to my mind confirms the correctness of the reading *da-θa-sa* is that it makes perfect sense. The wording of (*a*), 'as long as you are *da-θa-sa*', is given a possessive turn in (*b*): 'as long as to you is strength'. Hence *da-θa-sa* must mean 'strong' or 'vigorous', and *daθ*- is evidently the OPers. form of the Av. base *das*-. The latter is known from two abstract derivatives, *das-var-* and **dās-man-* (attested in *dāsmanī-* from

† Benv., *BSL*, 1951, 38, maintains that *taumā* does not here mean 'strength', but 'semen'. Why should sexual, rather than general strength, be the limiting factor in looking after inscriptions? Moreover, had 'semen' been intended, the Elamite version would surely in col. III, 86 have had the ideogram *NU.MAN* 'seed', which invariably occurs where the OPers. text has *taumā* in the sense of 'family'.

COMMENTARY [44.4–45.2

*dāsma(n)-nī- 'leading to *dāsman-', see Benv., *Origines de la formation des noms en indo-européen*, 22). Bth. translated both words by 'health', in view of the Pahl. transl. *drustīh*. Thanks to the equivalence of OPers. *da-θa-sa* and *taumā* we can now say more accurately that the meaning of *dasvar*- and *dāsman*- is 'physical vigour'.

The form of *da-θa-sa* is not less interesting than the meaning. We must evidently normalize the spelling as *daθa(n)s*, the OPers. equivalent of Av. *dasąs. Thus we meet for the first time with the OPers. nom. sg. masc. ending of -*ant*- participles, which, as was to be expected, is -*a(n)s*. In OPers. the base *daθ*- 'to be vigorous' was apparently still used as a verb. The active voice of the participle here used agrees with the voice of IIr. *jīv-* 'to live', which belongs to the same semantic sphere.

44⁵. The expression *pərəθu aipi* is similar to OPers. *dūraiy api* 'a-far', and thus literally means 'a-broad'; cf. *awrā́ pərəθu aipi vījasāitīš* 'rain-clouds which spread over a broad surface', *Yt* 8.40.

45¹. This is the usual translation of the beginning of the stanza, from which only Gdn. and Htl. diverge: 'Welchem Mithra dienstbare gehilfen (*asti*-) auf allen thürmen und an allen fenstern als späher sitzen' (Gdn.); 'in dessen Unterkommen die Gaben (sich befinden). Auf allen Gipfeln... sitzen Miθras Späher...' (Htl.). Gdn.'s interpretation breaks down over *ašta*, or even the variant *asta*. But the alternative suggested by Htl. cannot be rejected off-hand. There is no need to change *ašta* to *aštaya*, as Htl. proposed, since *ašta* could be the loc. of *ašti*-, parallel to the Gāthic loc. *aštō*, see note 138¹. The 'gifts' (*rātayō*) Htl. had in mind are the libations mentioned in st. 32. These, as we saw, Mithra takes to his home in Paradise, not to his earthly abode. One cannot, however, be sure that he does not also keep a store of 'gifts', whether libations or other votive offerings, on earth. My reason for preferring the traditional interpretation is that Htl.'s translation not only cuts off the first line of 45 from the remainder of the stanza, which is thereby deprived of the benefit of the connecting relative *yeṅhe* (to my mind *miθrahe* is objective, not possessive, genitive, and means 'contract', not 'Mithra'); it also makes of the first line an unsuitable conclusion of st. 44, where a different word for 'Unterkommen' (viz. *maēθanəm*) is used; had Htl.'s meaning been intended one would have expected to find **yaθra rātayō* 'where the gifts (are)'.

45². *yōi paurva miθrəm družinti*, Bth. (*Wb*. 871, line 11 from foot): 'die zuvor den M. belogen haben'; thus also Lo.; Gdn.: 'welche zuerst die treue brechen', similarly Wi.; Da.: 'qui font maint (?*pouru*-) mensonge à M.'; Htl.: 'welche von je den M. verletzt haben'. Those who respect the reading *paurva* have not done justice either to the present *družinti*, or to the context. Obviously what is meant is *qui primi* (=*ubi primum*) *Mithram decipiunt*. [See Addenda.]

There is another context in which Bth. wrongly accounted for *paurva*-:

45.2–45.3]   COMMENTARY

the dog is described in *Vend.* 13, sections 45 and 48 respectively, as having the character (*haēm*)† of a warrior and a child; by way of illustration he is said to be

  raptō paurvaēibya yaθa raθaēštā̊ in 45, and
  pairi.taxtō paurvaēibya yaθa apərənāyuš in 48.

Da. translated 'il marche en avant comme un guerrier', and 'il creuse la terre avec les pattes, comme un enfant'; Bth. 'indem er sich an die beiden Vorausgehenden hält, (ist der Hund) wie der Krieger' (replacing *raptō* by the variant *yatō*), and 'im Herumlaufen um die beiden Vorausgehenden (ist der Hund) wie ein Kind'.

In 48 (but not in 45) Da. had sensed what is obviously the right explanation of *paurvaēibya*. This dual occurs again in *Vend.* 8.37, where Bth. correctly understood (*Wb.* 870) that the adj. referred to an implied *paδaēibya: frā.mē gaδwa zazayąn nižbərəta nōiṯ ainižbərəta nižbərətāṯ hača paourvaēibya* 'let them fetch my dogs: these are to be brought out‡— they are *not* not to be brought out—by means of a bringing out by the front (legs)'. Accordingly *Vend.* 13.45 must be translated 'in attacking with his front paws (he is) like a warrior', and *Vend.* 13.48 'in running about on his front paws he is like a child'. By extruding *raptō*, loc. of **rapti-* 'attack', from the text, Bth. suppressed the only OIr. form with which Parth. *rf-*, MPers. *rp-*, 'to assail' (cf. Henning, *BSOS*, IX, 87) can safely be connected. [See Addenda.]

45³. *isənti* 'they seek out' gives the clue to an improved understanding of the difficult passage *Yt* 5.95. There the goddess Anāhitā gives the following answer to Zarathuštra's question 'what happens to the libations which daēva-worshipping owners of Falsehood offer to you after sunset?':

  nivayaka nipašnaka
  apa.skaraka apa.xraosaka
  imā̊ paitivīsənte
  yā̊ māvōya pasča vazənti
  xšvaš.satāiš hazaŋrəmča
  yā nōiṯ haiti vīsənti
  daēvanąm haiti yasna

† The stem of *haēm* is scarcely *haya-* as Bth. assumed, but rather **havya-*, lit. 'property, *suitas*', to *hava-* '*suus*'. For the phonology, cf. *kąsaēm*, nom.-acc. neuter of *kąsaoya-* (=**kąsavya-*), and *aēm* 'egg' (see Henning, *Asiatica, Festschrift Weller*, 291), which may, after all, represent *avyam, from *āvyam, with shortening of *ā* even before *u̯* followed by *y* (cf. the beginning of note 103¹). Thus interpreted, *aēm* would not exclude Da.'s interpretation of *apāvaya-* as 'deprived of testicles', which Henning, *ibid.* n. 17, disbelieves.

‡ I take *nižbərəta* and *ainižbərəta* as corruptions of (*ai*)*nižbarata* (cf. the variants with *-barət-*), caused by the following *nižbərətāṯ*. °*barata-* will be future passive particle, like *yazata-*. The double negation *nōiṯ ainižbərəta* has a parallel in *nōiṯ anaipi.pārəmnāi*, below, p. 247.

Bth.–Wo., accepting Gdn.'s emendation of *haiti* to *paiti*, translated: 'die Schaudererregenden, die Scheelsüchtigen, die Hohnsüchtigen, die Schmähsüchtigen stellen sich—sechshundert und tausend—bei diesen (Zaoθra's) ein, die hinter mir zugeführt werden; (die Zaoθra's), die nicht (von mir) besucht werden, (dienen) zur Verehrung der Daēva's'.

It seems strange that *haiti* should twice have taken the place of *paiti*; if, however, one of the two *haiti*-s, preferably the first, belonged to the text from the beginning, it would be understandable if a nearby *paiti* had been changed to *haiti* by a careless copyist. We should accordingly read

yā nōit̰ haiti vīsənti
daēvanąm paiti yasna,†

and understand *vīsənti* as consisting of *vi* + *isənti* (cf. OInd. *vīcchati* 'to seek for', to *iṣ-* + *vi*): 'chasing(?), heeling (??cf. *pāšna-*, Sogd. *pšn'*, "heel"), shuffling (cf. Skt. *ava-caskarire*, *apa-skirate*, "to scrape with the feet"), reviling, they (*scil.* the daēvas)‡ approach in large numbers (*lit.* 600 and 1000) these (libations), which fly (away) after me: (me) who (am) not being present they (viz. the libations) seek at the sacrifice to the daēvas'. Thus a reasonable picture of the situation is obtained: the daēvas eagerly approach the libations offered to Anāhita, but these, finding that the goddess has failed to attend the sacrifice, fly away in search of her.

45⁴. On *haiθīm.ašavajanasca* see note 38³.

46¹. *avā̊*, nom. sg. masc. of *avant-* (to *av-* 'to help'), with Da. and Lo., rather than of *ava-* 'he', with Bth., *Wb.* 166 sq.

46². *pavā̊*, nom. sg. masc. of *pavant-*. Attention was drawn in *JRAS*, 1952, 177, to the more plausible meaning we gain in *Y* 44.18 if we replace Gdn.'s reading *apivaitī* by the better attested *apavaitī*, and interpret the latter as acc. dual fem. of *apavant-* < **ā-pavant-* 'protecting':

kaθā ašā tat̰ mīždəm hananī
dasā aspā̊ aršnavaitīš uštrəmcā
hyat̰ mōi mazdā apavaitī haurvātā
amərətātā yaθā hī taēibyō dā̊ŋhā

'shall I obtain in accordance with Truth the salary—ten mares with a stallion, and a camel—as protecting which (= as protectors of which (*scil.* salary = animals)) thou, Mazdāh, hast given me, since they are thine, Haurvatāt (= water) and Amərətāt (plants)'. The water and plants, sustenance of horses and camels, are well described as their God-given protectors. For *pavant-* with *ā* cf. *ā-fəntō* and *ā-fyeiδyāi*, above, p. 174.

† H. Weller, *Anahita* (*Veröff. d. Orient. Sem. d. Univ. Tübingen*, IX, 1938), pp. 98, 122, more drastically changes the first *haiti* to *paiti*, the second to *hənti*, and translates: 'weil sie (die Opfergüsse) nicht (von mir) besucht werden, werden sie zu Opfern für die Abergötter'.

‡ Instead of adjectives the first four words may alternatively be nouns denoting four different types of daēvas.

46.2–48.2] COMMENTARY

Thieme's interpretation of *apivaitī* (*Asiatica, Festschrift Weller*, 661), confirms the impression that in the present Gāthic stanza this reading should be abandoned. [See Addenda.]

47[1]. *frasrūtəm zaranimnəm*, 'den berühmten, zornvollen' (Wi.); 'den berühmten den, (wenn) er zürnt...' (Wo., Lo.); 'welchen berühmten, goldenen...' (Spi.); 'l'illustre, au collier d'or' (Da.); 'den b., streitbaren' (Gdn.); 'dem weitb., dem zornigen' (Htl.); 'the praised in song, the furious' (Hzf., 434). In my opinion *frasrūtəm zaranimnəm* form a kind of hendiadys by which, of two asyndetically placed adjectives referring to the same noun, the second provides a qualification of the sense in which the first is to be taken. In such cases a *because* or *inasmuch as*, interposed between the two adjectives, will produce the required meaning. In the Mithra Yašt apart from the present case ('famous *inasmuch as* angry' = 'notorious in his anger') this device can be recognized with a fair degree of probability in $x^v ainisaxtəm\ pouru.spā\mathring{\delta}əm$, st. 109 (and 111) '(kingdom) beautifully strong (see note 109[1]) *because* provided with a numerous army', and in *dahma vīduš.aša*, st. 16, 'trained *inasmuch as* Truth-knowing' = 'trained by knowledge of Truth'. A possible case is also *yūxδahe pāirivāzahe*, st. 127, 'dexterous *insofar as* leaping' = 'dexterous in his leaps'.

48[1]. *aθra* has unnecessarily misled all translators into making the apodosis begin with this very adverb, although the parallel passage *Yt* 14.63 (quoted above, p. 194, n. †) makes it abundantly clear that the apodosis consists of the words *nōit̰ pāδa vīδārayeiti nōit̰ paiti.tavā̊ bavaiti*.

48[2]. Of the end of this stanza the following translations have been offered, none of which is entirely satisfactory: 'die Füße kann (jen)er nicht einstemmen; er vermag keinen Widerstand zu erheben. (Und so auch) diese Länder ⟨diese Widersacher⟩, wo er der weite Fluren besitzende Miθra vernachlässigt wird' (Wo.); 'ihre Füße erhält er nicht, er ist nicht ein Bestärker jener Länder, jener Kämpfer, welche mit Uebelwollen trägt Mithra der weitflurige' (Wi.); 'nicht schreitet dann einer mit den Füßen aus, keine Widerstandskraft haben(?) diese Völker und die Feinde, wo Mithra, der weite Triften hat, übel gehalten wird' (Lo.); 'plus ne tiennent de pied ferme, plus n'ont force de résistance ces nations, ces adversaires, quand les traite en mauvais traitement Mithra, maître des vastes campagnes' (Da.); 'und läßt ihre füße nicht mehr feststehen: wehrlos werden die völker und die feinde, weil der weitgebietende Mithra ihnen übel gewogen ist' (Gdn.); '...nicht ist er kräftigend in Bezug auf die Gegenden, auf die Feinde, die mit Uebelwollen trägt Mithra' (Spi., commenting: d. h. wohl in den Ländern, welche von Feinden besetzt sind macht Mithra seine Kraft nicht geltend); 'quando male ferentes (contumaces, duzhbereṯ, ipse male) perfert (non perfert, tolerat) Mithras' (Kossowicz, quoted by Spi., *Commentar*, II, 560); 'sie (welche) übel tragend trägt (Mithra, passivisch)' (Justi, *Handbuch* 158, s.v. *duzhbereṯ*).

COMMENTARY   [48.3-48.5

48³. As the subject of *paiti.tavā̊ bavaiti* cannot very well be Mithra, or the plurals *daiṅhāvō* and *haṃərəθā̊*, the sentence is necessarily impersonal (cf. Reichelt, *Aw. Eb.* §716, where the present examples and *nōit̰...ava.ašnaoiti* in st. 24 should be added); the preceding *nōit̰ pāδa vīδārayeiti* (cf. note 28¹) is best interpreted in the same way. Lo., who realized that the two verbs are impersonal when he translated the parallel passage in *Yt* 14.63 (quoted above, p. 194, n. †), was misled in the present context by the line *tā daiṅhāvō tē haṃərəθā̊*. The very fact that in *Yt* 14.63 the identical series ends with *bavaiti* is convincing proof that *daiṅhāvō* and *haṃərəθā̊* should be kept out of the *bavaiti* clause.

48⁴. *paiti.tavah-*, lit. 'whose strength is against', is the only bahuvrīhi quoted by Duch., *Comp.* 185, in which *pati* has adversative meaning. Among the compounds 'à premier terme régissant' Duch., *Comp.* 197, quotes *paitipa-* 'against the current'; we should add *paiti.biši-* 'which counteracts enmity' (correctly analysed by Bth. as containing **biši-* 'enmity'), and the Iranian source of Arm. *petmog*. The latter is the name of a Persian religious sect referred to by Eɫišē, which was last discussed by Zaehner, *Zurvan*, 29. The hitherto unexplained word is a slightly imperfect reproduction† of an Iranian term *patmōγ < **pati-maoγa-*, meaning '(directed) against heresy'. It would seem that the founders of the *petmog* denomination had made it their avowed task to refute various *ahramōγ* views (cf. Zaehner, *Zurvan*, 45).

48⁵. *dužbərəntō baraiti* must be the opposite of *hubərətō baraiti* in st. 112. Since in either stanza the only attested reading is *baraiti*, it was methodically wrong of Bth. to change the verb to *baraite*, especially as even so the syntax of both periods, as interpreted by him, remains faulty. If the verb is *baraiti*, with Mithra as only possible subject, then *dužbərəntō* and *hubərətō* are most likely the respective objects. In this case their endings can only be correct, and there is no reason to think that they are not, if the respective stems are *dužbərənt-* and *hubərət-*. The latter is without difficulty understood as a transitive synthetic compound meaning 'who treats well', cf. *huš.ham.bərət-* 'which collects well', etc., see Duch., *Comp.* 61 *sq.*, 80 *sq.* The strange stem *dužbərənt-* must accordingly have resulted from a contamination of the expected **dužbərət-* 'treating badly' and a synonymous participial compound *dužbarant-* (cf. *gāθrō.-rayant-*,‡ *ašaoxšayant-*) which is, in fact, preserved in the variant *dužbarəntō* of J₁₀. Kossowicz and Justi came very near this understanding of *dužbərəntō* (cf. note 48²), but failed to draw from it the appropriate

† Dr C. Dowsett, consulted on the explanation proposed above, kindly offered the suggestion (17 March 1955) that the original *patmog may have been affected by *mogpet*, which various scribes in any case have confused with *petmog*. Normally, of course, the Iranian preposition *pati°* turns up as *pat°* in Armenian, whereas *pati-* 'chief' gives Arm. *pet*.

‡ Connected by Henning with Sogd. *r'y-*, see *GMS*, §565, n.

48.5–50.1] COMMENTARY

conclusion for *hubərətō* in st. 112, which Justi quotes as a nom. sg. of *hubereta* 'wohlverehrt' (*Handbuch*, 329). On the syntax of this period see note 112[4], end.

50[1]. All translators render *upairi harąm* by 'on the Harā', except Wi. who has 'above the H.' On the other hand for *haraiθyō paiti barəzayå̄* Wi. and Spi. have 'an der hohen H.', Da. 'sur le H.B.', Bth. 'von dem H. Gebirge', Gdn. 'aus der hohen H.', Lo. 'auf der hohen H.', Htl. 'von der...H.'

It seems that in contrast with st. 44, where the reference to Mithra's 'abode' being spread throughout the material world is intended to emphasize the god's ubiquity (cf. notes 95[1], end, and 145[1]), the present passage introduces us to 'high-dwelling' (st. 140) Mithra's real 'home'. For its location three considerations must be borne in mind: (1) Mithra's abode is built by Ahura Mazdāh and the Incremental Immortals, whose own abode, according to *Vend.* 19.32, is the Paradise (*garō nmānəm*); (2) Ahura Mazdāh worships Mithra in Paradise (st. 123), and it is from there Mithra sets out on his journey in st. 124; (3) in st. 32 Mithra is said to deposit his libations in Paradise, in st. 136 to take them 'to his abode'. The conclusion that Mithra's abode is in Paradise seems obvious.

We can therefore only translate *upairi harąm* by 'on the Harā' if we assume that Mount Harā is part of the Paradise. This on the whole is unlikely. It is true that the Pahl. Commentary to *Vend.* 19.30 places the far end of the Činvat Bridge on Harā. But in the Book of Ardā Virāf that pious man after crossing the Bridge (v, 2) does not reach the Paradise proper, *Garodmān* (x, 1), until he has passed through the 'Region of the Mixed' (*Hamīstagān*, cf. Bth., *Wb.* 1187, Nyberg, *Oriental Studies...Pavry*, 346), and the three Halls of Good Thought (Av. *humata-*), Good Word (Av. *hūxta-*), and Good Deed (Av. *hvaršta-*). Even if we were to call 'Paradise', broadly speaking, the whole stretch from the Bridge at Harā to *garō nmānəm*, Mithra's abode, as we saw, must still be located in the *garō nmānəm* itself, the holiest part of 'Paradise' (cf. Bth., *Wb.* 513), and apparently well removed from the Harā. Moreover, the constant epithet of *garō nmānəm* (in Sogd. *rwxšn'γrδmn* it has even merged with it) is *raoxšna-* 'light, bright', which, if Mithra's abode is situated in *garō nmānəm*, agrees well with the statement in our stanza that there is no night or darkness where the abode was built. This statement could scarcely apply to the Harā, across which the sun is said elsewhere to rise (*Vend.* 21.5, cf. also *Yt* 10.118). Lastly, it is unlikely that a people familiar with mountains should have entertained the idea that no mists rise on the highest mountain of all.

We are thus justified in concluding that the Paradise, and with it Mithra's abode, is situated *above* the Harā (with Wi.), out of reach of the mists which rise *from* the Harā (with Bth.). The fact that almost in the same breath (st. 51) Mithra is said to survey the material world from the

Harā does not invalidate this conclusion. The Harā is a natural vantage-ground for Mithra to linger on in transit, cf. st. 13; being the highest mountain on earth it is also the nearest peak on which to step down from a residence situated above it. From an anticephalic (as against antipodean) point of view the Harā may be considered the upside down *bālā-xāna* or watch-tower of Mithra's heavenly palace.

As a secondary development, however, the impression may have gained ground even in ancient times, that Mithra's abode was *on* the Harā. This is at least arguable from the fact that when it was desired to find lodgings for the Zarathuštrian Sraoša, who in the process of 'Zoroastrianization' had become closely associated with Mithra (cf. above, p. 193), he was assigned a house *on* (*paiti*) Harā (*Y* 57.21); possibly, too, behind the wording, quoted *in calce*,† of *Yt* 12.23 (borrowed from the present stanza) and *Yt* 12.25, lies the idea that also Rašnu, Mithra's other close associate, lives on top of Harā, although no 'abode' or 'house' is mentioned. The notion attested in later Zoroastrian sources also has to be borne in mind, that Mihr, Sroš, and Rašn act as judges of the souls of the deceased at the Činvat Bridge, whose one end, as we saw, rests on Mount Harā.

50². *pouru.fraorvaēsya-* 'mit vielen Ausläufern (eig. Hervordrehungen)' (Bth., *Wb.* 868.901; followed by Htl.); 'vielbestiegen' (Wi.); 'sehr weithin reichend' (Spi.); 'vielbesucht' (Gdn.); 'aux révolutions nombreuses' (Da., thinking of *Yt* 12.25 quoted *in calce*).

51¹. Since Mithra's sister (above, p. 195) Aši is also their sister (see above, p. 168, n. ‡), Mithra is a brother of the incremental Immortals, who in st. 92 are said to consider him the *ahū* and *ratu* of the material world.

52¹. On Nairyō.saŋha replacing Rašnu in this stanza cf. above, p. 194. As a member of Mithra's escort he appears again, half disguised, in st. 66. The translation of the *hapax legomenon māyu-* by 'delightful, voluptuous' (as against Bth.'s 'geschickt, gewandt, findig', Lo.'s 'wunderkräftig', Da.'s 'obéissant', Wi.'s 'der Rufer', Spi.'s 'weise', Htl.'s 'listenreich', Gdn.'s 'wunderthätig') is suggested by the probable connection of the word with Av. *māyā-* 'delight', *mayah-* '!ust', Pš. *mīna* 'love', *mayan* 'lover'; on the quantity of *ā* cf. *GMS*, §124. Thus interpreted this epithet of Nairyō.saŋha fits in with the quality of 'un éphèbe dont la

† *Yt* 12.23: *yaṯčiṯ ahi rašnvō ašāum upa harąm bərəzaitīm paouru.fraourvaēsąm bāmyąm yaθra nōiṯ xšapa nōiṯ təmā*, etc. 'even when you, Truth-owning Rašnu, are on the high Harā, etc.' The use of *upa* instead of *upairi* in *Yt* 10.50 cannot affect our argument above since, for once, it is here a question of Rašnu, not of Mithra, and, above all, *upa* is the preposition mechanically employed all along from section 9 to 38 of *Yt* 12.

Yt 12.25: *yaṯčiṯ ahi rašnvō ašāum upa taērəm haraiθyā barəzō yaṯ mē aiwitō urvisənti starasča māsča hvarača...* 'even when you, Truth-owning Rašnu, are on the peak of Mount Haraitī, round which my stars, moon, and sun revolve'.

beauté juvénile exerçait une irrésistible séduction' (Cumont, *Le Mani-chéisme*, I, 61), which prompted Persian (occasionally also Parthian, see *infra*) Manichees to select Narisah to represent the Legatus Tertius in his role of decoy for the sensual appetites of the Archontes, cf. Henning, *Mir. Man.* I, 192, n. 6. Similarly in Zervanism it is Narsa whom Ohrmazd places behind Satan to incite the lust of the renegade women, cf. Zaehner, *Zurvan*, 184–6. Nairyō.saṃha's good looks are also conveyed by his epithet *huraoδa-* 'of beautiful aspect' in *Y* 57.3, *Vyt.* 40.

Nairyō.saṃha's association with Mithra has left a trace in Western Mithraism, where he is represented by Hermes–Mercury, not only, one may surmise, on account of his beauty, but also because he acts as messenger of Ahura Mazdāh (*Vend.* 19.34, 22.7) and the gods (*Gt. Bd.* 177.8, see Zaehner, *Zurvan*, 186, n. 6), cf. Cumont, *TMMM*, I, 145. On the other hand the interchange, in Parthian Manichean texts, of Narisaf and Mihr-yazd in the role of Legatus Tertius (see Introduction, p. 40), is hardly to be interpreted as revealing an intimate relation between the two gods in Parthian mythology. It is more likely that the Parthian Mani-cheans, after assigning the role of Legatus Tertius to Mihr, occasionally replaced the latter by Narisaf because this god was the representative of the Legatus in MPers. Manicheism.

52². 'Er tötet ihn, seis in der Schlachtreihe, seis im (Einzel)angriff (wörtl.: als einen beim...getöteten)'; thus rightly Bth. The other trans-lations, 'er tötet ihn in der Schlacht oder im Zweikampf' (Lo., following Spi.), 'il frappe un coup qui brise leurs bataillons, qui brise leur force' (Da.), 'schlägt ihn reihengeschlagen oder machtgeschlagen' (Wi., similarly Htl.), 'tué dans les rangs de l'armée (= dans la mêlée), tué par un assaillant' (Benv., *Vṛtra*, 15, n. 1), miss the point. The evil-doer is *amō.jatō*, 'killed in the process of attacking', if he 'hastens to the fore with fast step'; but even had he stayed with his regiment he would have been killed, *rasmō.jatō* 'killed (whilst being) in (the ranks of) his regi-ment'. The latter alternative had been dealt with by the poet repeatedly before (cf. sts. 36 *sq.*, 42 *sq.*, 48); he did not therefore enlarge on it in the present stanza of which, on the other hand, the former alternative is the main topic. Logically after *māyuš* the poet should have stated: *amō.-jatəmča dim jainti*, and then proceeded to his general conclusion: *rasmō.-jatəm vā.dim jainti amō.jatəm vā*. However, as the general conclusion the poet intended to pronounce in the last two lines included the *amō.-jatō* fate of the evil-doer who 'runs to the fore', he refrained from stating this fate in advance of the conclusion. Economy was thereby achieved, but not without some loss of clarity.

53¹. *bāδa* 'at times' is correlated to *bāδa* in st. 73, where Mithra, instead of complaining, speaks joyfully. The meaning of *bāδa* is given by Sogd. β'δ...β'δ 'tantôt...tantôt', cf. Benv., *Essai de grammaire sog-dienne*, 174. A series of such *tantôt*-s is found in sts. 83 *sq.* The Av.

COMMENTARY [53.1–60.1

superlative of *bāδa* is *bāiδištəm*, to be translated as 'at all times' (cf. Bth., *ZWb.* 196), to which belong Parth. *b'dyst'n*, Man.MPers. *b'yst'n*, Pahl. *bāstān*, see Henning, *TPS*, 1944, 111. As a frequentative verbal particle (cf. MPers. *b'*, see Henning, *ZII*, IX, 247) *bāδa* occurs in st. 68.

54¹. To *hvapah-* may belong, apart from Pers. *xūb*, Sogd. *xwp*, also Parth. *zxrwb* < **zrxwb* 'gold-worker', see Henning, *BSOAS*, XI, 723. *hvapō* is taken as nom. sg. by most translators, as voc. referring to Ahura Mazdāh by Bth. and Htl.; Lo. hesitates between the two views.

55¹. The words from *frā* onwards were translated as follows by Bth.-Wo.: '(so) würde ich mich—eignen sonnigen unsterblichen Lebens—mit dem Alter der bestimmten Zeit zu den ašagläubigen Männern aufmachen; mit (dem Alter) der bestimmten (Zeit) würde ich hinzukommen'. Similarly Gdn., Lo., and Da. The same sentence occurs, apart from st. 74, also in *Yt* 8.11, where after *jaymyąm* the following additional words occur: *aēvąm vā avi xšapanəm duyē vā pančāsatəm vā* 'for one night, or two, or fifty'. This addition suggests that *θwaršta-* refers to a *limited*, rather than an *appointed* time. Semantically either interpretation is compatible with the basic meaning of *θwaršta-*, which is 'cut'. As to *upa.θwarštahe*, which appears to be a compound constituting the verbal part of the genitive absolute clause (lit. 'my own...life being interrupted'), the meaning here assumed for it is based on that of *upa.-θwaršti-* and *upa.θwarəsa-* '(artificial) cutting, breach, opening (of a dike, respectively house)'. Some doubt on the interpretation here proposed must, however, remain, for the words *xᵛahe gayehe xᵛanvatō aməšahe* occur alone at the end of *Y* 9.1, where they seem to form a genitive absolute clause of their own. [See Addenda.]

60¹. I can make no connected sense of these six words. Wi. translated: 'nicht verletzend den Bauer, der da schaltet über seine Stätte unbedrängt' (emending to *vasō.yaonəm ainitəm*); Da.: 'qui ne fait point de mal au laboureur...?...'; Htl.: 'der den nach Wunsch seine Bahn ziehenden, nach Wunsch Rinderweide besitzenden, den viehzüchtenden Weidebesitzer hier, nicht überwältigt (= nicht vernichtet), ihn, der nach Wunsch seine Stätte sucht (oder: wählt)' (separating *vasō* from **yaonă-yəntəm*(?) taken as denominative); Wo. and Lo. leave blanks; Gdn. operated with drastic emendations. On *fšuyant- vāstrya-* 'pasteur du gros *et* du petit bétail' see Benv., *BSOS*, VIII, 407 *sq*. For *vasō.yaona-*, an epithet applied to the Fravašis in *Yt* 13.34, Benv. has proposed the translation 'cheminant librement', *Vrtra*, 53. To the internal evidence adduced by Benv. that *yaona-* means 'course', outside proof can now be added from Khotanese and Pašto. In Khot. Bailey has pointed out that *gyūna-, jūna-*, means 'way of moving, gait' (*TPS*, 1954, 138); in Pš. the corresponding word is *yūn* 'gait, movement, step', plur. *yāna* (on the analogy of *špūn:špāna* 'shepherd', etc., cf. Trumpp, *Grammar of the Paštō*, 61.111), of which Morg., *EVP*, 100, rightly doubted the previously

60.1] COMMENTARY

suggested affiliation with OIr. *yāna-*; Pš. *-ūn-* is from *-aun-* also in *γūna* 'hair' (Av. *gaona-*).
In this connection two words may be mentioned which only in appearance belong to *yaona-* 'course', viz. *yaona* in *Yt* 5.87, and *θrāyaone* (dat.) *Yt* 4.9, 14.46, *θrāyaonō* (nom. pl.) *Yt* 5.86. The context of *yaona* is as follows:

Yt 5.87   *θwąm kaininō vaδre yaona*
             xšaθra hvāpå jaiδyånte
             taxməmča nmānō.paitīm

Bth.–Wo. translate: 'dich sollen heiratsfähige (*vaδrya-*) emsige Mädchen... (*yaona*?; vll. "in der Heimstätte") um [gute?] Herrschaft bitten und um einen heldenhaften Hausherrn'; Da.: 'les jeunes filles au sein stérile te demanderont de bonnes maîtrises...'; Lo.: 'dich sollen unberührte Jungfrauen, die schöne Arbeiten machen(?), um Besitztum bitten...'; Weller (cf. p. 201, n. †), p. 119: 'Mädchen (Frauen) mit unfruchtbarem Schoß sollen um ein recht fleißiges Feld, d. h. um einen recht fruchtbaren Mutterleib, bitten'. The interpretation 'stérile' is based on OInd. *vádhri*, on which see Wn., *KZ*, 46 (1914), 267, n., Wn.–Debr., II, 2, 859. But 'filles au sein stérile' is hardly what one expects to find in this passage. Benv. accordingly assumed a compound *vaδri.yaona- 'qui s'achemine au mariage' (*Vṛtra*, 51, n. 3). To this one may object that girls so described would no longer need to ask Anāhita for a husband. In any case *vaδri-* as 'marriage' would be hard to justify. It is easier to maintain Gdn.'s reading *vaδre* and interpret it as acc. pl. n. of *vaδrya-*, an adjective derived from *vaδra- 'wedding'. *vaδre* will then agree with *yaona*, plural of a neuter derivative of *yu-* 'to join', the *vaδryam yaonam being literally the 'wed-lock'. *hvāpå*, defined by the instr. sg. *xšaθra*, can be taken as a further neuter pl. agreeing with *yaona*: 'it is you the girls ask for wedlocks liberal in estate, and for a strong husband'. On *hvāpah-* see below, note 92[4]; *xšaθra-* means 'estate' also in *Yt* 5.130 (*yaθa azəm...masa xšaθra nivānāni* 'that I may acquire large estates') and *Yt* 17.7 (*tē narō xšaθra xšayente aš.baourva...* 'these men are powerful by means of an estate† in which there is much eating, etc.').
[The interpretation of *xšaθra-* as 'estate' is unexpectedly confirmed by the meaning *šahr* has in the dialects of Bašāgird (referred to as Baškard by the natives), a region in which I had occasion to travel after the above lines had been written. *šahr* there denotes throughout the cultivated oasis, as distinct from the village or cluster of huts in which the cultivators and their families dwell. These oases usually consist of one or

---

† Bth. and Lo.: 'rule over estates'. The interpretation here proposed of *xšaθra* as a singular is favoured, as my student David Street points out, by the words *yahmya starətasča gātuš* 'and in which a couch (sing.) is spread', in the next line but one.

208

COMMENTARY   [60.1

more date groves surrounded by a few small maize or wheat fields. The huts are often situated at a distance of half a mile or more from the *šahr*, and the natives made it quite clear that this term did not include the village. Such Baškardīs as have any notion of towns refer to them as *šahrestān*-s.]

If now we turn to *θrāyaonō/e* we must compare *Yt* 5.86:

> θwąm āθravanō marəmnō
> āθravanō θrāyaonō
> mastīm jaiδyąnte spānəmča

(according to Wo.: 'dich sollen die betenden Priester, die *θrāyavan*-(?) † Priester um Wissen und um Heiligkeit bitten') with *Yt* 4.9 and the almost identical phrasing of *Yt* 14.46: *zaraθuštra aētəm mąθrəm mā fradaxšayō* (*Yt* 14.46 *fradaēsayōiš*) *anyāṯ piθre vā puθrāi brāθre vā haδō.- zātāi āθravanāi vā θrāyaone*. This means according to Lo. (who here follows Bth. and Gdn., *Drei Yasht*, 80): 'O Zarathuštra, lehre diesen Zauberspruch niemand anderem als dem Vater oder dem Sohn oder dem leiblichen Bruder oder dem...(?) Priester'. It is at once obvious that this translation does not account for the distribution of the disjunctive particle *vā*, and that Da. was right with his rendering 'O Z., ne laisse enseigner cette Parole que par le père à son fils, par le frère à son frère né du même sein, par le prêtre à son élève'. Bth. probably discounted Da.'s translation because he did not consider *fradaxšaya*- and *fradaēsaya*- to be causatives; but Da.'s view of the situation can be upheld even if we translate: 'do not teach this Divine Word except either to the father for (transmission to) the son, or to the brother for (transmission to) the uterine brother, or to the priest for (transmission to) the disciple'. The meaning 'priestly disciple' is thus assured for *θrāyaone* by the context.

For a closer definition of this disciple we must turn to the question and answer in *N* 11: *čvaṯ nā aēθrapaitim upōisāṯ? yārə.drājō; θrizarəmaēm xratūm ašavanəm aiwyåṅhaṯ*, which Bth. (*Wb.*, s.v. *aēθrapati*-) renders: 'Wie lange soll man einen Lehrer besuchen? Ein Jahr lang. Drei Jahre hindurch soll man dem Studium der frommen Weisheit obliegen', adding the comment: 'd. h. man soll zu drei Lehrern je ein Jahr lang gehen'. Such being the practice of priestly study, the *θrāyavan*- will be a student 'attending a **θrāya*-', and the **θrāya*- (lit. 'triad', neuter of **θrāya*- 'triple') a triple course of studies, involving three years and three teachers. The Av. adj. **θrāya*-, accordingly, stands to OInd. *trayá* 'triple' (cf. *trayī́ vidyā́* 'the triple science') as Av. *θrāyō* 'three' to OInd. *tráyas*. In *θrāyavan*- we thus find Avestan confirmation of Bth.'s opinion that the long *ā* of *θrāyō* 'three' is not a secondary development of the Younger Avestan language, see *GIP*, I, p. 131. A free translation of

---

† Weller, *loc. cit.*: 'Besitzer von Schutzmitteln, d. i. Zaubermitteln'.

θrāyavan- which may commend itself to readers familiar with present Cambridge University regulations, is 'a Tripos student in Divinity'.

61¹. On *zaēnahvant-* and cognates see F. W. Thomas, *JRAS*, 1946, 10 *sqq.* (where *huzaēna-* should perhaps be added, see below, note 143⁶); cf. also Henning, *BSOAS*, XI, 716 *sq.*

61². *ərəδwō.zənga-*, the posture of the watchful servant, cf. Ved. *ūrdhva- sthā-*, Oss. *urdugistæg*, etc., most recently discussed by Bailey, *TPS*, 1954, 129. There is, however, an ingenious alternative interpretation suggested by Gdn., which Bth. was perhaps wrong in condemning. In *Y* 62.5 (*nairyąm...hąm.varəitīm ərəδwō.zəngąm axᵛafnyąm āsitō.- gātūm jaγāurūm*), as Gdn. pointed out, the contrast of Manly Valour being 'awake' although 'lying on a couch' suggests that the two preceding words express a similar contrast, 'not sleeping' although 'in a position of rest (*lit.* with drawn-up shanks)'. Accordingly Gdn., referring to the parallel Skt. expression *ūrdhvajānu(ka)* 'raising the knees (=sitting)', translated our passage 'den...auch wenn er ruht, wachsamen'.

61³. *frat̰.āpəm* 'filling the water(s)', followed by *tat̰.āpəm* 'causing water (rain) to fall' (on which cf. Lo., *Asiatica, Festschrift Weller*, 407, n. 5), provides the clue for an understanding of the mysterious epithet *frāpa-* of Satavaēsa, who in *Yt* 13.44 is also said to be *tat̰.āpō*. Bth. left *frāpa-* untranslated; Duch., *Comp.* 186, suggested 'ayant l'eau prête', thinking of the preverb *frā̆* as Da. ('qui pousse les eaux', *ZA*, II, 326; 413) and Justi ('zum Wasser gehörig') had done before. It is, however, clear that *frāpəm < *fra-āpəm* (base ²*par-* 'to fill') is merely a doublet of *frat̰.āpəm*, the two words standing in the same relation to each other as *barō.zaoθrā-* and *barat̰.zaoθrā-*, or Khot. *barbīra-* and *baravira-* (cf. *GMS*, pp. 251 *sq.*). On Satavaēsa's survival in Manicheism as a rain-god see Mary Boyce, *BSOAS*, XIII, 909. It may be noted, against Bth., that not only the stem of *tat̰.āpəm* is thematic (as Duch., *Comp.* 34, rightly has it), but in all probability also that of *frat̰.āpəm*, considering that its doublet has the thematic gen. *frāpahe* in *Yt* 8.0. [See Addenda.]

61⁴. *karšō.rāzah-*, which is also used as an epithet of *āsna- frazainti-* 'noble progeny', means according to Bth. 'der dem Gau die Gesetze vorschreibt'; Wi., followed by Lo., has 'der die Furche richtet', Htl. 'der die Äcker furcht', Spi. 'der einen Kreis (Versammlung) veranstaltet', Da. 'gouverneur de la terre', B. Geiger (*WZKM*, 41 (1934), 124) 'dem Gau die Richtung weisend, ihn leitend'. My own preference goes with Gdn.'s 'richter der linie, s. v. a. ordner, schalter oder in specie ein feld-, grenzscheider', which makes of Mithra and the 'noble progeny' arbiters in boundary disputes affecting neighbouring states and estates. *karša-* as 'Gau' is extremely doubtful; the well-attested meaning is 'line drawn on the ground', including a 'furrow'.

61⁵. *dāmi.δātəm*, 'created by the creator', that is, by Ahura Mazdāh,

COMMENTARY [61.5

who is referred to as *dāmi-* elsewhere, see above, p. 169. The epithet, which is well attested with this meaning, confirms Ahura Mazdāh's statement in st. 1 that it was he who created Mithra. Nevertheless Nyberg, *Rel.* 59, and Widengren, *Hoc.* 108, 129, have inferred from *Vend.* 19.15, 35, where, according to Bth., *dāmi.δātəm* refers to Ahura Mazdāh and means 'creator of the creation', that Mithra is in our stanza invoked as a δημιουργός, 'who creates the creation'. In the two paragraphs of *Vend.* 19 the sentence *nizbayemi ahurō mazdā̊ ašava dāmiδātəm* occurs, which in Bth.'s translation is made to appear altogether illiterate: 'ich rufe herab den ašaehrwürdigen Ahura Mazdāh, der Schöpfung Schöpfer (acc. of *dāmi.dāt-*)'. Even if we were to accept this translation, Nyberg and Widengren's interpretation of *dāmi.δātəm* in *Yt* 10.61 flies in the face of the available evidence: on the one hand it is Ahura Mazdāh, not Mithra, whom Bth. supposed to be described as a creator in *Vend.* 19.15, 35; on the other hand st. 1 of our Hymn clearly states that Mithra is a creature, not a creator.

Nyberg's and Widengren's theory is nevertheless so serious an obstacle to the correct understanding of the Avestan Mithra, that it will repay us to examine more closely Bth.'s interpretation of *Vend.* 19.15, 35, on which this theory solely and precariously rests. In §§ 13–14 of *Vend.* 19 Ahura Mazdāh instructs Zarathuštra to 'call down', in the following order,

(*a*) the Religion,
((*b*) the Aməša Spəntas; the wording, borrowed from *Y* 57.23, jars with the context of *Vend.* 19.13; we may assume that this item should be struck out, having been introduced under the influence of *Yt* 10.92 = *Y* 57.24, where the Aməša Spəntas are mentioned together with the Religion)
(*c*) Θwāša, Zrvan, Vayu,
(*d*) Vāta and, probably, Ārmaiti,
(*e*) the Fravaši of Ahura Mazdāh 'whose soul is the Ma̧θra Spənta', and
(*f*) *imaṱ dąma yaṱ ahurahe mazdā̊* (last words of § 14) 'this creation of Ahura Mazdāh'.

In §§ 15–16 Zarathuštra is said to agree (see above, note 25[1], (1)) to Ahura Mazdāh's suggestion. He calls down successively

(*A*) *ahurō mazdā̊ ašava dāmiδātəm*,
(*B*) Mithra,
(*C*) Sraoša,
(*D*) the Ma̧θra Spənta,
(*E*) = (*c*),
(*F*) = (*d*), and
(*G*) = (*a*).

Even assuming that (D) is intended to correspond to (e), where the Maθra Spənta is at least mentioned, Zarathuštra, or rather the author of this story, appears to have ignored (f), and added on his own initiative (B) and (C), in Bth.'s interpretation also (A). That Zarathuštra should flout, or be represented as flouting, Mazdāh's command, nobody will earnestly believe, and a glance at the position of (a) = (G) in the two lists shows that in fact this is not the case. For, once we realize that Zarathuštra takes last what in Mazdāh's list comes first, we also see that Mazdāh's last item is the one Zarathuštra takes up first. Da. was therefore right with his translation of (A): 'd'Ahura Mazdāh la sainte création', which grammatically, too, is far more convincing than Bth.'s; instead of three wrong endings it implies one or none, according as we consider *ahuro* a mistake, or, with Da., the first half of an Av. equivalent of the OPers. juxtaposition *aʰuramazdāh-. dāmiδātəm* in (A) is then the acc. neuter of the usual adj. *dāmi.dāta-*, and means 'what is created by the creator' = 'creation'. Once this is understood, the reading *dąma dātəm* 'created creation', of Gdn.'s text in §15, which Bth., probably rightly, replaced with the variant *dāmiδātəm*, turns out to be fully justified as an alternative way of referring to *dąma* in (f).

Now we are in a position to account for the presence of Mithra and Sraoša in §15. As they are not visible in §§13–14, one of the items (a)–(f) must be capable of implying them. This is evidently (f) = (A), as we know from *Yt* 10.1 that Mithra is Mazdāh's creature, and from *Yt* 17.16 that Sraoša is Mithra's brother (cf. above, p. 195). That (B) and (C) serve as illustration of (A) is confirmed by the parallel structure of *Vend.* 19.35, where after the words *nizbayemi ahuro mazdå ašava dāmi.dātəm* we read: *nizbayemi ząm ahuraδātąm āpəm mazdaδātąm urvarąm ašaonīm...zrayō vourukašəm*, etc. 'I call down the Ahura-created earth, Mazdāh-created water, Truth-owning vegetation, the sea Vourukaša, etc.' Here the list of items created by Ahura Mazdāh, of which the first two are expressly stated to be his creation, is obviously intended to illustrate the preceding neuter *dāmi.dātəm*.

Thus *Vend.* 19.15, 35, far from being capable of serving as the one Avestan proof that Mithra was regarded as 'creator of the creation', does in fact make it clear that its author, just as the author of our stanza, thought of Mithra as 'created by the creator'.

64^1. Wi.: 'in welchem das Verständniß der reinen, breithin nützenden Lehre das große, mächtige niedergelegt ist; in welchem der Same ausgebreitet ist auf die sieben Karšvare's'. Wo.: 'in den machtvolle Erhabenheit zur Ausbreitung der schönen Religion, der weithin sich verbreitenden, niedergelegt; des Antlitz nach allen den sieben Erdteilen gerichtet (ist)'. Da.: 'qui a saisi la belle Religion, au loin épandue, d'une prise grande, forte et profonde (*litt.* en qui, dans la prise (*vyāne*) de la Religion, dépôt (*niδātəm*) grand et fort); et dont le regard s'étend sur les sept Karshvares'. Lo.: 'in dessen Seele für die schöne, weitverbreitete Religion (machtvolle Größe?) niedergelegt ist, dessen(?) Same (?= Nachkommenschaft; oder: "Antlitz"?) über alle Erdteile verteilt ist(?)'.

64². *vyāna-* 'soul', cf. Benv., *Inf.* 38, Bailey, *Zor.* Prob. 106, n. 4.

64³. *maza* and *amava* are nom. sg. n. of *mazant-* and *amavant-* respectively, built on the nom. sg. masc. *maza* and *amava* (on which see Bth., *GIP*, I, p. 220) as if these belonged to *-an-* stems. On the supposed Av. neuter *mazaṯ* see note 44³.

64⁴. *niδātəm* is best taken with Da. as a neuter noun meaning 'deposit', but in the special sense of 'pledge'.

64⁵. *yahi* may be a corruption of *yahmi*, which is the reading of H₃ and J₁₀; but it is hard to see why such a common and unmistakable form should have been distorted into an unknown word. We must take care not to consign to oblivion what may in fact be a valuable relic. Considering that Av. *yeṅhe* (**yahyā*) and OInd. *yásyām* have the secondary loc. fem. affix *-ā(m)* of the nominal inflection (on which cf. Wn.–Debr., III, pp. 43*e*, 120 *sq.*, 505), it may well be that *yahi* is the genuine pronominal loc. sg. fem., representing IIr. **ya-si̯-i*. Throughout the Younger Avesta fem. forms tend to be used for the neuter, cf. *čiθrå̄*, next note. On the sporadic pronominal neuters *yå̄*, *yōi*, and *aētaṇhąm* cf. Bth.'s notes to *GIP*, I, §§416a, 417a. *yahi*, having become redundant as a fem. beside *yeṅhe*, may have been set aside to distinguish the neuter where necessary, seeing that the normal masc.–neut. *yahmi* was ambiguous. The present stanza, if correctly interpreted by us, provides just such a case where the distinction is desirable, since a masc. *yahmi* occurs in the first relative clause. Syntactically we may in any case take the second relative pronoun as being governed by the following *paiti*, and referring to the neuter *niδātəm*.

64⁶. An elegant explanation of the coexistence of the meanings 'appearance' and 'race (etc.)' in Av. *čiθra-* is advocated by Duch. in *Anthropologie religieuse* (Supplement to *Numen*, II, 1955), p. 98, n. 6, who compares the semantics of Lat. *species*. It may be worth taking stock of the various forms and meanings attested.

In Avestan, *čiθra-* is adj., '*manifestus*' (cf. OInd. *citrá* 'brilliant'), and substantive with two meanings: (1) 'appearance' (with various shades of meaning, cf. also *hučiθra-* 'beautiful', and *duš.čiθra-* 'ugly'), and (2) 'species, race, progeny, clan, family'. Meaning (2) is attested for the sg. *čiθrəm* in *Y* 32.3 and *Yt* 13.89,† as well as *Yt* 13.87 (*čiθrəm airyanąm*

† From *Y* 32.3 we learn that *čiθrəm* can be construed with the ablative of the progeny's source: *yūš daēvā vīspåṇhō akāṯ manaṇhō stā čiθrəm* 'all you daēvas are the progeny of (*lit.* from) the Evil Mind'. This construction supplies the clue for the understanding of two generally mistranslated lines in *Yt* 13.89: (*zaraθuštrahe...*) *yō paoiryō čiθrəm urvaēsayata | daēvāaṯča haotāṯ mašyāaṯča*. According to Bth. this means 'der zuerst sein Gesicht abkehrte von dem Daēva- und

daḥyunąm 'the family of Iranian nations'), for the plur. čiθrǎ or čiθrå̆ (fem. ending, cf. *GIP*, I, p. 233, n. 3c, where Bth., instead of 'nach der s-Klasse' should have printed 'nach der ā-Klasse', as in his following note d) in st. 112 of our Hymn and in Y 44.16 (*yōi hənti čiθrā mōi dąm* 'who are the offspring in my house', see *JRAS*, 1952, 176). In addition, Bth. rightly postulated for *čiθra-* the meanings 'seed' (in compounds, possibly also in the present stanza), which is a development of 'race', and 'origin', which is a development of 'seed'.

Outside Avestan we find in conformity with the Av. meaning (1): Pahl. *čihr(ak)* and *čihrīk* 'visible', *čihr* 'visible form', *čihrēnītan* 'to give visible form to', *čihrak* (NPers. *čihr(a)*) 'countenance', see Bailey, *BSOS*, VII, 281, Zor. Prob. 91 sq.; Man. MPers. *cyhr* 'face' (*Mir. Man.* II), plur. *cyhr"n* 'Gestalten' (*BBB*), *cyhr'wynd* 'beautiful' (see Henning, *Sogdica*, 37); Parth. *cyhr(g)* 'form, appearance' (see Mary Boyce, *BSOAS*, XIII, 912, n. 1), *hwcyhr* 'beautiful', *dwrcyhr* 'ugly', *h'mcyhrg* 'homomorphic' (Henning, *Asia Major*, N.S., III, 194, n. 54, Mary Boyce, *MHCP*, 171, n. 2); Pš. *cēr* 'similar', *bar-cēr* 'clear, manifest', *cēra* 'picture' (Pers. LW), see Morg., *EVP*, 17; Gāndhārī *cirorma* 'face-covering', see Bailey, *BSOAS*, XI, 794.

With the Av. meaning (2) the following agree: OPers. *ariya čiθ^{r}a* 'of Iranian lineage', Man. MPers. *cyhr* 'essence, origin, offspring' (*Mir. Man.* I); Man. Parth. *cyhrg* 'essence, nature' (*Mir. Man.* III, 876.23, 886.16; Mary Boyce, *BSOAS*, XIII, 915, line 13; XIV, 438, verse 17, and *MHCP*); Inscriptional Parthian |*kē cyhr* |*ac yazdān*, translated by ἐκ γένους θεῶν;† Pazand *čihara* 'essence, origin, source'; NPers. *čihr(a)* 'origin, essence'; Khowar *puši-žeri* 'kitten', see Morg., *BSOS*, VIII, 660.

With the adj. Av. *čiθra*, OInd. *citrá* may go the OPers. name Τισσα-φέρνης if = 'having splendid *Farnah*' (Justi). [A connection of Arm. *čitak* 'necklace' with OIr. *čiθra-* is suggested by Bailey, *TPS*, 1956, 107.]

64⁷. *vīδātəm* 'distributed' is from *dā-*, in contrast with *vīδātəm* 'set',

Menschengezücht'. This is clearly unsatisfactory because (1) the statement makes poor sense, (2) *urvaēsaya-* is causative, and (3) one would expect a preposition (*hača*) to take charge of the ablative. Spi., *Commentar*, II, 611 sq., tried to meet these objections by treating the abl. as depending on *paoiryō* in comparative function: 'welcher das (Himmels) Rad (Var. *čaxrəm*) früher drehen ließ als die Daewas und geholfenen (Var. *aotāt̰*) Menschen (da waren)'. This interpretation, too, satisfies neither text nor context. If, however, *čiθrəm*... *haotāt̰* is taken, in accordance with Y 32.3, to mean 'progeny of the evil breed', everything becomes clear: '(Zarathuštra...) who put to flight the progeny of the evil breed of both daēvas and men'. The caus. *urvaēsaya-* means 'to turn away, *fugare*' also in Yt 5.131. On Zarathuštra putting to flight the daēvas cf. Yt 19.81 and Y 9.15 (disguised as men).

† See M. Sprengling, *Third Century Iran, Sapor and Kartir*, Chicago, 1953, p. 7.1 and plate 10.1. Cf. Hzf., Zor. II, 431 [and Otakar Klíma in *Archiv Orientální*, 24 (1956), 292 sq.].

see note 44². The pledge to the Religion which Mithra has stored in his soul, is that he will bring about the conditions of peace which will favour the spreading of the Religion. This one infers from *Yt* 13.94 *sq.* (quoted above, p. 27), where the spreading of the Religion over the seven climes is synchronized with Mithra's and Apam Napāt's pacification of countries in turmoil and consolidation of the central authorities (*frətəmatātō*, see note 145¹). The pledge is in keeping with Mithra's office of executive chief and 'purifier' of the Religion, which he appears to hold in st. 92 (cf. note 92¹, last para. but one). In return for these services the Religion paves Mithra's paths (st. 68), and Čistā, her Double, escorts him on his journeys (st. 126).

65¹. For Henning's etymological connection of *arədra-* with Sogd. *i 'rδwky*', *'yw'rδkw*, see *GMS*, §154.

65². On *fraxšti.dā̊* see *BSOAS*, xiv, 487.

65³. *āzūiti-* 'fat', cf. Lo., *NGGW*, 1935, 141 *sq.*; on Nyberg's translation 'sprinkling (with cow urine)' cf. Henning, *Zor.* 12 n.; on the Gāthic passages where *āzūiti-* occurs (*Y* 29.7, 49.5) cf. *JRAS*, 1952, 178.

66¹. Bounty (*Pārəndi*), here 'swift-charioted', in *Visp.* 7.2 simply 'swift' (*rəvī-*), is a frequent companion of Aši. She appears with Mithra and other divinities also in *Yt* 8.38 (see below, note 68²), and is found in *Visp.* 7.2 in the company of Justice (*Arštāt*, see below, note 139²), the Religion, Rašnu, and Mithra. [See Addenda.]

66². *raoraθa-*; on *rao < ra(γ)u-* see Duch., *Comp.* 21. In *BSOAS*, xiv (1952), 483, n. 2, a warning was uttered against believing O. Szemerényi's derivation of Oss. *ræuæg* 'light, quick' from **rabaka-*; this is an obvious impossibility, since intervocalic *b* results in Oss. *v*, not *u*. Nevertheless *ræuæg* was uncritically entered by Pokorny under IE **rebh-* (*IEW*, 9. Lieferung (1955), p. 853). I have no doubt now, in view of the identity of meaning, that Benv. was right in connecting *ræuæg* with Av. *raγu-* (*JAs.* 1936, 200 *sq.*). As we must start from **raγuaka-*, Ossetic evidently shares the Av. development of OIr. *-gu-* to *-u-*, although so far no other Oss. example has been noticed. To *raγu-* belong, in addition to the Iranian words quoted by Benv., also Khot. *rraysga- < *rajuka-* (S. Konow, *NTS*, xi, 74), Waxi *rānjk* (Morg., *IIFL*, ii, 537), and Sogd. *ryncwk < *ranjuka-* (Henning, *BSOAS*, xi, 482, n. 5), which go with the stem *ra(n)j-* of the Av. comparative *rənjyah-*, superl. *rənjišta-*. [Addenda.]

66³. *θwāša-*, translated with Gdn. as 'Firmament', on the strength of Zaehner's remarks, *Zurvan*, 89.

66⁴. All translators, except Htl., co-ordinate *yasča...haθrākō* with *uγrā́sča...fravašayō*, taking *haθrāka-* to be a member of Mithra's escort. Wo.: 'und der Einiger (Wi. 'die Versammlung', Spi. 'wer ein Versammler ist', Gdn. 'die Einigkeit', Lo. 'die Gemeinschaft') der vielen ašagläubigen Mazdāhanbeter'. Da. has 'et Celui qui tient réunis les... adorateurs de Mazda', with the contradictory note: 'il s'agirait de Mithra même'. Htl. thought of resolving this contradiction by referring *yas(ča)* to Mithra: 'der die vielen vereinigt...die Mazdāh-Opferer'. Before taking our choice we must try to determine the meaning of *haθrākō*, which is more likely to be a masc. epithet of a god than an abstract 'Gemeinschaft'. However, the causative meaning which Bth. and others attribute to this adj. is not what one would expect from a *-ka-* derivative of the adverb *haθrǎ* 'simultaneously, in one place, together'. The comparable adjectives listed by Wn.–Debr., II, 2, 519 *sq.*, are all intransitive, cf. OInd. *ánuka* 'dependent (*lit.* what, who, is *anu*)', *ápāka* 'removed, distant', etc. Accordingly the *haθrākō* should be 'one who *is* together, simultaneous, with'. The gen. governed by *haθrākō*, instead of the expected instr., can be understood if the construction of the synonymous compounds with *hǎmō-* (discussed at the end of note 109[6]) is borne in mind: 'being in the same place as many Mazdayasnians' might be expressed in Av. by *hāmō.gātuš pourunąm mazdayasnanąm*, 'being at the same time as (= simultaneous with) many M.' by **hamō.ratuš pourunąm mazdayasnanąm*. *haθrāka-*, which here very likely means both 'being in one place with' and 'being simultaneous with', could easily have borrowed this construction.

Assuming that such is the meaning of *haθrākō...mazdayasnanąm* one is at first tempted to follow Htl. and identify the *haθrākō* with Mithra. The present two lines would then be a further statement of Mithra's ubiquity, on which see note 95[1], end. To adapt our translation to this view it would be sufficient to replace the comma after 'Truth' by a semicolon, and suppress the pronoun 'he': '...of Truth; and who shares place and time...'.

On the other hand, if this were what the poet had in mind, he would have to be charged with unwonted ambiguity, for usually he refers back to Mithra by a simple *yō*, even after one or more preceding *yō*-s, cf. sts. 65, 69, 98, 103, 143. Moreover, in *Vend.* 19.34 we read: *narō ašavanō ham.bavainti nairyō savhō ham.bavaiti* 'die ašagläubigen Männer vereinen sich, Nairyō.saṃha vereinigt sich (mit ihnen)' (Wo.). Although these words appear to have been torn out of their original context, the description *pourunąm haθrākō ašaonąm mazdayasnanąm* would seem to be applicable to Nairyō.saṃha in the circumstances contemplated in this fragment. We may perhaps amplify the two passages as follows:

Vend. 19.34, ⟨wherever and whenever⟩ Truth-owning men assemble, Nairyō.saṃha is with ⟨them⟩;

Yt 10.66, ...and ⟨Nairyō.saṃha,⟩ who is together with Truth-owning Mazdayasnians ⟨, wherever and whenever there are⟩ many ⟨of them⟩.

Nairyō.saṃha is, of course, sufficiently accredited with Mithra to be acceptable as a member of his escort in the present stanza, cf. above, note 52[1].

COMMENTARY   [67.1–68.2

67[1]. That *arəzahī-* is the eastern (and consequently *savahī-* the western) *kišwar* was made clear by Henning, *Sogdica*, 28 sq.

67[2]. The word in question is either *čixra*, a *hapax legomenon* rendered by Bth. on etymological grounds as 'energy', or, with two MSS., *čaxra*, or, with one MS., *čaθrahe*. Gdn.'s *čiθra*, which he himself abandoned in the Addenda to his edition, is an emendation. Lo., following Wi., translated 'mit passendem Rad', Da. 'avec la roue qui roule régulièrement'. But *čixra* as *lectio difficilior* deserves preference, and there is no point for the present in departing from Bth.'s guess at the meaning.

Wo. took the following *vərəθrayna* as the abstract noun, Da. as the name of the god. Either interpretation is possible, since on the one hand we have *zaēna hačimnō* 'equipped with a weapon' in st. 141, on the other hand *hačimnō..rašnu* 'accompanied by Rašnu' in *Yt* 13.3 (see above, p. 37). Considering that Vərəθrayna the god comes into his own very shortly (st. 70) I have opted for his presence being anticipated at this juncture only through the quality he represents. This view is recommended by the comparison of *hačimnō x^{v}arənaṇha* in the present st., with *x^{v}arəna* (thematic) *hačimnō* in st. 141, which, coming five lines after *zaēna hačimnō*, apparently means 'equipped with fortune', not 'accompanied by Fortune'. We are thus relieved of the necessity of speculating whether *čixra-*, too, instead of a quality, should be taken as the name of the divinity representing this quality.

68[1]. *hangrəwnāiti* 'guides'; thus with Spi., Gdn., Bth., Htl. The most accurate rendering is Wi.'s 'mitergreift'. Da. has 'soulève', Hzf., 434, 'with whom drives in his chariot'.

68[2]. Here we find Mithra's two sisters (cf. above, p. 195), Aši and the Religion, in action. Aši appears implicitly as his charioteer in sts. 76 and 143.† Elsewhere (*Yt* 8.38) she and Bounty (*Pārəndi*, cf. note 66[1]) are associated with Mithra, Ahura Mazdāh, and the Incremental Immortals (cf. note 51[1]), in assisting the arrow which was shot by the legendary archer Ɣrəxša.‡ We shall see below, p. 228, that in *Y* 60.7 Aši and Mithra together are singled out as bestowers of fortune, riches, and noble progeny, to the house of the faithful. The association of Aši and Mithra in the minds of early Zoroastrian priests is also indicated by what is apparently a gloss added to *Yt* 17.2: *uta.hē...ava.baraiti...uta.hē... jasaiti avaṇhe yō ašīm yazāite zaoθrābyō [hō miθrəm yazāite zaoθrābyō]*

---

† Aši drives her own chariot in *Yt* 17.17. Her charioteering propensities make it likely that her epithet *x^{v}anaṭ čaxra-*, which she shares with Drvāspā, means not 'deren Räder sausen', as Bth. has it, but 'impeller of wheels' (base ²*hav-*). [On *čanaṭ.čaxra-* cf. Bailey, *TPS*, 1956, 105 sq.]

‡ On the other hand *Yt* 8.7 names as supporters of the arrow Ahura Mazdāh, Mithra, the Waters, and the Plants; the last two also accompany Mithra in st. 100 of our Hymn.

'she (Aši) brings...to him, and comes to the help of him...who worships Aši with libations [he will worship Mithra with libations]'.† It is thus very likely that Cumont was right in identifying with Aši the *Fortuna* who appears on Mithraic inscriptions, see *TMMM*, I, 151.

On the give-and-take arrangements between Mithra and the Religion see note 64⁷.

68³. *xᵛīte*, dative of *xᵛīti-*, cf. Benv., *JAs*. 1936, 199.

68⁴. *frādərəsra-* elsewhere qualifies the star Sirius, the sky, the soul of Ahura Mazdāh, and, in *Vend.* 22.1, his house, that is, according to Da., the Paradise. Bth.'s translation 'hervorleuchtend, strahlend' ill agrees with the meaning of the base *darəs-* 'to see', while Da.'s and Lo.'s 'visible afar' is scarcely applicable to soul and Paradise. I take *frādərəsra-* as meaning primarily 'transparent, *perspicuus*'; for the preverb cf. Oss. *ræsug* 'transparent, clear' < **frasauka-*, to Av. *suka-* 'sight, light'; for the suffix see Wn.–Debr., II, 2, 853. Secondarily, as in many other languages, from 'transparent' the meaning 'clear = spotless' developed. In describing the horses as *asaya-*, 'shadeless', the author of our passage (which also occurs in *Y* 57.27) probably thought of *frādərəsra-* as meaning 'transparent'.

68⁵. *yaṯ...bāδa irinaxti* 'as often as he launches', lit. 'as at times he launches', see note 53¹.

68⁶. *hu.irixtəm...irinaxti* 'launches well-launched', lit. 'lets go, releases, well-released', implies that Dāmōiš Upamana is situated *behind* Mithra's chariot, cf. notes 9⁴ (p. 168) and 127¹. Gdn.'s translation 'hält stets die bahn frei' ('während ihm D.U. recht freie Bahn macht', *Lesebuch*) is wrong, and fails to account for *dim*.

68⁷. *varənya-*, which Bth., Lo., and still Christensen, *Essai sur la démonologie iranienne*, 9 *sq*., translate 'aus Varəna stammend', can hardly have anything to do with that country, in which Henning has recognized the modern Buner (*BSOAS*, XII, 52 *sq*.). Already Da. (*ZA*, II, 373 *sq*. (n. 33), followed by Jackson, *GIP*, II, 660, 663) had seen the obvious connection of *varənya-* with the demon of concupiscence called *Varan* in Pahlavi, but confused the understanding of the adjective by bringing in the name of the country Varəna. On the demon Varan cf. Zaehner, *Zurvan*, pp. 124, 174.

68⁸. All translators render stanza 69 (and the parallel sts. 98 and 135) as if its words were pronounced by the worshippers with apotropaic intent. They are more likely the utterance of the terrified daēvas and owners of Falsehood, who have been watching with dismay the arrival of Mithra and his escort. Otherwise it would be a strange coincidence that,

† The usual translation 'who worships Aši with libations worships Mithra with libations', apart from being nonsensical, fails to account for the second subj. *yazāite*, and the enclitic *hē* of the *ava.baraiti* clause. Note that the gloss is missing in two MSS.

COMMENTARY   [68.8–71.1

each time the exclamation *mōi* (*mā*)...*jasaēma* occurs, it is immediately preceded by the mention of *daēva*-s and *drvant*-s.

Lo.'s interpretation of *fratərəsa*- in sts. 68, 97, 99 and 134 as 'to flee', which is based on the meaning of Pš. *tṣ̌əl*, *taṣ̌tēdəl*, is not to be recommended, as it does not apply to the OIr. forms of this base without preverb, from which the Pašto verbs derive.

70¹. On Vərəθraγna's role, or one of his roles, as Mithra's assistant cf. *Yt* 14.47 (quoted above, p. 193).

70². *hū kəhrpa varāzahe*, cf. the boar visible next to Heracles = Vərəθraγna on a Mithraic relief, Cumont, *TMMM*, I, 143; II, 344, perhaps also II, 266 and 339 (Mon. 239, *a*). See also above, p. 62, n. †. On the god's incarnation in a boar see Benv., *Vṛtra*, 35, 69, 73.

70³. The *s* of *tiži.dạstra*- is more readily explained as due to the analogy of the synonymous *tiži.dạsura*- (cf. also *karətō.dạsu*- 'biting with knives (=spurs)'), than to that of a theoretical nom. **tiži.dạs* of **tiži.dant*- as Bth. thought.

70⁴. I am indebted to H. W. Bailey for the correct explanation of *asūra*- in *tiži.asūra*-. The word is connected with Sogd. *'ns(')wr* in *VJ*, hitherto wrongly translated as '*chose inappréciable*'; in reality it means 'tusk (of an elephant)', as shown by Oss. (Dig.) *ænsur(æ)*, (Iron) *ssir*, 'tusk'. The Av. initial *a*- therefore represents IE *n̥*-.

70⁵. Wi. rendered *paršv-anika*- by 'angesicht-triefend', cf. Skt. *parṣati* 'to sprinkle' (NPers. *pāšīdan*, Orm. *prusnaw*- 'to sprinkle', Morg., *IIFL*, I, 404) and Av. *āpō*...*paršuyå̊* ('channel-water' according to Henning, *BSOS*, IX, 91). Other translations include 'von struppigem aussehen' (Gdn., suspecting 'zusammenhang mit skr. *parṣa*, *parṣin*'), 'mit triefendem Maule' (W. Geiger, *OK*, 158), 'cherchant la bataille' (Da.), 'mit gesprenkeltem Gesicht' (Bth., referring to OInd. *pṛṣat* 'spotted'; approved by Duch., *Comp.* 161), 'mit (borstigem?) Gesicht' (Lo.). Bailey supports the translation of *paršu*- as 'spotted' by comparing Georgian *pharšamangi*, MPers. *frašēmurv* 'peacock', see *BSOS*, VI, 596 *sq.*

71¹. *frạštačō hamərəθāδa*, Wi. 'der hervorstürzt vom Gegner gefolgt'; Gdn. (followed by Bth. and Lo.) 'den feind im laufe überholend'; Da. 'poursuivant...l'ennemi'; Spi. 'welcher vorwärts stürzt nachdem er an den Feind sich gehängt hat'; Htl. 'der, dem Gegner entgegenlaufend, sich an ihn macht'. For the meaning assumed by Gdn. there is a parallel in Arəzō.šamana's epithet *afrakatačim* (*Yt* 19.42), for which Bth.'s translation 'not running forward (from the cover he had taken)' is unconvincing; one expects a positive, not a negative quality.† Gdn.'s

† The epithets which precede *afrakatačim* in *Yt* 19.42 include *frāzuštəm* [...]*uštəm* (the latter probably a compound of °*zuštəm* with a preverb other than *frā̆*, cf. Bth., *Wb.* 420), which have been assigned by Bth. to the base *zuš*- 'to like'; Wo. accordingly translates 'den geliebten'. One would rather expect the

interpretation, *Drei Yasht*, 26, n. 4, is therefore to be preferred: he took *afrakatak-* for a bahuvrīhi containing **fraka-tak-* 'Zuvorkommer', hence meaning 'who has no *Zuvorkommer*, cannot be caught up with'.
71². *stija*, Wi. 'verwundend' (comparing Gr. στίζω, στίγμα); W. Geiger, *OK*, 447, n. 1, 'spitze Waffe' (comparing Ved. *tigmá*, etc.), a view which was revived by Duch., *BSOS*, IX, 865; Justi, Spi., Gdn., Da., Bth., Lo., Htl., 'im, mit, Kampf', comparing NPers. *sitēz*. Justi's interpretation deserves preference over Geiger's, inasmuch as a 'pointed weapon' is unwanted: the bestial treatment meted out to the victim clearly shows that the attacker is the boar, not Vərəθrayna, let alone Mithra, as might appear from some translations; the boar's fangs and

opposite meaning, and seek a connection with Oss. (Digor) *zust* 'hart, finster, grob' (Miller-Freiman's Dictionary, I, 617), provided, of course, the latter is not from **a-zušta-* (cf. Morg.'s explanation of NPers. *zōš* 'vehement, violent, wicked' as from **a-zauša-*, *EVP*, 106); cf. Middle Irish *guss* < **ĝhustu-s*, 'force, violence, anger', Pokorny, *Indogermanisches Wörterbuch*, 448? On the other hand, for the epithet *barō.zušəm* (Bth. 'wearing a jewel' (?)), which immediately follows *afrakatačim*, a connection with Pš. *zwaž* 'noise, clamour' may be suggested, which Morg., *loc. cit.*, derives from **zauša-*; 'raising a noise', with *bar-*, is reminiscent of the common Av. expression *vāčim bar-*, see Bth., *Wb.*, 935 *sq.*
There is a different **zuš-*, from which Morgenstierne derived Sanglechi *zōl* (etc.) 'sleeve', see *IIFL*, II, 424. This can be recognized in Av. *zuša* in the *Frahang i oīm* ('jewel' according to Bth.), and in Av. *frazuš-*, epithet of *aδka-* 'coat' in *Yt* 5.126 (quoted note 124²), which is generally thought to mean 'precious'. As Skt. *pra-hasta* means 'long-handed', cf. Wn., II, 1, 284, so *frazuš-* could mean 'long-sleeved'; cf. A. Waag, *Nirangistan*, 134, who sought in *aδka-* a coat with long sleeves. The Pahl. transl. of *frazuš-*, which also occurs in *N*, looks like *fr'c xw'stk*, that of *zuša* in *F* like *xwstwk*. Bearing in mind Yidγa *avlăsto* 'sleeve' < **upa-dastā-* (Morg., *IIFL*, II, 194), and the Pahl. word which Nyberg reads **apaδast* and translates 'Handschutz' (*Hilfsbuch des Pehlevi*, II, p. 11), one might read **'wstnk* instead of Pahl. *xwstwk*, and, with a slight emendation, **'wystk* instead of *xw'stk*. *'wyst-* and *'wst-* could represent an older **abi-sta-*, amplified by *-ka-* and *-na-ka-* respectively. The development will be as in **abi-štāta-* > MPers. *'wyst'd* > NPers. *ōstāδ*, *ustāδ* 'teacher', or **abi-stāna-* > MPers. *'wyst'm*, Arm. *ostan*, > Pers.-Ar. *ustān* 'province', see Henning, *ZII*, IX, 195, 224 *sq.* As to **abista(na)ka-*, this may be a reduction of **abi-d(a)sta-* (*na)ka-*, comparable to that which according to H. W. Bailey (personal communication) took place in Oss. *fīston*, Dig. *festonæ* 'handle' < **pati-dastāna-*, cf. Khot. *daśtānya* 'handle', *Asia Major*, N.S., II, 30 (48.1). Perhaps NPers. *āstīn* 'sleeve' similarly goes back to **ā-d(a)st-*. As to *zuš* in *Yt* 5.7 (*frā srīra zuš sispata urvaiti bāzu.staoyehi*) one would agree with Bth. that it ought to be an epithet of Anāhitā in the nom. sing., if one did not thereby have to attribute to the goddess also the strange epithet 'stouter than the arm', which is scarcely explained by the assumption that Anāhitā is here a river. The epithet would, however, suit wide sleeves, and the ending *-i* may be of the neuter dual. Has then *zuš* replaced an earlier *zuši*, possibly because as an attribute of Anāhitā, *srīra.zuš* 'having beautiful sleeves', was sought in this line? If so, translate 'she, the beautiful, shows off her flowing sleeves which are ampler than her arms', the comparative deriving its point from the earlier description of Anāhitā's arms as being 'stouter than horses' (*aspō.staoyehīš*).

tusks having been mentioned, there is no room for further 'pointed weapons'. On the other hand 'battle' or 'quarrel' in the case of a fleeing opponent is also questionable. An alternative is to see in *stija* a tossing motion of the boar's head, having regard to the method of attack of this beast, which consists in ramming the opponent with lowered tusks and lifting the head, so as to thrust them upwards into his body (cf. Alfred Brehm, *Tierleben, Säugetiere*, IV, p. 10). Thus interpreted *stig-* will belong to Dig. *tehun*, Iron *tīhĩn* 'to sift', whose original meaning is probably 'to shake, push, swing', judging by Dig. *ræ-tehun* 'to swing', *ba-ræ-tehun* 'to push, swing'. The initial is as in Av. (*s*)*taēra-*, etc.

71³. *naēδa manyete*, lit. 'he does not at all think', see note 19¹.

71⁴. The correct explanation of *yənąm* and *ava.dərənąm* (*Vend.* 18.19), by imitation also *vyusą* (*H* 2.7, 25), before *saδayeiti*, is due to Benv., *Inf.* 18 sq. (who, however, unnecessarily regarded the present line *naēδa...saδayeiti* as an interpolation, *op. cit.* 16 sq.): they are nom. sg. masc. participles in *-ąs*, which lost their *-s* before the initial *s* of *saδayeiti*, after which final *-ą* was liable to be represented by *-ąm*; cf. *hubərətą(m)* below, note 112⁴. Hence *yənąm saδayeiti* = 'he seems hitting'.

71⁵. *čim* has been consistently misinterpreted; cf. Benv., *Inf.* 16, who thought it wrongly stood for *čit̰*. It is left untranslated by everybody, unless Gdn. meant to refer to it when he wrote 'und nicht glaubt *genug* gemordet zu haben'. Actually *čim* is the acc. sg. masc. of the indefinite pronoun *kasčit̰*; this would be *kəmčit̰* or **čimčit̰* in positive sentences, but °*čit̰* is bound to be absent after a negation, see Bth., *Wb.* 426, sect. 4. Hence *naēδa čim* means 'and not anybody'; cf. also *naēδa.čiš* in *Vend.* 18.11 (quoted above, note 19¹), where *čiš* is nom. sg. fem. agreeing with *āfritiš* (*čiš* is fem. also in *H* 2.10: *čišča čarāitiš ahi*; on *čiš* as nom.-acc. sg. n. see Bth., *Wb.* 427, n. 2).

71⁶. OPers. **mrdu-*, corresponding to Av. *mərəzu-*, survives unrecognized in Pahl. *mrd'*, which translates Av. *manaoθrī-* 'neck', see Bth., *ZWb.* 205 sq. The Pahl. ending compares with that of *b'z'*, Sogd. *β'z'* from OIr. *bāzu-* (cf. *GMS*, §971), or of *ns'* from *nasu-*. It is noteworthy that *-ă-* stems from OIr. *-u-* stems are found also in Pašto, cf. *zana* 'chin' from **zanu-* (see Henning, *Sogdica*, 49 sq.), *ōṣ̌a* 'tear' from *asru-*, *aca* 'thigh-bone' from *asču-* (Morg., *NTS*, XII, 262). The origin of *-ă* is no doubt to be sought in forms like Av. *nasāvō*, *bāzāuš*, OPers. *dahyāuš*, on which cf. Meillet–Benveniste, *Grammaire du vieux-perse*, pp. 165 sq. For other forms belonging to Av. *mərəzu-* 'peg, vertebra', see Henning, *JRAS*, 1942, 241 sq., *BSOAS*, XIII, 642, n. 2. [See Addenda.]

71⁷. On *uštāna-* cf. Bailey, *TPS*, 1954, 135 sq.

75¹. *tē* is by most translators regarded as personal pronoun: 'Seien wir Dir Felderbeschützer, nicht seien wir Dir Felderverderber...' (Wi., similarly Spi.); 'Wir wollen auf deiner flur wohnen bleiben, nicht deine flur verlassen' (Gdn.); 'Wir wollen dein Land in Schutz nehmen, wir wollen (dein) Land nicht im Stich lassen' (Wo.). However, no explanation has come forth as to what is to be understood by 'your šōiθra-'. Only Wi. appropriately mitigated the awkwardness of this interpretation by taking tē as a *dativus ethicus*. But Da.'s understanding of tē as demonstrative pronoun is much to be preferred: 'Puissons-nous garder nos terres! Ne pas quitter nos terres!'; the worshippers are more likely to be concerned with their own šōiθra-s than with Mithra's.

75². At the end of the series of five compounds with °*iričō* the predicate most easily supplied after *māδa* is ⟨*buyāma taṯ.iričō*⟩, *taṯ* being, as often, implied in the rel. pron. *yaṯ* which here is the object of *nivānāṯ*: 'and ⟨may we⟩ not ⟨become such as have to abandon that⟩ which the strong-armed shall guard for us...'. The request that strong-armed Mithra may guard the property of the worshippers is telescoped with the prayer that the property may not have to be abandoned by them. Other renderings differ: 'und nicht (soll das geschehen), damit der starkarmige (Mithra) uns berge(nd schütze) vor den Feinden' (Wo.); 'dann nicht, wenn der Starkarmige uns beschützen wird' (Lo.); 'et que le Dieu aux bras redoutables ne nous écrase pas devant nos ennemis!' (Da., following Wi.); 'nie so lange nur er der starkarmige...' (Gdn.). In *nivānāṯ*, which Bth. entered under his base ⁴*van*-, Lo. (quoted by Henning, *ZII*, IX, 177, n.) wanted to find a late form of *nipānāt; however, both the unusual present stem thereby assumed for *pā(y)*- 'to protect', and the 'Middle Persian behaviour' of Av. *p* which this suggestion entails, make it inadvisable to rely on it.

76¹. Bth. replaced *hurāθyō* by the better attested reading *hurāθvō*, standing for *hurāθivō 'mit schönem Wagenzeug versehen'. Gdn. translated 'du bist...ein guter wagenstreiter', Da. 'tu as...de belles voies'. Clearly we cannot translate 'having a good chariot' as Wi. and Duch. (*Comp.* 33 (§57), 189 *sq.*) did, since *rāθi-, as a vṛddhi-derivative of *raθa*- 'chariot', should not also mean 'chariot'. After the reference to Mithra's good horses, it is more likely that by *rāθi- his charioteer (viz. Aši, cf. sts. 68 and 143) is meant than his 'Wagenzeug'; cf. Ved. *sárathi* 'charioteer', and Wn.–Debr., II, 2, 303 *sq.*

77¹. I take *zbayāi* as introducing direct speech from *āča.nō* to *hufraborəitiča*. Other translators assume a single sentence, disregarding the alternation between 2nd (*θwā*) and 3rd person (*jamyāṯ*); Da.'s and Lo.'s remedy consists in treating *āča.nō jamyāṯ avaiṅhe* as a parenthesis.

78¹. *tā...yā* = '*ita...ut*'. Less well Bth.: 'then...when'. The other translators, disregarding correct grammar, take *tā* as acc., *yā* as nom. pl. fem.: 'those countries which strive'.

78². This *hubərəti*- 'good care' is the abstract belonging to *hubərət*- 'treating well' and *hubərəta*- 'well treated' (see notes 48⁵, 112⁴), and must apparently be distinguished, as by Bth., from *hubərəti*- 'good offering', which occurs several times in a series with *ušta.bərəti*- and *vanta.bərəti*-.

However, Spi., followed by Lo. and Htl., has 'welche sich guter Darbringung befleißigen für den Mithra'. [See Addenda.]

79[1]. Lit. 'the judge (*rašnuš*) who renders (*daiδe*) gaining-in-prominence (*manavaintīm*, adj.) the abode [for or of him] to whom ([*ahmāi*] *yahmāi*) Rašnu has given [it] for long succession'. This translation of a most intriguing stanza has two advantages over the other renderings quoted below: it requires no emendations, and offers a logical antecedent to what follows in st. 80. It is based on three assumptions:

(1) That it is Rašnu who provides the faithful with an abode which Mithra makes prosper. Such generosity on Rašnu's part is not met with anywhere else, but may be linked with the activity of Rašnu's and Mithra's brother (cf. above, p. 195) Sraoša, who, according to *Y* 57.10, 'after sunset builds a strong house for the poor man and the poor woman' (*driyaošča drīvyåsča amavaṱ nmānəm hąm.tāšti pasča hū frāšmō.dāitīm*). That Mithra makes 'gaining-in-prominence' the abode built by Sraoša and given by Rašnu would agree with the care Mithra is said elsewhere to take of the house (cf. sts. 28, 30, 80, and below, p. 229, on his office of $x^v\bar{a}\theta r\bar{o}.disya$- of the house). Thus, if our interpretation of the present stanza is correct, Mithra, Rašnu, and Sraoša would appear to have been *inter alia* a triad of tutelary genii of the house.

(2) That the first *rašnuš* is not the name of the god but the common noun which occurs in the plural in *Visp.* 16.1, where 'fire-born *rašnuš*' are worshipped between 'fire-born gods' and Fravašis.† Bth. translated *rašnu-* by 'just', but one wonders if 'the just' would have been mentioned in one breath with gods and Fravašis. Moreover, *-nu-* adjectives formed from the base normally have active voice, see Wn.–Debr., II, 2, 741. Hence *rašnu-* ought to mean 'ruling', or perhaps 'judging', when substantivized 'ruler' or 'judge'. The latter meaning is preferable, since Mithra, to whom *rašnuš* here stands in apposition, is said in a Pahlavi passage to be a judge (*dātaβar*) (see Zaehner, *Zurvan*, 101, n. 5); cf. also his position of *ahū-* and *ratu-* in st. 92.

(3) That the *hapax legomenon manavaintīm* can be explained by combining Wi.'s reference to Arm. *manavand* 'more, rather, better', *manavandzi, manavandt'ē* 'above all, chiefly', with Da.'s to Av. *hąm. vaintī-*, the former suggestion to be used as a guide for the meaning, the latter by way of formal analogy.‡ According to Bth., *Wb.* 1352, *hąm.vaintī-* means 'victorious', and is derived by haplology from **hąm.vanant-*. Similarly we may postulate an adjective *manavantya-* (nom.–acc. neut.

---

† *ātarš.čiθrəsča yazatō yazamaide, ātarš.čiθrəsča rašnušča yazamaide, ašāunąmča fravašayō yazamaide, sraošəmča yim vərəθrājanəm yazamaide*, etc.; Bth. was no doubt right in following the string of MSS. which omit °*ča* after *rašnuš*.

‡ [The comparison of Arm. *manavand* will have to be abandoned if Bailey's analysis of this word as from **nāmǎvant-* (*TPS*, 1956, 107) should prove to be correct. Our interpretation of *manavaintīm* can nevertheless be maintained.]

manavaintīm) from **mana-vanantya-* 'apt to gain eminence (or sim.)'; this will be a *-ya-* extension of **mana-vanant-* 'gaining eminence' (cf. the reading *manavaintəm* in the Apparatus to st. 81), as Ved. *sahantya* 'prevailing, conquering' of the participle *sahant-*. **mana-* 'eminence', the noun from which Av. *mainya-* 'authoritative' was derived (on which cf. Henning, *BSOAS*, XII, 309), will then belong to Bth.'s base ³*man-* '*pro-mineo*', the only finite form of which we have endeavoured to eliminate (see note 20²), but which is also attested in *mati-* '*promunturium*' and *framainya-* 'eminence'. Substantially we thus find ourselves in agreement with Gdn.'s etymological interpretation, though not with his understanding of word and stanza: 'accusativ eines themas *manavainti* oder *manavaintya* hausbesitz zu wz. *van. mana* ist das grundwort des § 137 fl. uns begegnenden *mainya*'.

If the above assumptions are granted, *manavaintīm* can be taken as predicative adj. agreeing with *maēθanəm*, an antecedent *ahmāi* or *ahe* being implied for *yahmāi*. *daiδe* will then have Bth.'s meaning V (*Wb.* 716 *sq.*) 'einsetzen als, machen zu'. When so used *dā-* is usually active, but the middle is found in *Y* 58.1 *tat sōiδiš tat vərəθrəm dadəmaidē hyat nəmə̄* 'this, the prayer, we make (our) weapon and shield' (cf. Benv., *Vṛtra*, 12 *sq.*), in *Y* 55.6 *frašəm vasna ahūm daθāna* 'rendering the existence extraordinary† at will', and in *A* 3.7 *sqq.*

Not surprisingly, other translations differ considerably from the above and from each other: 'welcher die wohnung des Rashnu teilt, welchem Rashnu zu dauernder gemeinschaft den (mit)besitz des hauses einräumte' (Gdn.); 'welchem R., als er sich ein Haus baute, zu dauerndem Bund die Hausfreundschaft anbot' (Gdn., *Lesebuch*, 20); 'qui à R. (*ou* à l'homme véridique) a donné une demeure et à qui R. a apporté en longue amitié toute la force de son âme' (Da.); 'der von R. die Wohnung empfangen hat, dem R. (sie) zu langer Gemeinschaft... übertragen hat' (Wo.); 'durch welchen R. ein Haus erhalten hat; welchem R. für lange Freundschaft (Huldigung?) darbrachte' (Lo.). Most translators replace the first

† Of all the explanations of Av. and OPers. *fraša-* listed by Bailey, *TPS*, 1953, 29 *sq.*, the one which recognizes it in the Arm. loanwords *hrašk'*, *hrašakert* still seems to me the most convincing. Bailey's connection of *fraša-* [still upheld in *TPS*, 1956, 100 *sqq.*] with Ved. *pṛkṣd* (which implies an IE **pṛksó-*) meets with some difficulty on the OPers. side: as shown by Bailey, *pṛkṣá*, said of a bull and of horses, means 'strong'; this meaning very much favours Justi's explanation of the name Πρηξάσπης as containing the OPers. equivalent of Ved. *pṛkṣá* (*Iranisches Namenbuch*, 255); the compound thus means 'having strong horses'. Whether Πρη- here represents OPers. *pṛ-* (Av. *pərə-*) or *fra-* (in which case Πρη(ξα)- would stand to Ved. *pṛ(kṣá)* as Av. *fra(ša)-* was supposed to stand to it in Bailey's interpretation), one would normally assume that *-ξa-* stands for OPers. *-xša-* (IE *-qso-*), and cannot therefore be connected with OIr. (*fra*)*ša-*; the expected 'Greek' form of the latter appears in Φρασαόρτης. The name Ξέρξης < *xšayārša* can scarcely be invoked to support the assumption that ξ represents š in Πρηξα-, since the second ξ is obviously due to the influence of the first.

rašnuš by the variant *raš⟨n⟩əuš* of the second *rašnuš*, which, however, no MS. carries in the parallel stanza 81.

After this note was written, R. C. Zaehner's discussion of the present stanza appeared in *BSOAS*, XVII, 247. The author states that 'it is more than likely that originally the term *rašnu-* (meaning, presumably, "the righteous, the director, ordainer, or builder") was simply an epithet of Mithra'; *Yt* 10.79 is accordingly translated: 'who as Rašnu built a house to which as Rašnu he brought *manavainti-* for long association'. That *rašnu-* is an epithet which originally belonged to Mithra is an inference Zaehner draws from the present passage relying on his theory that 'Rašnu and Sraoša, when associated with Mithra, are merely aspects of him'. This theory, in its turn, Zaehner is content to base on his suspicion that in a certain myth, related in the Pahlavi Rivāyats, Rašnu 'is merely the representative of Mithra'. The myth runs as follows in Zaehner's translation: 'After the resurrection of the dead, Rašn Činvand (the separator or avenger), son of Vivanghān,† offers those men who sawed Yam (in two) up in sacrifice. They all die and lie dead for three days'. It is hard to see why Rašnu, a well-defined personality in Iranian mythology, should not be given leave to perform this rather unimportant sacrifice in his own rights, but must be held to be here acting as Mithra's 'representative'. If in addition we remember that there is Avestan evidence which refutes Zaehner's other theory on Rašnu, according to which he is Mithra's 'sinister' companion (see the Introduction, p. 70), we shall again be free to look at st. 79 without the preconceived view that Rašnu is an 'aspect' of Mithra.

As soon as we do so Zaehner's treatment of the stanza reveals two shortcomings. First, no notice is taken of the existence of a common noun or adj. *rašnu-*, which reduces the identification of the god Rašnu with Mithra to an unnecessary and, seeing that there is no other support for it, unlikely hypothesis. Secondly, as far as I can see, wherever else a *nmānəm* or *maēθanəm* is mentioned in the Avesta, it is defined as belonging to somebody, usually by means of a pronoun referring to the owner (cf. sts. 28, 30, 38, 44, 50, 137, *Y* 57.10 (above, p. 223), *Yt* 13.107 (below, p. 228, n. ‡), *Y* 60.7 (below, p. 228), *Y* 57.21, etc.); occasionally a local adverb replaces the pronoun (*Vend.* 3.32), or the identity of the owner is clear from his appearing in the same sentence (*Vend.* 11.10). If then, contrary to our opinion, *yahmāi* should *not* refer to the owner of the

---

† The Pahlavi text, as printed by Zaehner, has ¹*avēšān* ¹*kē-šān Yam kirrēnīt Rašn Činva⟨n⟩d ⟨i⟩ Vīvanghānān yašt-ē* ¹*bē* ¹*kunēδ*. One wonders if Zaehner's restoration of the Iḍāfat is compelling. Surely *Vīvanghānān* ought to refer to *Yam* from which it is separated by only three words. It would not be beyond a Pahlavi scribe to misplace a patronymic, but perhaps we need not even go thus far. Could not the sentence mean: 'Rašn Činvand offers up in sacrifice to the son of Vivanghān (viz. Yam), those men who sawed Yam (in two)'?

abode, the stanza would presumably be on a par with *Yt* 10.77 and *Yt* 16.2, where the owner is the subject of the sentence. But against the *maēθanəm* in question being Mithra's own abode stands the fact that he has already two abodes (sts. 44 and 50 *sq.*), neither of which can be meant here, since the one on earth, being the whole earth, scarcely requires 'building', whilst the one in Paradise was built by Ahura Mazdāh and the Aməša Spəntas, not by Rašnu. If in our stanza a third abode were meant, it would be strange that our author does not define it against the other two, by mentioning its location or purpose. As to *manavaintīm*, Zaehner tentatively changes the word to **manō.vaintīm*, translates it by 'she whose victory is in, or over, the mind', and argues on the strength of *Yt* 17.6 (quoted at the beginning of the next note) that by this epithet Aši is meant, who is here introduced as Mithra–Rašnu's consort. Although Aši is indeed connected with the house, as the next note will show, the verb (*fra*)*bar-* seems little indicated to take Aši as object, whether the subject is Mithra or Rašnu. It must be admitted that in view of its position at the end of the stanza, *manavaintīm* is more easily taken as a noun, with Zaehner and other translators, than as an adjective. However, in the absence of a convincing explanation of such a noun, the alternative interpretation of the word as an adjective deserves consideration.

79². *haxəδra-* is generally thought to mean 'association, friendship'; it is here under the care of Rašnu, elsewhere of the Fravašis (see *infra*), or of Rašnu's sister (above, p. 195) Aši. In *Yt* 17.6 Aši safeguards *haxəδra-* for the house of the pious Zoroastrian, as Rašnu does in the present case: *hubaoiδiš baoδaite nmānəm | yeṅhe nmāne ašiš vaṅuhi | sūra pāδa nidaθaite | āgrəmaitiš darəyāi haxəδrāi* 'well-scented is the house (of him) in whose house the good Aši strong(ly) sets her feet, watching† over long *haxəδra-*'.

† *āgrəmati-*, as Hzf. recognized in *Ap. I.* 60 *sq.*, belongs to the OPers. word in *Beh.* 1, 21, the correct reading of which has now been established as *āgariya*, see Benv., *BSL*, 1951, 32 *sq.* M. Benveniste translates both words as 'consentant, bien disposé', and refers to his note in *JAs.* 1934, II, 179 *sq.*, where he had attractively suggested for the verb *abi-gar-* the meaning 'to take, adopt', connecting it with OInd. *hárati*. (However, in his *Noms d'Agent*, 21, 25, M. Benv. himself reverts to the old translation of *aibijarətar-* as 'Lobpreiser'.) The extension of Benv.'s comparison of *hárati* to OIr. *ā-gar-*, and the consequent translation of the latter by 'to adhere', is due to Duch., *Comp.* 74. What is true of *abi-gar-* need, however, not, and evidently does not, apply to *ā-gar-*: from the use of Akkad. *pitqudu* 'watching, looking after' (cf. Hzf., *loc. cit.*) to translate OPers. *āgariya*, we can safely conclude that the latter belongs to *gar-* 'to wake' and means 'watchful, solicitous', as already Weissbach had seen, see *Keilinschriften der Achämeniden*, p. 12, n. *a*. The same meaning can fittingly be assigned to Av. *āgrəmati-*. Before the reading of the OPers. cognate became clear, Bailey, *Zor. Prob.* 4 *sq.*, had connected *āgrəmati-* with the base *gram-*. [The base *gar-* 'to take' has now been recognized by Bailey also in Av. *niγrāire*, see above, note 40¹.]

COMMENTARY [79.2

In the identification of the meaning of *haxəδra-* insufficient attention has been paid to *Yt* 6.5; there the word occurs in a sentence which *happens* (see *infra*) to be preceded by an invocation of Mithra and his mace:

    yazāi miθrəm vouru.gaoyaoitīm
    hazaŋra.gaošəm bāevarə.čašmanəm
    yazāi vazrəm hunivixtəm
    kamərəδe paiti daēvanąm
    miθrō yō vouru.gaoyaoitiš
    yazāi haxəδrəmča yat̰ asti
    haxəδranąm vahištəm
    antarə mā̊ŋhəmča hvarəča.

'I will worship grass-land magnate Mithra, thousand-eared, ten-thousand-eyed! I will worship the mace which is well brandished on the evil head(s) of the daēvas (by him) who (is)† grass-land magnate Mithra! I will also worship the *haxəδra-* between moon and sun, which is the best of the *haxəδra-*s.' The last three lines were translated as follows by Wo.: 'ich will auch die Gemeinschaft (mit Mithra [see *infra*]) verehren, die die beste unter (allen) Gemeinschaften ist zwischen Mond und Sonne'. The objection to this translation is that whatever the meaning of *haxəδra-*, a number of *haxəδra-*s can scarcely have been situated between moon and sun. One would find more acceptable Lo.'s (and Da.'s) rendering '... das Bündnis (Da. 'amitié'), das das beste Bündnis ist, (das) zwischen Mond und Sonne', if either friendship, or communion, or a pact, could easily be imagined to have existed between moon and sun. 'Pact', moreover, takes us too far from the meaning of *hak-* 'to follow', of which *haxəδra-* is an abstract noun;‡ this obvious etymology makes it clear that the relation here envisaged between the two luminaries is that of mutual 'succession'. 'Succession' in the sense of 'line of successors, heirs, descendants' is also the immediately convincing meaning of *haxəδra-* in all other passages, and one which satisfactorily accounts for its frequent epithet *darəya-* 'long'. This meaning, which implies a collectivity of successors for each house which is thus to be blessed, also accounts for the use of the plural in *Yt* 13.30, where the Fravašis are said to be 'the best to be dwelt with

---

† Gdn. did well to reject the inferior reading *vouru.gaoyaoitōiš*, whose accidental origin is shown by the fact that the MSS. which carry it have *miθrō yō*, not *miθrahe yat̰*. Evidently we have here an ellipse of the demonstr. pronoun which should have provided the grammatical agent to *hunivixtəm*. The unabridged sentence should have run: ... *vazrəm hunivixtəm ... ahmāi yō miθrō yō vouru.gaoyaoitiš. ahmāi* was left out in accordance with the practice mentioned above, p. 161, n. †; the first *yō*, a proper relative pronoun, was omitted because the second *yō*, in addition to its usual Iḍāfat function, was capable of fulfilling the relative function required in the context. The construction *past participle* + *agent introduced by the rel. pron. in the nom.* is found also in st. 143, see note 143[10].

‡ In *Yt* 17.6 also Lo. translates *haxəδra-* by 'Freundschaft'.

for long successions' (*darəyaēibyō haxəδraēibyō upa.šitəe vahištå*).† How suitably the Fravašis are invoked in connection with 'succession' can be seen from *Yt* 10.3, where they 'give noble progeny'.

By using *haxəδra-* with the first of two connotations which the word has in common with Engl. *succession* ((1) a following in order, (2) collectivity of successors), while in all other Av. passages the second connotation is found, the author of *Yt* 6.5 enables us to recognize the meaning of the word. A misunderstanding can now be dispelled, whose origin is to be traced to popular etymology: the Pahlavi commentators, imagining that *haxəδra-* belonged to *haxi-* 'friend', translated both by the same word *hambrāt*. When they came to interpret the context of *haxəδra-* in *Yt* 6.5, they avoided the impasse arising from their wrong notion of the word, by dragging in Mithra, whose name happened to precede immediately the *haxəδra-* passage. Accordingly *haxəδrəm* was translated by **hambrātīh ⟨i⟩ Mihr-yazd*, if we accept Schaeder's emendation *ap.* Widengren, *Hoc.* 99. The arbitrary combination of *haxəδra-* and Mithra in *Yt* 6.5, which was probably prompted by the correct notion that *haxəδra-* (as 'line of successors') does fall within Mithra's competence (cf. his epithet *puθrō.dā-* in st. 65), led to the strange conception of Mithra 'going between sun and moon', as the NPers. commentator has it (see Da., *Et. Ir.* II, 290, Widengren, *loc. cit.*). Instead of taking this theory for what it obviously is, a late upshot of Zervanite speculation about Mithra's mediating propensities (on which cf. Zaehner, *Zurvan*, 101), some scholars actually imputed it to the author of *Yt* 6.5. Hence we have Wo.'s unwarranted 'Gemeinschaft (mit Mithra)', quoted above, and the misleading statements about an Avestan Mithra 'Mesites' standing between sun and moon, of Cumont, *TMMM*, I, 122, 174, Widengren, *loc. cit.*, and Duch., *Orm.* 129, n. 1.

With the passage quoted at the beginning of this note, in which Aši looks after *haxəδra-* in the house,‡ Bailey, *Zor. Prob.* 5, n. 6, compared *Y* 60.7: *mā yave imaṭ nmānəm x^vāθravaṭ x^varənō frazahīṭ, mā x^vāθravaiti ištiš, mā x^vāθravaiti āsna frazaintiš, x^vāθrō.disyehe paiti ašōišča vaŋhuyå darəyəm haxma* 'may neither comfort-giving fortune, nor c.-g. riches, nor

---

† Here Wo. and Lo. were driven to translate *haxəδraēibyō* as if it stood in the singular, 'Gemeinschaft' and 'Freundschaft' respectively.

‡ For 'Aši in the house' cf. further *Yt* 13.107: *yeŋhe nmāne ašiš vaŋuhi | srīra xšōiθni fračaraēta | ... | yō azgatō arəzyayå | havaēibya bāzubya | tanuye ravō aēšištō | yō azgatō arəzyayå | havaēibya bāzubya | hamərəθəm paiti yūiδištō* 'in whose house beautiful majestic Aši would walk around, ... (he is) he who in the unbearableness of battle (is) most able to make room for himself with his arms, (he is) he who in the unbearableness of battle (is) most able to fight the opponent with his arms'. Both Bth. ('der unbezwingliche der (=in der) Schlacht') and Lo. ('der nicht wankend(?) in der Schlacht') took *azgatō* as a nom. sg. masc., although the gen. *arəzyayå* clearly does not permit this. It is obvious that *azgatō* is the loc. sg. fem. of *a-zgati-* (to Ved. *sagh-* 'to be able to bear').

c.-g. noble progeny, ever fail this house, while there is long *haxman-* with (*or* between) the "comfort-assigner" and good Aši'. There can be little doubt that the meaning of *haxman-* is 'association', as is generally assumed (cf. Ved. *sákman* 'association, attendance'), and no doubt that *haxman-*, like *haxəδra-*, belongs to *hak-*. It does not follow from this relationship that *haxəδra-* has the same meaning as *haxman-*. It will be noted that while the long *haxman-* of Aši and the 'comfort-assigner' is a *condition* for the presence in the house of fortune, riches, and noble progeny, the long *haxəδra-* is the *object* or purpose for which Aši or the Fravašis associate with a given house.

As regards the 'comfort-assigner' ($x^v\bar{a}\theta r\bar{o}.disya$-), in whom Bth. did not know whether he should recognize—on equally weak grounds—Aša or Ahura Mazdāh, his identity is scarcely open to doubt. He is a god who either shares with Aši the prerogative of 'long association' with a house, or is himself enjoying 'long association' with Aši. On Aši's close connection with Mithra we have dwelt in note 68[2]. Mithra's care of the house is explicitly stated in st. 30. The title 'comfort-assigner' suits no Avestan god better than Mithra, for not only does he bestow 'comfort' in sts. 5 and 33 (where instead of $x^v\bar{a}\theta ra$- its synonym *havaŋha-* is used; cf. also *havaŋhō.dā-* in st. 65, and *īštīm pouruš.$x^v\bar{a}\theta ram$...baxšāni* in st. 108), but he dispenses the very favours which in Y 60.7 are said to be $x^v\bar{a}\theta ravant$- 'comfort-giving', viz. fortune (sts. 16, 108), riches (sts. 33, 108), and noble progeny (st. 108, cf. also *puθrō.dā̊* in st. 65). We may therefore confidently equate the $x^v\bar{a}\theta r\bar{o}.disya$- with Mithra.

80[1]. Bth. treated *paiti niš.harəta* as an abbreviation of *paiti(š.harəta) niš.harəta* 'du bist der Be-, der Über-wacher des Gemeindebezirks...'. Benv., *Mages*, 9, n. 3, regarded *paiti* as either a corruption of *pāta*, due to the influence of the following *θwā paiti*, or a form of *pati-* 'master'; he translated: 'toi le maître(?) de la communauté, tu es le veilleur des non-impies'. Similarly Lo. ('du, o Herr der Gemeinde, ein Wächter bist du derer, die nicht lügen') and Wi.; Gdn. translated as if the text had *pāta*. The difficulty can be avoided by taking *paiti.niš.harətar-* as the *nomen agentis* of *har-* with two preverbs. This is probably what Da. had in mind when he wrote: 'c'est toi qui conserves le clan de ceux qui ne mentent pas'. *paiti* and *ni* are used together also in *paiti nisrinuyāṭ*, *Vend.* 3.20, whose meaning does not differ appreciably from that of the simple *nisirinuyāṭ* in *Vend.* 14.2; cf. also the unclear *aēvō bayəm paiti nidaδāiti* in *Vič.* 2.

80[2]. To interpret the difficult last four lines it is best to take as subject of *daiδe* an implied *varəzānəm*, to be understood from what precedes; to the same *varəzānəm* the rel. pron. *yahmi* can then be referred. Otherwise either *yahmi* cannot be accounted for, or one has to put up with the strange idea that the 'divinatory trial' is apparently organized by 'victoriousness': 'mit dir als Herrn (Lo.: 'durch dich, o Herr',) erwerbe ich mir ja die beste Gemeinschaft und den ahurageschaffenen Sieg, bei dessen Gericht die miθrabetrügenden Menschen in Massen erschlagen liegen' (Wo.). Others take *paiti* as postposition: 'denn mit dir habe ich das treuste siegbringende bündniß, durch das die treubrüchigen menschen hinterrücks erschlagen in massen zu boden liegen' (Gdn.); 'car avec toi il (= Rašnu) a formé la plus excellente des amitiés, et avec Vərəθraγna, créé par Ahura; toi devant qui tombe maint Miθrô-druj, frappé devant l'eau qui sait' (Da.); '...Denn durch dich gewinne ich die beste Bundesgenossenschaft und den gottverliehenen Sieg [—says Rašnu].—Vor welchem die mithrabelügenden Menschen bei dem Gottesgericht erschlagen in Massen daliegen' (Gdn., *Lesebuch*, 20). On *vīθiši* 'un procédé divinatoire', see Benv., *TPS*, 1945, 75 *sqq.*, who refers *yahmi* to Mithra: 'Miθra par(?) le *vīθiš* duquel gisent en masse, abattus, les parjures'.

84[1]. None of the translations suggested for *dvācina piθe hačimna* inspires confidence: 'die Thürgeherin vom Topf gefolgt' (Wi.); 'die beiden Aeltern(?) vereint' (Spi.); 'zwei auf dem wege sich begegnende' (Gdn.); 'l'homme hésitant entre deux chemins' (Da.: 'sorte de duel absolu, "deux chemins étant là"'); 'irgend zwei, (die) sich zu (gegenseitigem) Schutz zusammentun' (Wo., followed by Lo.); 'irgend zwei, die eine Fehde haben' (Kern, accepted by Benv., *Inf.* 49; *piθe* = instr. sg. of *piθya-).

The new solution here proposed, although also based on guess-work, has the advantage of fitting the circumstances of *karde* 22 (sts. 83–7) more closely than is the case with previous attempts. In st. 86 the cow stretches out her 'hands' to Mithra as she is being led away in circumstances similar to those described in st. 38. In st. 85, to go by our interpretation, the lamenting voice of both cow and pauper is said to reach the heavenly lights. It would accordingly be natural enough to find that the apparently female suppliant referred to as *dvācina* is the cow. That she is thus made to invoke Mithra twice, in st. 84 and again in st. 86, is scarcely a valid objection, since the two supplications concern different aspects of the situation: it is because she is being driven to the estate of Falsehood (st. 86) that the cow is deprived of her milkmaid (st. 84), in which respect her plight is comparable to that of the poor man who is deprived (*apayatō*) of his rights (st. 84).

That *dvācina* is a nom. sg. fem., and not a dual or plural, is suggested by the 3rd sg. *zbayeiti*. Admittedly *ustānazastō* remains incorrect, but the word occurs with the same ending in st. 86, where there is no doubt that it refers to the cow. *dvācina* 'who longs to be milked', lit. 'who longs for the milkmaid', can be analysed as a thematization of *dvā-činah-, cf. the thematic *tạθrō.činō*, quoted below, p. 255, and *haomō.čanəm* 'desiring haoma-'. *dvā*- may be the fem. of a compound **dug-(g)va*- 'who milks cows', the inversion of Ved. *godúh*, with the simple base in front, as happens with *isə.xšaθra-* 'desirous of power', or *vītar-azah-* 'overcoming anxiety', cf. Duch., *Comp.* 199. Thus interpreted, *dvā-* bears close semantic and formal resemblance to the name of Zarathuštra's mother *duγδō.vā-* < **duγδa-gvā-* 'by whom cows are milked = milkmaid', a name from which one gathers that it must have been common for the milking to be done by women. The disappearance of *-ug-* has a parallel in *drvant-* 'owner of Falsehood' < **drugvant-*, and the long *ā* was transferred to the compound from the nom. sg. fem. **dvā*, whose final length is characteristic of monosyllables (cf. Bth., *GIP*, I, p. 178, A 1). [See Addenda.]

hačimna might be translated by 'beset', as is done by us, perhaps somewhat freely, in the case of *varāzahe...yō...upa.haxtō†ā.manaŋha*, st. 71. But we should also bear in mind the peculiar use of Av. *hak-* in the sense of 'having, being with'. *hačaiti dim aspahe aojō* (*Yt* 19.68) means, in practice, 'it has the strength of a horse'; the corresponding participial clause **aspahe aojaŋha hačimnō* will mean 'having, being with, the strength of a horse'. Similarly *miθrəm...hačaite...xratuš* in st. 107 can be turned round to **miθrō xraθwa hačaite* (or *hačimnō*) 'Mithra has (*or* who has) insight'. It may therefore be possible to interpret *piθe hačimna* as either 'beset by **piθya-*', or 'having, being with, **piθya-*', **piθya-* being the substantivized neuter of an adjective derived from **piii-* 'swelling', abstract of Av. *pay-*, Ved. *páyate* 'to swell, overflow', to which also belongs Av. *payah-*, Ved. *páyas*, 'milk'. **piθya-* thus means 'connected with, or caused by swelling', in particular the swelling of milk in the udder, and the neuter **piθyam* may be thought of as the technical term for the discomfort arising from this swelling if the udder is not milked in time. *piθe* would be, of course, a regular Av. development of **piθya*, instr. sg. n. of **piθya-*.

84². On *driyu-* see K. Barr in *Studia Orientalia Ioanni Pedersen... dicata*, 21 *sqq.*, whose conclusions need to be revised in the light of the evidence provided by the Sogd. derivative of this word, δrγwšk- (žwxšq-, jwxšq-).

85¹. Lo.'s translation of *yaṭčiṭ nəmaŋha vāčim baraiti* by 'wenn er im Gebet die Stimme erhebt' fails to account for °*čiṭ*. The limiting nature of this clause is confirmed by its absence in the parallel passage of st. 89, where Haoma is expressly stated to have chanted 'with loud voice'. There is a noteworthy touch of warmth in the thought that the *poor* man's voice reaches Heaven even if owing to his weakness it is uttered softly. The other translators regarded *yaṭčiṭ* as co-ordinated with *yaṭ* (*gaoš*)*čiṭ* (see next note).

† [*upa.haxta-*, lit. meaning 'followed, accompanied by, associated with', provides the obvious etymology of Khot. *vahī*, which M. J. Dresden, *Jātakastava*, 29 r 2 translates 'escorted by'.]

85². The last two lines and the beginning of st. 86 have been rendered as follows: 'mag er laut die Stimme erheben oder ins Ohr. Die in die Irre geführte fürwahr ruft...' (Wi.); 'ob er demutsvoll, ob er mit lautem rufe seine stimme erhebt. Welchen (die kuh) die...getrieben wird...ruft' (Gdn.); 'soit qu'il prononce sa prière à voix basse ou qu'il lève la voix (*yaṯ gaoščiṯ* "ou en son retentissant"?). La vache, emmenée captive, l'invoque...' (Da.); 'sei es daß er betend seine Stimme erhebt oder (laut) rufend. Das als Beute fortgeschleppte (Rind) ruft...' (Wo.). All these translations, which (apart from Wi.'s) Bth. justified by postulating for the sake of *gaoš*° a stem ³*gav*- 'rufend, schreiend', seem unrealistic. Is it likely that *gaoš* does *not* mean 'cow' when the initial *yā* of st. 86 so obviously refers to the cow, who in the usual interpretation remains unmentioned?

More promising is the line taken by Lo. and Htl.: 'dessen Stimme nach jenen Lichtern emporgelangt..., wenn er mit Verehrung, wenn die Kuh die Stimme erhebt, die gefangen...' (Htl.); '...wenn er im Gebet die Stimme erhebt. Den auch die Kuh fürwahr, die gefangen fortgeführt wird,...' (Lo.). However, *gaoš* cannot be nom., and *yaṯ* referring to the masc. Mithra, as Lo. has it, is awkward. There is at least one convincing example (cf. Bth., *Wb.* 1260) of *yaṯ* being used as a comparative conjunction, viz. *Yt* 17.61: *ana θwā yasna yazāne, ana yasna frāyazāne, yasə.θwā yazata vīštāspō* 'mit diesem Opfer will ich dich verehren, mit diesem Opfer will ich (dir) huldigen, wie dir Vištāspa... opferte' (Wo.). This opens up the possibility of taking *gaoš* as the gen. of *gav*- 'cow' (cf. Bth., *ZWb.* 222, n. 3), with °*čiṯ* in the usual meaning of 'also, too': 'the voice of the poor man reaches..., makes the round..., pervades..., as (that) of the cow, too, (reaches..., makes the round..., pervades...)'. It will be noted that as in st. 84 the °*čiṯ* of *driyuščiṯ* links the pauper's invocation to that of the cow, so in st. 85 the °*čiṯ* of *gaoščiṯ* links the pervasiveness of the voice of the cow to that of the pauper's voice.

86¹. *aršan*-, basically 'male', sometimes 'hero', see Benv., *BSL*, 1949, 101 *sq*. There is no justification for Lo.'s translation 'Mithra the bull', which Htl. and Hzf., p. 426, adopted.

86². *kaδa.nō...apayāṯ* 'wann wird unser Held...die Rinderherde einholen' (Bth., since *apayeinti* in *Y* 57.29 clearly has this meaning); Lo. (following Wi., Gdn.,† Da.): 'wann wird...Miθra...uns die Herde erreichen lassen'. The latter alternative obviously makes better sense; there is no reason why a causative present stem *apaya*- should not have existed beside the intensive stem *apaya*- (< *apāya-, with formative as in OPers. *grbāya*-; on *ǎ* cf. *mǎyā*-, note 52¹ above), as we have *pataya*- with both intensive and causative value, see Bth., *GIP*, 1, §§145, 151.

† In *Lesebuch* Gdn. changed over to Bth.'s interpretation.

86³. There is a slight inconsistency between *kaδa.nō fraourvaēsayāiti* 'when will he divert *us*', and *azəmnąm* '(*me*) who am being driven', due to the poet's envisaging both the one cow that speaks and her unfortunate companions. The same inconsistency is found in st. 38. It is scarcely advisable to take *azəmnąm* with Bth. as representing *azəmną*, acc. plur. masc. (*Wb.* 223): a worse inconsistency would arise, that of the speaking cow mistaking her own sex.

86⁴. I take *vaēsma-*, which only occurs here, as meaning 'a conglomeration of *vīsō*', understanding by *vīs-* in this case 'a house with its yard', cf. Bth., *Wb.* 1456, 1(*a*) and 2(*a*); Wi., Gdn., Lo., and Htl. have 'Haus', Bth. 'Wohnung', Da. 'repaire'.

87¹. *daiṅhusasti-* 'command of countries'; *sasti-* means both 'praise' and 'command, order'. Wi. (followed by Bth. and Lo.) translated 'Landes Ruhm', Gdn. 'landesgebieter' and (*Lesebuch*) 'Reichsgewalt', Da. 'empire', Htl. 'Herrschaft des Landes'. The implications of the translation we have adopted are discussed in note 145¹ below.

88¹. For recent views on *frašmi-* see Henning, *Sogdica*, 24 (who suggests 'messenger'), and Bailey in *TPS*, 1953, 32 (where the translation 'invigorating' is upheld). The *Dēnkart* sentence quoted by Bailey (*kaδ-aš fratom frāšm vistarīhēt* 'when its (the sun's) *frāšm* first is spread-out'), settles the approximate meaning which the most probably related *frāšma-* has in the Av. idiom for 'sunset', *hū frāšmō.dāiti-*, 'the setting of the sun's *frāšma-* (rays, *or* glow)'. Benv.'s connection of *frāšma-* with *'dyšm* 'moon' in the Pahlavi Psalter, < **adi-šma-*, *JAs.* 1936, 230 *sq.*, is therefore very attractive; whether, as Benv. suggested, Parth. *n'šmy* (in his opinion 'obscurity') and Sogd. *nšmy* (reputedly 'west', according to Henning, *loc. cit.*, perhaps 'south') also belong here, is uncertain. Since only nominal forms are attested, *m* need not belong to the base, but may be suffixed. As an alternative to Bailey's base **fraš-* (IE **prek̑s-*) 'to invigorate', which he also finds in Av. and OPers. *fraša-* (cf. above, p. 224, n. †), one may therefore think of IE **k̑es-* (OInd. *kṣā́yati* 'to burn'), and interpret *frāšma-* as *frā-š-ma-*, with *š-* < **k̑s-*. At all events, the meaning 'glow' can be assigned to *frāšma-* simply on Avestan and Pahlavi contextual evidence, independently of etymological considerations.

frāšmi- is then easily understood as a derivative of *frāšma-*, meaning 'possessed of glow, glowing'. The Haoma (Ved. Soma) plant is described as *zairi.gaona-* in Avestan, *hári* in Vedic, 'gold-coloured, yellowish'; if, as is likely, the plant is the rhubarb (see Morg. in *Sarūpa-Bhāratī, Dr Lakshman Sarup Memorial Volume*, pp. 30 *sqq.*), its tawny or golden stalks might well have seemed to worshippers of Haoma to be 'glowing' in the sun. Bth. and Lo. left *frāšmi-* untranslated, Gdn. has 'genießbar' (base *as-*); Da. anticipated Bailey with his 'invigorant'. [Cf. now also *TPS*, 1956, 103 *sq.*, where Bailey further discusses his interpretation of *frāšm°* as 'energy'.]

88². Literally 'on the highest peak on Harā the high'; cf. st. 90.

90¹. While all translators except, apparently, Gdn. agree in taking *yim* in st. 89 as referring to Haoma, there are strange vagaries in respect of the relative pronouns in 90. Bth., who rightly had referred *yō* (*Wb.* 719, line 7) and *yahmāi* (*Wb.* 917 bottom) to Haoma, unaccountably took *yeṅhå* (which stands for the masc. *yeṅhe* by attraction to the gender of *kəhrpō*) as referring to Mithra (*Wb.* 1213, line 15). This contradiction is not apparent in Wo.'s translation, who presumably thought of Haoma throughout the stanza, as apparently did Lo.; Wi. hesitated over referring *yō* to Haoma or Mithra, but seems to have assumed that the same god was meant throughout the stanza. Gdn., followed by Htl., referred all relative pronouns in 90 to Mithra, assuming that st. 91 reports in direct speech the homage of the sun announced at the end of st. 90. Moreover, to judge by the interpunction, Gdn., but not Htl., also referred the initial *yim* of st. 89 to Mithra. Spi. found Mithra throughout st. 90, as well as, apparently, in st. 89. Da. differs from everybody else in that he refers all rel. pronouns in 89–90 to Haoma, but translates from *yahmāi* onwards: 'lui pour qui le soleil aux chevaux rapides éveille de loin les hommages (de Mithra): "Hommage à Mithra, etc."'.

It is not surprising that, amidst such confusion of those who should have known better, Cumont, *TMMM*, I, 197, could teach the heresy: 'l'Avesta nous apprend que ce héros (= Mithra) est le premier qui ait préparé le Haoma destiné au sacrifice dans un mortier céleste émaillé d'étoiles (*Yasht* x 23 [*sic*, read 120?], 90)'!

That *yim* in st. 89 refers to Haoma should be clear from what precedes in st. 88. It is because Haoma is a divine *zaotar*- (89) that he worships Mithra in 88 with *barəsman*-twigs, as the human *zaotar*- does in st. 137. Being a priest, one would think Haoma is primarily a priest concerned with the *haoma*-sacrifice, in fact the first *haoma*-priest; consequently to refer the initial *yō* of st. 90 to Mithra means to fly in the face of all likelihood. If *yō* refers to Haoma, then *yeṅhå* can only refer to Mithra if *yahmāi* does likewise. It is, however, hardly Mithra whom the sun would revere 'from afar'. Both st. 13 and st. 142 seem to indicate that Mithra precedes the sun only by a short distance. Moreover, as they both travel in the same direction, we should be confronted with the undignified situation that Mithra 'perceives' the sun's reverence from behind. On the other hand, 'perceiving' is a singularly apt expression if the Haoma-plant is meant, as it waits in the rarefied cold mountain air (cf. *Y* 9.26; 10.3, 4, 12) for the first rays of the sun to warm it.

The immediately following *nəmō* of st. 91 should not deceive us into interpreting that stanza as quoting the sun's words. The device of our poet of resuming at the beginning of a stanza one or more words he had used at the end of the preceding stanza, is also met with in sts. 30 *sq.*, 99 *sq.* (*dašinəm*), and 126 *sq.* (*upamana*-). There is no difficulty in under-

COMMENTARY [90.1–92.1

standing st. 91 as an invocation of Mithra on the part of the worshipping poet (cf. sts. 29 *sqq.*, 77 *sq.*, 80, 93 *sq.*). If, however, it is thought that the words *must* have a divine speaker, the obvious candidate is Haoma, on the assumption that the relative clauses of sts. 89–90 constitute a long parenthesis, after which the text goes on to quote the *vaγžibyō* announced at the end of st. 88.

90². This interpretation of the three lines as constituting one closely-knit sentence (lit. 'whose well-grown body the Incremental Immortals were praising [of him the well-grown body] Ahura Mazdāh praised') is preferable to the one generally adopted, which co-ordinates *bərəjayən* with *bərəjayat̰*: 'es ehrte der Weise Herr, es ehrten die Heiligen Unsterblichen seine wohlgebildete Gestalt' (Lo.). On *yeṅhå* cf. note 143¹.

92¹. This is the only stanza in the Hymn where I think an editor is justified in making a slight change in the order of lines. The line *frā. hē aməšå spənta*, redundant after the enumeration of the individual Aməša Spəntas, and forcing a disturbing plural subject on the singular *vərənta*, is missing in the parallel passage *Y* 57.24: *aya daēnaya fraorənta | ahurō mazdå ašava | frā vohu manō | frā ašəm vahištəm | frā xšaθrəm vairīm | frā spənta ārmaitiš | frā haurvatās frā amərətatās | frā āhūiriš frašnō* ('the questioning of Ahura') | *frā āhūiriš t̰kaēšō* ('the doctrine taught by Ahura').†

On the other hand, the translations hitherto proposed of the second half of the stanza (the last five lines), as far as they adhere at all to the text, offend the requirements of elementary grammar in two respects: (1) both *hē* and *θwā* are referred to Mithra, despite the discrepancy of person, and (2) *yōi* is referred to *gaēθanąm*, despite the discrepancy of gender. From the point of view of sense, the poet is made to state a platitude: Mazdāh having conferred on Mithra the chieftainship of people, people consider him their chief. Moreover, according to the traditional interpretation the stanza contains no reference to any relation between Mithra and the Religion, so that the Religion would seem to have been introduced at this point to no purpose.

Grammar being the only reliable criterion in such a difficult context, we must needs find the masc. plural which *yōi* takes up. As the five lines in question do not contain any, and the line *frā. hē aməšå spənta* is redundant in the first half of the stanza, it is hard to escape the conclusion that this line was moved up from the second half of the stanza by a scribe who thought that it ought to close the list of the individual Aməša Spəntas. Support for this conclusion is provided by the enclitic *hē* after *frā* in line 8, which is out of tune with the *hē*-less *frā*-s of the five preceding lines, but agrees with *frā. hē* in the next line but one. Since the latter is followed

† Lo., *Die Yäšt's*, p. 77, n., misleadingly pretends that *Y* 57.24 also has *frā. hē aməšå spənta*.

235

by a verb in the singular, the obvious place for our dislocated line is after *gaēθanąm*, so that the same verb by implication supplies the predicate of which *aməšā̊ spənta* is the subject, and *frā* the preverb:

> *frā haurvata aməṛtāta*
> *bərəja vərənta daēnayāi*
> *frā.hē mazdā̊ hvāpā̊*
> *ratuθwəm barāṯ gaēθanąm*
> *frā.hē aməšā̊ spənta*
> *yōi.θwā vaēnən* etc.

With this restoration we gain as masc. plur. subject of *vaēnən*, represented by *yōi*, the Aməša Spəntas with or without Ahura Mazdāh. Moreover, in the first half of the stanza the 3rd sing. *frā...vərənta*, which is a repetition of the initial *fraorənta*, becomes understandable: it represents a kind of group inflection, *frā vohu manō, frā ašəm, frā xšaθrəm, ...vərənta*, standing for *frā v.m. vərənta, frā ašəm vərənta, frā xšaθrəm vərənta*, etc.

The whole stanza thus acquires syntactical smoothness and unity of sense, and the situation is plausible: Mazdāh and the Immortals, having invested the Religion with the jurisdiction (*ratuθwəm*) of living beings, consider Mithra her executive agent (*ahū-* and *ratu-*) and 'purifier' (cf. below, notes 92[6-8]). Mithra's co-operation with the Religion, which thus discloses itself as the main topic of the stanza, lends poignancy to the reference to the 'two lives' in the following stanza 93.

The previous understanding of this stanza is best seen in Wo.'s conscientious rendering, which essentially agrees with everybody else's, except Da.'s: '...es bekannten sich dazu die Aməša Spənta's nach dem Brauch der Religion. Es übertrug ihm (Miθra?) der wohlwirkende Mazdāh die Ratav-schaft über die Menschheit, (auf daß) sie dich (Miθra?) unter den Geschöpfen für den Ahū und Ratav der Menschheit ansehen, für den Vervollkommner dieser besten Geschöpfe (oder "den besten Vervollk. dieser Gesch."?)'. Da. differs: '...l'ont professée les Amesha-Spentas, dans le désir de la religion; l'a professée le bon Mazda. [Mithra] conférera la maîtrise du monde à qui reconnaît en lui le Maître et le Seigneur du monde, le purificateur de cette excellente création'.

92[2]. After Henning's convincing identification of Ved. *bráhman* with OPers. *brazman-* and MPers. *brahm*† (*TPS*, 1944, 108 *sqq.*), there is no justification for maintaining Bth.'s translation of Av. *bərəg-* as 'rite', which was based on his etymological connection of this word with Ved. *bráhman*. On the other hand, since Pahl. *ārzū(k)* 'desire', by which Av. *bərəg-* is translated, belongs etymologically to the Av. present stem *bərəjaya-* 'to praise' (which in Khot. gives *bulj-* 'to praise'), as Henning pointed out in *BSOAS*, XI, 487, n. 2, there can be little doubt that *bərəg-* is simply the root noun of the base of *bərəjaya-*, with the meaning of Pahl.

† For which Bailey, *JRAS*, 1951, 194, has a different etymology.

COMMENTARY   [92.2–92.6

ərzū < *ā-bərəǰ-u-. The close semantic relation of 'desire' and 'praise' is also met with in derivatives of the IE base *lubh-, cf. Engl. *love*, Germ. *Lob*, Lith. *liaupsė* 'praise'. The translation 'desire', which already Spi. and Da. had adopted for *bərəg-*, suits all passages except *A* 3.4, where *saškuštəma ašahe bərəjō* means 'the most learned pursuers (*lit.* desirers) of Truth'; here either *bərəg-* is a noun of agent, cf. Wn.–Debr., II, 2, 9 *sq.*, or a synthetic compound **aša-bərəg-* was resolved into *ašahe bərəg-* on the analogy of the common expression *ašahe bərəja* 'out of desire for Truth'. *yō ašāi bərəjyąstəmō* in *N* I is 'he who most intensely longs for Aša'. [See Addenda.]

92³. This is the meaning of *ašəm vahištəm* (also occurring in st. 33), not, of course, 'the best Truth', which would imply a plurality of Truths.

92⁴. *hvāpah-* is an epithet of Ahura Mazdāh, *āsna frazaintiš* 'noble progeny', and, according to our interpretation above, p. 208, *vaðre yaona* 'wedlock'. It is translated on several occasions by Pahl. *x^vāpar*, which also translates *x^vāpara-* its own Av. forbear. *x^vāpara-* is found as epithet of Aši, the Fravašis, *frazaintiš* (without *āsna*), and *ząˊ* 'the earth'. Its meaning may thus be near to that of *hvāpah-*, and Bth.'s tentative etymological connection of the two words is likely to be correct. In the *Škand Gumānīk Vičār* Pazand *x^vāwar* occurs as opposite of *anāmurž* 'merciless' in XII, 12, and is followed immediately by *awaxšāišn* 'mercy' in X, 23, XI, 4; P. de Menasce accordingly translates it by 'clément, généreux', and the abstract *x^vāwarī* by 'compassion'. The meaning 'liberal, generous' would fit the Av. contexts of both *hvāpah-* and *x^vāpara-*, and 'forgiving' in Middle Iranian may be understood as a semantic development of 'generous'. If this analysis is correct, *hvāpah-* is not likely to be related to OInd. *ápas* as Bth. thought. Perhaps one should rather think of Ved. *āpí* 'friend, ally', Greek ἤπιος 'benevolent, friendly'; we should then have an -*ah-*:-*ar-*:-*i-* alternation °*āp-ah-*:°*āp-ar-*:*āp-i-*; *x^vāpara-* would be a case of compound thematization, comparable to *frayara-* and *uzayara-* from *ayar-*.

92⁵. *frā...barāṭ*, taken as preterite with Bth., cf. *GIP*, I, p. 57.

92⁶. Unless the last three lines have been transferred here from another passage as Bth. suspected (*Wb.* 283, n. 5), *θwā* can only refer to Mithra. Bth.'s reason for doubting the genuineness of these lines was presumably that in *Yt* 13.92 (quoted at the beginning of note 92⁸) it is Zarathuštra who is said to be *ahūm ratūmča gaēθanąm*. The conclusion we shall arrive at in note 92⁸ that the *θwā* in question, like Zarathuštra, is a 'purifier' of the Religion, at first sight seems to confirm Bth.'s suspicion. However, in principle we should resist the tendency to declare an Avestan passage spurious merely because it does not on first consideration appear to say what we expect. This especially applies to the present Hymn, in none of whose other stanzas is there reason to assume an intrusion of the kind suspected by Bth. The alternatives are two: either, despite the appearance, the poet does suddenly address Zarathuštra in the second person; or we must take the three lines at their face value and accept the information that the titles in question belong to Mithra as well as to Zarathuštra. It is difficult to resign oneself to the first alternative, since a vocative *spitama* would seem to be essential, and altogether in this Hymn Zarathuštra is addressed only by Ahura Mazdāh (sts. 1 *sq.*, 119, 121 *sq.*, 137 *sq.*, 140 *sq.*, 143) and Mithra (st. 118). In favour of the second alternative the following may be said.

In *Y* 29, despite immense difficulties, this much is clear: in st. 2 Truth and other Entities (cf. *JRAS*, 1952, 174 *sq.*) are asked by the Creator of the cow whether they have a *ratu-* for the cow; in st. 6 Ahura Mazdāh states that neither an *ahū-* nor a *ratu-* has been found; however, in sts. 7–8 Zarathuštra is declared to be the one who will look after the cow for men, and in st. 9 the cow herself describes him as a 'powerless protector'. Zarathuštra's expressed view was, accordingly, that because no divine being would act as *ahū-* and *ratu-* of the cow, he, Zarathuštra, had been appointed her protector, as a *substitute* for an *ahū-* and *ratu-*. The reference to 'men' in st. 7 suggests that the *ahū-* and *ratu-* Zarathuštra had in mind was to have looked after the interests not only of the cow but also of the cattle-breeding community, and fulfilled more or less the function Zarathuštra himself was later credited with fulfilling, that of an *ahu ratušča gaēθanąm*. It would seem from *Y* 29 that Zarathuštra thought it proper for such a post and title to exist, *and be held by a divine being*. May it perchance have existed, but have been held by a divinity Zarathuštra was not prepared to acknowledge? In the end Zarathuštra does not allow the title to any divine being, but undertakes to fill the gap as best he can by his own efforts. Once he had died, his followers, unhampered by the sense of his own inadequacy he seems to have entertained, may well have quoted this very Gāthā in support of the title *ahu ratušča gaēθanąm* which they gave him. With regard to Zarathuštra's surprising hesitation to award the title to a divine being, we may then conjecture, on the strength of *Yt* 10.92, that up to his time the title was held by Mithra, whom the prophet wished to keep out of his religious system. [See Addenda.]

There are two supporting reasons for this conjecture. One, Mithra's Sogdian title δ'm'γγt, will be discussed in the next note. The other is that Zarathuštra shares the title *ahu(ča) ratušča dāmanąm* with no less a partner than Ahura Mazdāh, the god being according to *Visp.* 2.4 the *ahuča ratušča* of the supernatural (world of) creatures (*mainyavanąm dāmanąm*) and the supernatural existence (*mainyaoyā̊ stōiš*), while Zarathuštra is the *ahuča ratušča* of the empirical (world of) creatures (*gaēθyanąm dāmanąm*) and the empirical existence (*gaēθyayā̊ stōiš*): *ahmya zaoθre barəsmanaēča θwąm ratūm āyese yešti yim ahurəm mazdąm mainyaom mainyavanąm dāmanąm mainyaoyā̊ stōiš ahūmča ratūmča; ahmya zaoθre barəsmanaēča θwąm ratūm āyese yešti yim zaraθuštrəm*

COMMENTARY   [92.6

spitāməm gaēθīm gaēθyanąm dāmanąm gaēθyayå stōiš ahūmča ratūmča.
'By means of sacrifice (*yešti-*, see Benv., *Inf.* 31) I invoke upon libation and *barəsman-* you the supernatural *ratu*, Ahura Mazdāh, *ahū-* and *ratu-* of supernatural creatures, of the supernatural existence; by means of sacrifice I invoke upon libation and *barəsman-* you the *ratu-* pertaining to nature, Zarathuštra the Spitamid, the *ahū-* and *ratu-* of empirical creatures, of the empirical existence.'† The apportionment to Zarathuštra of the *gaēθyanąm* share of the title *ahūča ratušča dāmanąm* is an obvious innovation on the prophet's own views, since, as we saw, he did not even consider himself an *ahu ratušča*. However, the other share of the title, the *mainyavanąm*, so obviously suits Ahura Mazdāh that it very likely was his before Zarathuštra was joined to him in partnership. Who then was Zarathuštra's predecessor, Ahura Mazdāh's original partner in charge of the *gaēθyanąm* share? It clearly cannot have been a human being. Among divinities the candidate who immediately springs to one's mind, even apart from the evidence of *Yt* 10.92, is Mithra, because the division of the *dāmanąm ahū-ratu*-ship in two halves, one of which is held by Ahura Mazdāh, strongly recalls the dual partnership which is advertised, as it were, in the dvanda *miθra ahura*, formerly **miθrā vourunā* (see above, p. 47), the name of the ancient firm which Zarathuštra's intervention all but dissolved. However, if we are right in surmising that the absorption of *Vouruna by Ahura Mazdāh was initiated by Zarathuštra (see above, pp. 46 *sqq.*), the god who in and before Zarathuštra's time was Mithra's partner in charge of the *mainyavanąm* share must have been *Vouruna. The situation we obtain is then as follows. One of *Vouruna and Mithra's joint functions was the *ahū-ratu*-ship of the world, *Vouruna being in charge of the supernatural, Mithra of the empirical world. Zarathuštra would not hesitate to pass on to Mazdāh *Vouruna's *ahū-ratu*-ship, but found himself altogether unable to name a divine *ahū-ratu* in charge of the empirical world, because that should by rights have been *Vouruna's partner Mithra. After Zarathuštra's death he himself was appointed by his disciples to fill the vacancy, while the worshippers of Mithra naturally continued to assign the *ahū-ratu*-ship of the empirical world to Mithra. That of the two shares Mithra should be assigned the *gaēθyanąm*, the sway over the empirical world, is in full agreement with the contents of our Hymn, according to which he is constantly and ubiquitously hovering over this earth.

As regards the 'purification' of the Religion, one might argue that this being eminently Zarathuštra's task (cf. *Vend.* 10.18 *sq.*, quoted note 92[8], p. 242), a well-meaning scribe might have added the last line of the stanza

† Wo. oddly translated *mainyaom mainyavanąm dāmanąm* by 'den (im höchsten Maße) geistigen unter den geistigen Geschöpfen', and *gaēθīm gaēθyanąm dāmanąm* by 'den (im höchsten Maße) stofflichen unter den stofflichen Geschöpfen'. I am following Da. in the translation of these words.

239

under the misapprehension that the preceding ones referred to Zarathuštra. However, there are, as we saw, special bonds between Mithra and the Religion (cf. note 64[7]), and it would be rash of us to decide that Mithra cannot also 'purify' the Religion, when we have only the haziest notion of what this occupation entails.

92[7]. For Mithra to be considered the *ahu ratušča gaēθanąm* '*in creaturis (dāmōhu)*' may mean either that out of the creatures he, as one of them, had been selected to be the *ahū-ratu* of living beings, or else that he exercises the *ahū-ratu*-ship of living beings in the (world of) creatures. The second alternative is to be preferred, as it is the one which more easily accounts for *yazatəm sūrəm dāmōhu səvištəm* in st. 6: 'the god (who is) strong, strongest in the (world of) creatures'. Accordingly, *dāmōhu ahūm ratūmča gaēθanąm* is synonymous with *gaēθyanąm dāmanąm*... *ahūmča ratūmča* in *Visp*. 2.4 (quoted in the preceding note), which, as its own context reveals, is synonymous with *gaēθyayā̊ stōiš ahūmča ratūmča*; in practice, therefore, *dāmōhu ahūm ratūmča gaēθanąm* means 'the temporal and religious judge of the living world'.

Unexpected confirmation that this is indeed an ancient Iranian title of Mithra comes from the hitherto obscure Sogdian expression δ'*m'yγtyh 'myδry* in the *VJ* passage quoted above, p. 35. Gauthiot's translation was 'dieu des créatures', which Benv. replaced by 'le Mithra des créatures'. This unusual way of referring to Mithra is based on the analysis of δ'*m'yγtyh* as plural oblique case of *δ'*m'yγ*, a supposed extension of δ'*m* 'creature(s), world' (<*dāman-* 'creature, creation') by an otherwise unknown suffix -'*yγ*. I should prefer to see in δ'*m'yγtyh* the singular oblique case of a compound δ'*m'yγt*, pronounced δāmext or· δāmexat, and consisting of δ*āme*, the oblique case of δ'*m* 'world' and the Sogdian word for 'judge'. The latter in Buddhist texts appears as '*γt'w* (pronounced *əxtu*), in Manichean texts as *xtw* (pronounced *xatu*). *xatu*, being a light stem, though of an unusual type (cf. *GMS*, § 1190), may be expected to behave like other light stems in composition. As the light stem βγγ, βγw, βγ', etc. (βaγi, βaγu, βaγa, etc.) loses its vocalic endings in the heavy compound "δβγ 'chief god' (see *GMS*, §498), so *xatu*, in composition with the heavy δ*āme*, was almost bound to be reduced to -*x(a)t*. The superfluous final -*h* of δ'*m'yγtyh* may be compared with that of *z'ktyh* in *VJ*, 97[2].

The title 'judge of the world' not only suits Mithra in the *VJ* passage and in general extremely well, but in view of Mithra's Avestan title under discussion, whose affinity with the Sogdian title jumps to the eye, confirms a suspicion on the etymology of Sogdian *xatu* which W. B. Henning imparted to me at least twelve years ago. In Henning's opinion *xatu* is the Sogdian representative of Av. ²*ratu-* 'judge', influenced in its initial by *ahū-*, the cause of the strange development being differentiation from the homophone Av. ¹*ratu-* 'time', Sogd. *rtw* 'moment'. With this explanation, convincing despite the irregularity involved, Sogd. *xatu* becomes the representative of both Av. *ahū-* and *ratu-*, and Mithra's Sogdian title δ*āmext* an almost exact replica of Av. *dāmōhu ahūm ratūm(ča gaēθanąm)*. [See Addenda.]

[The recent attempt to explain δ'*m'yγtyh* '*myδry pt'ych* as 'chieftain of the local genii' (δ'*m'yγ-* = δ*āmēx-* '*naivāsika*', '*myδry pt'ych* = Agnean *mṣapantin-*, H. W. Bailey, *BSOAS*, XVIII, 35, XIX, 49 *sqq*.), is open to the following objections: (1) The meaning of Agnean *mṣapantin-* is unknown, as the essential part of the Skt. equivalent ⟨*senā*⟩*dhipati* has to be restored by conjecture; in the series 'father's wife, *mṣapantin-*'s wife, kinsman's wife' one would rather expect *mṣapantin-* to refer to family relationship than to a 'Heerführer'. (2) The *n* of °*pantin-* would seem to preclude the assumption of a loan from °*pati-*; in any case there is nothing to show that *mṣapantin-* is a loanword. (3) It is awkward that, on the one occasion **miša-* is supposed to occur in Sogdian, it is spelled not *miša-* but *miθra-* by 'inverse spelling'. (4) Even less reassuring is the assumption that *pt'ych* is here not the familiar *patič* 'in the presence of' (which in the context makes excellent sense), but a *hapax legomenon*, feminine of *pati-* 'master', and this despite the expectation, shared by Bailey, that such a feminine should turn up as °*pan-*. (5) The Khot. fem. *jašti* does not provide a valid parallel for the supposed fem. suffix of Sogd. *pt'ych*; the Khot. word is fem. because it *translates* the Skt. fem. abstract noun *devatā* 'divinity', but Sogd. **mišipati-* '(male) commander of troops', not being a translation of a noun whose gender is fem., should not take on a fem. ending. (6) Conversely, if -'*ych* in *pt'ych* is not a fem. ending, it becomes impossible to connect the word with *pati-* 'master'; the final *-h* is a most unreliable reason for postulating the existence of a Proto-Sogdian word **patīčā-* meaning 'mistress'; gratuitous *h*-s were commonly added at the end of a line (cf. *pt'ych* 'in the presence of' in the same text, line 359), and from there were occasionally imported into the interior; possibly, too, our MS. was copied from one where the present *pt'ych* stood at the end of a line; cf. the meaningless *h*-s in the interior of *VJ* lines with "*prynh* 51 e, "*z'wnh* 19, *prmh* 1171, *ptškw'nh* 372, *pwny'nh* 906, *pyδ'nh* 227, *t'yw*"*kth* 333, 1024, *tntr'kh* 629, *wyr'ych* 36 e. (7) The difficulty of the Sogdian expression does not lie with '*myδry pt'ych* 'in the presence of Mithra', but with the *hapax legomenon* δ'*m'yγtyh*; here Bailey's proposal carries no conviction: one does not see why a derivative of *dāman-*, a word meaning 'creation, world, creature', should take on the same meaning ('local genius') as a derivative of *nivāsa* 'abode'; moreover, no suffix *-ēx* is known in Sogdian: of the two alleged parallels *mazēx* 'great, big' represents the OIr. comparative *mazyah-*, see *GMS*, § 403 (on the positive meaning cf. Wn.–Debr., II, 2, 460), while *tm'yγ* is merely a slip of Benveniste's (*Grammaire sogdienne*, p. 95) by which I too was once misled

(*GMS*, § 1106); the text has the expected *tm'yk*. In conclusion we may say that Gauthiot's, Benveniste's, and Henning's (*BSOAS*, XI, 484, n. 5) understanding of '*myδry* as 'Mithra' remains unshaken, and is confirmed by the analysis of δ'*m'yγtyh* as a compound meaning 'judge of the world', to which Bailey could not yet refer.] [See Addenda.]

92⁸. Neither of Wo.'s two alternative interpretations of *vahištąm* (see note 92¹, end) is satisfactory, and no other has been suggested, apart from Gdn.'s suppressing the word altogether. To agree with *yaoždātārəm* the ending would have to be amended to (*vahišt*)*əm*; as a gen. pl. *vahištąm* is questionable as regards both form and sense. But there is no reason why we should not let ourselves be guided by what *vahištąm* appears to be, viz. an acc. sg. fem. referring to *daēnā-*. Of *daēnā-* we read in *Y* 44.10 that she is *hātąm vahištā*, which Bth. translates 'für die Seienden die beste', Lo. 'die beste von allen', Duch. 'de toutes les choses existantes la meilleure'. Equally to the point is *Yt* 13.92, a passage to which we referred at the beginning of note 92⁶ because it says of Zarathuštra very much what is said of Mithra in the present stanza: (*zaraθuštrahe*...) *yim isən aməšā̊ spəntā...ahūm ratūmča gaēθanąm, staotārəm ašahe yaṯ mazištaheča vahištaheča sraēštaheča, paiti.fraxštārəmča daēnayāi yaṯ haitinąm vahištayāi.* '(...of Zarathuštra...) whom the Incremental Immortals desired (to be) the *ahū-* and *ratu-* of living beings, the one who would pray to Truth (*or*: would pray the *Ašəm vohū* prayer) (which is what is) greatest, best, and most beautiful, the one who would be the interpreter (cf. *paiti.paršti-*, note 33²) of the best Religion there is (*lit.* of the Religion which (is) the best of the existing (religions))'. Accordingly *åŋhąm dāmanąm vahištąm* (*daēnąm*) may be either 'the best (Religion) for the(se) creatures', or, with gender attraction, '(the Religion, which is) the best (creature) of the(se) creatures'. In the latter case we may remember that *daēnā-* is a 'creature' inasmuch as she is Aši's (etc.) sister (see above, p. 195), and therefore Ahura Mazdāh's daughter.

As was pointed out at the end of note 92⁶, the fact that Mithra thus becomes a *yaoždātar-* of the Religion should not deter us from the proposed interpretation of *vahištąm*. In considering the possible meaning of *yaoždā-* in this connection, we must bear in mind that *daēnā-* in Zoroastrianism is both the Religion in general and the individual belief each man holds. The 'purified'† individual belief is identical with the 'pure' and 'purifying' Mazdayasnian Religion. *Vend.* 10.18 *sq.* illustrates the interplay of 'purification' and the two aspects of *daēnā-*: *yaoždā̊ mašyāi aipi.ząθəm vahišta hā yaoždā̊ zaraθuštra yā daēna māzdayasniš yō hvąm daēnąm yaoždāite humatāišča hūxtāišča hvarštāišča. daēnąm ərəzvō yaoždaiθīša aθa zī aiṅhā̊ asti yaoždāitiš kahmāičiṯ aṅhə̄uš astvatō havayā̊ ərəzvō daēnayā̊ yō hvąm daēnąm yaoždāite humatāišča hūxtāišča hvar-*

† 'Perfected' according to Bth., 'consecrated' or 'hallowed' according to Lo.

štāišča. 'The best (scil. Religion) who purifies man's future birth (quoted from Y 48.5) is this purifying one, O Zarathuštra, who (is) the Mazdayasnian Religion (of him) who purifies his individual Religion by means of good thoughts, words, and actions. You (=Zarathuštra) should indeed purify the Religion, for thus takes place the purification of her (who is) truly the individual Religion (belonging) to each (man) of the material existence who purifies his individual Religion by means of good thoughts, words, and actions.' Wo. understood the words *daēnąm ərəzvō yaoždaiθīša* to mean 'you should purify your individual *daēnā-*', but this is to reduce the whole of what follows to a pointless tautology. The subject of *yaoždaiθīša* is Zarathuštra, and what he 'purifies' is obviously the religion of which he is the founder. Zarathuštra himself states in *Y* 44.9 that he 'purifies' the *daēnā-*:

 kaθā. mōi yąm yaoš daēnąm dānē
 yąm hudānaoš† paitišə sahyāt xšaθrahyā
 ərəšvā xšaθrā θwāvąs asīštiš‡ mazdā
 hadəmōi ašā vohučā šyąs manaŋhā?

Here most translators take *hudānaoš* as depending on *daēnąm*. But one wonders how Zarathuštra would set about purifying someone else's *daēnā-*. In the light of the *Vend.* passage we have examined I ventured (*JRAS*, 1952, 176) to translate the Gāthic stanza as follows: 'is the *daēnā-* which (*yąm...daēnąm* by case-attraction, instead of *daēnā yąm*) I purify for myself the *asīšti-* ('promised reward') of the clear-sighted, which (*yąm* in line 2, referring to *asīšti-*) the Lord of Power (*paitišə... xšaθrahyā*)—such as Thou art, Mazdāh—dwelling in one house with Truth and Good Mind, may decree (*sahyāt*) through his high power?' Thus the *daēnā-* which the prophet 'purifies for himself' is the Zarathuštrian Religion, which for those who are clear-sighted enough to accept it is the reward decreed by Mazdāh.

In what exactly Zarathuštra's or, in the present case, Mithra's 'purification' of the Religion consisted, we cannot tell; Lo.'s translation of *yaoždā-* by 'to hallow' is only a palliative, since it does not apply to the numerous passages in which the object of *yaoždā-* is something impure, or not yet pure.

95[1]. *marəzaiti va karana* 'streifend berührt er die beiden Enden' (Wo.); 'berührt er beide Ränder' (Lo.); 'der so breit wie die Erde herzuschreitet... der fegt die beiden Enden dieser Erde' (Spi.); 'caresse les deux extrémités' (Da.). In st. 99 Mithra flies forth *dašinəm upa karanəm ainhå zəmō*, which is generally translated in non-committal fashion 'over the right-hand edge, *or* end, of this earth'. Only Htl. ventured to explain that the 'right-hand edge' was the western end of

 † See Bth., *ZWb.* p. 223, bottom.
 ‡ Gdn., Bth.: *asīštīš*; see the variants.

the earth. Herein he relied on Bth.'s opinion that where *dašina-* refers to a direction it means 'west'. However, W. Geiger (*OK*, 141 *sq.*) and Lo. (*ZII*, 2, 206) were no doubt right in defining the direction *dašina-* as 'south', cf. also Sogd. δšny w't, probably 'south wind', Henning, *BSOAS*, xi, 729. In the 'two edges of the round earth', we may then see the southern and the northern semicircles, which together form the circumference of the earth. As Mithra in returning from the West after sunset (see Introduction, p. 39) is said to be *zəm.fraθå*, 'as broad as the earth', he is bound to sweep or touch *simultaneously* the northern and the southern 'edges' of the earth, as well as the space they enclose. It will then be seen that even if Av. *dašina-* should not normally refer to the South, it must do so in the present case, since Mithra in travelling from West to East necessarily has the South on his right.

In what follows after st. 95, the poet, unable to pursue in one breath Mithra's all-embracing sweep over the entire surface of the earth, in my opinion describes in two stages what is really a single performance of the ubiquitous god. The second stage begins at *fravazaite* in st. 99, and continues to the end of st. 101. Here Mithra is seen flying over the 'southern edge of the earth', that is, within the southern semicircle. This is the half which is inhabited by men, as can be seen from Mithra's striking the anti-Mithrian countries he comes across on his way. Accordingly from st. 97 to *fravazaite* in st. 99 the first stage can be recognized, that is Mithra's eastward flight across the northern half of the earth. In these lines the words 'left-hand (= northern) edge of the earth' do not occur, but the location of the area in question can be guessed from the exclusive mention of frightened demons: for the home of the daēvas lies in the North, as appears from the Avestan passages quoted by Bth., *s.v. apāxtara-*.

Other passages in the Hymn on which one might have hoped to be able to test this theory neither confirm nor contradict it. In st. 67 Mithra comes from the East towards X'aniraθa, the inhabited central continent of the earth; the daēvas are terrified at the approach of the god and his awe-inspiring escort. We cannot tell where the daēvas are at that moment, whether in X'aniraθa or, as one may imagine, in the northern half of the earth, from where they would be just as able to espy the approaching danger. In sts. 133 *sqq.* the situation is even less clear: *after* smiting both demons and men Mithra flies over all seven continents, and the demons, who have just been 'smitten', are 'afraid'.

Such unenlightening statements can be excused if we bear in mind the practical difficulty which faced the poet: in his earth-wide sweep across the earth Mithra is at all times everywhere; any attempt to depict him in this fluidity of time and space was bound to result in chronological and topographical contradiction. The poet, however, will scarcely have been worried by such contradiction, which in his eyes will merely have served

COMMENTARY [95.1–97.1

to emphasize the god's ubiquity; elsewhere this ubiquity becomes apparent from the unlimited size of Mithra's abode on earth in st. 44 (cf. note 50[1], second para.), and possibly his title *vīspanąm dahyunąm daiṅhupaiti-*, on which see note 145[1].

96[1]. *satō.dārā-*; *dārā-* 'blade' is now attested also in Khot. *dairä* (see Bailey, *BSOAS*, XV, 538), and Chr. Sogd. *xyry d'r* 'sword-blade', *AAWL*, 1954, 843.36, 37, where Hansen translates 'mit dem Schwerte'; the Syriac version has the first time merely *b-syf'* 'by the sword', but the second time *b-pwm' d-ḥrb'* 'by the mouth of the sword', the sword's 'mouth' being its blade, as the Indo-Iranian arrow's 'mouth' (see note 129[1]) is its point.

96[2]. Bth.'s connection of *zaēna-* with OInd. *hetí* was challenged by Morg., *EVP*, 28, who connects the latter with Pš. *γəšai* 'arrow', thus attributing to it an IE initial aspirate velar.

96[3]. *zaranyehe*. Bth. was no doubt right in emending this form to *zarənyehe* in accordance with what the best MSS. have in st. 132 and *Yt* 5.78. But surprisingly he was content with assigning to *zarənya-* the same meaning as to *zaranaēna- / zarənaēna-*, viz. 'made of gold'. If the word has anything to do with 'gold' it can only mean 'gilded', since a mace 'cast in yellow iron' is evidently not made of gold. This was noticed by Spi. who here translated 'goldfarbig', but ʌas 'golden' in *Yt* 5.78, where *zarənya aoθra paitišmuxta* can very well mean 'shod with gilded shoes'.

97[1]. The meaning of *pəšō.tanū-* is generally agreed to be 'whose body is forfeited', as first suggested by Hübschmann, cf. *Arm. Gr.* 228. The compound is paraphrased in *F* 25 *b* by *tanūm piryete*, which Hübschmann translated 'he forfeits his life through guilt'. *pəša- = pr̥ta-* for Hübschmann meant 'schuldig (= both "guilty" and "owing"), verwirkt'. Bth. also translates *pəša-* as 'verwirkt', but his treatment of the present stem *pirya-* differs from Hübschmann's: it is, in his opinion, the passive of a base [3]*par-* 'to condemn', to which also the present stem (*fra*)*pərənu-* belongs. In practice, however, Bth. has to translate *tanūm piryete* by 'he is *condemned to lose* his body', and *ratuš ratunaēm frapərənaoiti yātəm gaēθanąm* by 'the ratu *condemns* the layman *to lose* a share of his property'. Strictly speaking *pəša-* should then mean 'condemned to be lost'. Yet derivatives from this base in other Iranian languages, including some which like Av. *pəša-* represent OIr. **pr̥ta(ka)-*, show no trace of the allegedly basic meaning 'to condemn': Arm. *partk* 'debt, what is owed, guilty', *part* 'due', *partakan* 'obliged, due, guilty', *partim* 'to owe, become indebted, be bound, ought, must', etc. (see Hübschmann, *loc. cit.*); Sogd. *'prtk* 'guilty', *ptyqn* 'owing', *ptkntw* 'owing', *pwrc* 'debt' (see Henning, *BBB*, p. 89, cf. *GMS*, §487); Sogd. *p'r(h)*, Pš. *pōr*, 'debt' (see Henning, *BSOAS*, XII, 607, n. 2); MPers. *'pwrdg* 'guilty'. Even Av. *pāra-* means 'guilt', not 'condemnation'.

Accordingly we should revise Bth.'s interpretation of his base ³*par-*. *tanūm piryete*, in the light of the above MIr. words, need mean no more than 'he owes his body', and *pašō.tanū-* can be understood as 'one whose body is engaged, due, who owes his body'. The individual so described is no longer free to dispose as he likes of his body or person, but is bound to submit it to the authorities for punishment. No doubt Bth. correctly interpreted the legal background: the *pašō.tanū-* had been condemned to 'owe his body', as a result of his criminal or irreligious behaviour, and has, in consequence, forfeited his body. But the verb *pirya-, prt-*, cannot mean everything at once; in determining its real significance our only safe guide is the semantic range of its MIr. cognates.

The greatest difficulty lies, of course, with *frapərənaoiti*. It occurs three times in *A* 3.10 *sqq*. Because the *ratunaya-*, 'layman', fails in his religious duties, he is punished by the *ratu-* three times with increasing severity:

...*ratuš*...*ratunaēm vārəmnəm staorəm frapərənaoiti*.
...*ratuš*...*ratunaēm yātəm gaēθanąm frapərənaoiti*.
...*ratuš*...*ratunaēm āhūirīm tkaēšəm frapərənaoiti*.

The third item is particularly troublesome; Bth. translated 'der Ratav verurteilt den Laien zum Verlust des ahurischen Richters (d. h. es wird ihm das Recht entzogen, einen Richter anzurufen)'; Da.: 'le Ratu déclare le disciple...exclu de la loi d'Ahura'. It would seem that, whether *tkaēšəm* here means 'doctrine' (as it ought to, in view of its epithet), or 'teacher', or 'judge', it can only stand in the same relation to *frapərənaoiti* as *vārəmnəm staorəm* ('a choice head of cattle') and *yātəm gaēθanąm* ('a share of possessions'), if all three are accusatives of relation, to be translated 'in respect of a head of cattle, a share of property, the doctrine'.

If the past participle *prta-* means 'owed, engaged, due', the transitive present *frapərənaoiti* may mean 'he makes liable, engages'.† When the *ratu-* 'engages' the layman in respect of a head of cattle or a share of his possessions, the layman becomes obliged to hand over the property in question. What his position is when the *ratu-* 'engages him, lays him under obligation, in respect of *āhūirīm tkaēšəm*', is difficult to say; from what follows in *A* 3.13, it seems that he becomes an outlaw. Bearing in mind that Iranian *prta-*, like Germ. *schuldig*, means both 'owing (a debt)' and 'guilty', we can perhaps say that the effect of *frapərənaoiti* in the first two cases is to declare the layman a debtor ('schuldig') in respect of property, whereas in the third case he is declared 'guilty ("schuldig")' in respect of *āhūirīm tkaēšəm*', and thereby possibly excommunicated. If we are right in using the term 'engaging' in this connection, it may be possible to relate *frapərənaoiti* to Skt. *ā-pṛṇoti, vyāpṛṇute*, 'to be occupied, *engaged* in'.

† 'To engage' here in the antiquated sense of 'to lay under obligation'.

COMMENTARY [97.1

The Pahl. translation of *pəšō.tanū-* is *tanāpuhl* which goes back to Av. *tanu.pərəθa-*, a synonym of *pəšō.tanū-*. But Pahl. also has a noun *puhl* 'corporal punishment',† which corresponds to an Av. word *pərəθa-* different from the second member of *tanu.pərəθa-*, cf. Bth., *ZWb.* 9, 193. To this *pərəθa-* belong Av. *an-āpərəθa-* 'that cannot be atoned', *dərəzānō.pərəθa-* 'requiring heavy(?) punishment', and *āpərəti-* 'atonement through corporal punishment'. These words Bth. assigned to a base ¹*par-* 'gleich machen', which Henning, *BBB*, p. 89, proposed to merge with Bth.'s ³*par-* (thus also Walde–Hofmann, *LEW*, s.v. *pār*).

Of Bth.'s two entries under ¹*par-* the first one, *pairyete*, is so uncertain that nothing can be learned from it.‡ The second entry, though unsatisfactorily translated by Bth., does seem to reveal the existence of a present stem *aipi-pāra-* which in the middle means 'to atone': *hā hē asti čiθa hā hē asti āpərətiš aipi.pārəmnāi iδa ašaone nōit anaipi.pārəmnāi drujō nmāne haiθyā aŋhən astayō* (*Vend.* 8.107). Bth. translated: 'das ist dafür die Sühne für den Gläubigen, der hier Buße tut; für den (aber), der nicht Buße tut (, gilt der Spruch§): im Haus der Drug sind das die rechten Gesellen'. Gdn., who has the same interpretation, commented (*Studien*, p. 6, n. 1): '*nōit* oder *an-* ist zu viel'. We must obviously give preference to Da.'s solution, which requires no alteration of the text: 'telle est... l'expiation qui dégage le fidèle qui se soumet à l'expiation; non celui qui ne s'y soumet pas: celui-là certainement ira habiter la maison de la Druj'. For *nōit anaipi.pārəmnāi* so taken there is a close parallel in *nōit ainižbərəta* (*Vend.* 8.37, quoted above, note 45²).

I find it difficult to derive from a single base the meanings 'atonement' on the one hand, and 'debt, guilt, obligation, etc.' on the other, and would therefore adhere to Bth.'s distinction of ¹*par-* and ³*par-*, but define the details differently: ¹*par-*, from which the doubtful *pairyete* should be excluded, means 'to atone', ³*par-* in the active (°*pərənu-*) 'to engage, make liable', in the passive (*pirya-*, *pəša-*, *pərəθa-*) 'to be engaged, obliged, owe'.

The reason why in this stanza, and only here, Aēšma is called *pəšō.tanū-* is not clear. It may be that this demon 'owes his body' to Mithra 'the Judge' (cf. note 79¹, point (2), and note 92⁷, last para.), who according

† Now also attested in the Kartīr Inscription on the *Kaʿba i Zardušt*, see Sprengling (above, p. 214, n. †), p. 47, line 13, last word.

‡ *kva tā dāθra pairyete*, *Vend.* 19.27. Bth. translates (*Wb.* 733): 'wo werden die Buchungen (mit einander) verglichen?' If *pairyete* is the correct reading the present stem *pairya-* could equally be the passive of ²*par-* 'to fill'; one might then translate with Gdn., *Studien*, p. 5, 'wo werden die Rechnungen ausgefüllt?' But the better attested reading is *pārayeinti*, which could be understood as an impersonal plural with *dāθra* as object: 'where does one (*lit.* do they) put across (Bth.'s ⁴*par-*) the accounts?' (the missing answer, to judge by what follows in the text, would be: 'one puts them across the Činvat Bridge').

§ *drujō nmāne*, etc. being a quotation from *Y* 49.11.

247

to the *Bahman Yašt* ultimately routs him (cf. Zaehner, *Zurvan*, 102, Bailey, *BSOS*, VI, 588 *sq.*). In the *Mēnōk-ī-xrat* it is Sraoša who smites Ēšm (see Zaehner, *op. cit.* 369), and in *Y* 57.10 Sraoša inflicts a wound on him.

97². The explanation of *Būšyąstā-* as 'procrastination' was first proposed by Wi. in his commentary to this passage. It was repeated by Bth. in *KZ*, 29 (1888), p. 547, n. 2, and Benv. in *Rev. hist. relig.* 130 (1945), 16.

98¹. Bth.–Wo., taking the whole of st. 98 (and st. 135) as an utterance of the worshippers, suppressed *as*: '...nicht sollst du...auf uns einschlagen, o Miθra, ...der—der stärkste der Yazata's, ...der—der siegreichste der Yazata's auf dieser Erde hier auftritt: er, der weite Fluren besitzende Miθra'. All other translators, excepting Lo., replaced *as* by *aš* (which occurs as a variant in st. 135, and in *Y* 9.15, on which see *infra*), and rendered *aš. vərəθrająstəmō* by 'most superlatively victorious' (or sim.). Lo. took *as* as 2 sg. inj. of *ah-* 'to be': '...der du der stärkste der Götter, ...der siegreichste der Götter bist.—Er tritt auf der Erde hier auf, Mithra der weite Triften hat'. I have followed Lo. in adhering to the *lectio difficilior as*, but not in his interpretation of the word, which makes of it a questionable *hapax legomenon* and entails the awkward separation of the last two lines as forming a separate sentence. Seeing that the first four lines are most likely an utterance of the daēvas (see above, note 68⁸), *as* can be fitted in as 3 sg. inj. of *ah-*,† in which function it occurs elsewhere in Avestan, either by considering lines 5–11 as no longer forming part of the daēvas' speech, or by assuming that the daēvas have reverted to speaking of Mithra in the third person, as they did in the first two lines.

A similar formula occurs in *Y* 9.15:

> *tūm zəmargūzō ākərənavō*
> *vīspe daēva zaraθuštra*
>
> . . . . .
>
> *yō aojištō yō tančištō*
> *yō θwaxšištō yō āsištō*
> *yō as vərəθrająstəmō*
> *abavat̰ mainivā̊ dāmąn.*

Here the translators who ignore *as*, or read *aš*, have an unfair advantage, cf. Wo.'s 'du, Zaraθuštra, hast bewirkt, (daß) die Daēva's sich alle in die Erde verkrochen..., (du), der der stärkste, der (du) der heldenhafteste, ...der (du) der siegreichste aus der Schöpfung der beiden Geister wurde(st)'. Lo., on the other hand, had to fall back on the poorly attested

---

† Mirza's suggestion (quoted by Tavadia, *Indo-Iranian Studies*, II, 58) to take *as* as nom. sg. masc. of Bth's ²*a-* (*Wb.* 11) is improbable. One would expect *ə̄ or *ō̄, cf. Bth., *GIP*, I, pp. 16 (§39, 3*a*), 180 (§304, II, 3).

COMMENTARY [98.1–101.1]

variant aša: '...(du)...der durch die Wahrheit (?) Siegreichste... wurde(st)'. If one must emend, I would rather suspect abavaṯ than as. Seeing that in Yt 13.76 we have

tā̊ zī hənti yāskərəstəmā̊
avayā̊† manivā̊ dāmąn

'they (= the Fravašis) are the bravest (cf. Henning in GMS, §396) of the creation of both spirits', one may well wonder if abavaṯ in Y 9.15 is not a mistake for avayā̊. To perceive how such a mistake may have come about, one may remember that instead of abavaṯ we have avavaṯ in Yt 5.120. abavaṯ is then perhaps the final outcome of successive attempts to find a form of bav- in avayā̊, on the part of scribes who, strangely enough, failed to understand as. If, however, abavaṯ is correct, it is advisable to interpret it in the sense of 'came into being, turned up, came to exist' (cf. Bth., Wb. 927 sq.), taking dāmąn as a locative: 'he who is (as) the strongest,...the most victorious, made his appearance in the creation (= world) of the two Spirits'. This would be a eulogistic resumption, in the third person, of the statement in Y 9.13: 'you, Zarathuštra, were born to Pourušaspa'.

99¹. On the meaning of dašinəm karanəm see note 95¹.

100¹. On Rašnu flying to the left of Mithra, as against his position in st. 126, see Introduction, p. 39.

100². The waters and plants accompany Mithra also in Yt 8.7 (cf. above, p. 217, n. ‡); they are objects of Mithra's care in st. 61. The Fravašis escort Mithra also in st. 66; they give 'noble progeny' to Mithra's worshippers, and are connected with 'long succession' in Yt 13.30, see above, pp. 227 sq.

101¹. Wo.: 'an sie verteilt er, (der) die Herrschaft führt, gleichmäßig (seine)...Pfeile. Und wenn er fahrend dahin kommt, wo die Miθrafeindlichen Länder (liegen), schmettert der als erster die Keule auf Roß und Mann nieder'. Lo.: 'auf jene (Mithra-feindlichen Völker) schleudert (comparing Ved. ava-bhṛ, abhi-pra-bhṛ) er—(denn) er vermag es—auf einmal seine...Pfeile, dann, wenn er dorthin geht, wo die m. Völker sind; er als erster schleudert...'. Wo.'s translation agrees with Spi.'s and Da.'s; the other translators take avi.dīš to mean 'against them', in support of which interpretation Gdn. refers to st. 37. But on the one hand what precedes avi.dīš are Mithra's assistants, not his opponents, on the other hand the very use of avi-bara- in st. 37 (where Lo. translates 'bringt er auf sie') goes against Lo.'s interpretation of this verb as 'schleudern' in the present case. The stanza is therefore best understood as contrasting the fighting weapon used by Mithra with those he distributes among his assistants.

† Thus with the best MSS.; Gdn.'s edition has vayā̊.

101². On the order of words in the dvandva *aspa vīra*, and in other combinations of these two nouns, see Benv., *BSOS*, VIII, 406, and Hzf., *Ap. I*. 95.

102¹. *parō.kəvid-*, which Duch., *Comp.* 75 (following Bth.) ranged with *aś∂mnō.vid-* (on which see note 39¹) as containing *-vid-* 'finding', and translated 'qui atteint le lointain', is better analysed as a compound with *vid-* 'to shoot', which is attested in Av. *vaēδa-* 'javelin', and in later Iranian forms quoted by Morg., *EVP*, 93, and Henning, *JRAS*, 1942, 234, n. 5.

103¹. Bth. guessed for *fravi-* the meaning 'prosperity', without being able to offer an etymology. The word is best understood as going back to an earlier **frāu̯i-*, abstract of *fra-av-*, cf. Ved. *prāvati* 'to favour, promote'. With long *ā* this noun survives unrecognized in Khot. *hāva-* 'profit, blessing'. The shortening of Av. *ā* before antevocalic *u̯* (also in *vavazān∂m*, note 124², before *-u̯ya-* possibly in *avya, p. 200, n. †) is familiar in *navāza-* (see Henning, *TPS*, 1942, 50, *GMS*, §§ 123–5), and in *yavaē°* 'always', against Parth. *y'wyd('n)*, MPers. *j'yd'n*, NPers. *jāvēd*. As was briefly indicated in *OIr. Lit.* § 10, the OPers. form *yāvai* 'always' can be restored with reasonable certainty in two much-discussed lines of the Beh. Inscription, col. v, 19 and 35. The identical text as read by Cameron was restored and translated by Kent as follows (*Journ. of Cuneiform Studies*, v, 56): *hya*: *Auramazdā*(line 19)*m*: *yad*[*ātai*]*y*: *yā*[*nam* : *avahyā*] : *ahatiy* : *utā* : *jīva*[*h*](line 20)*yā* [:] *u*[*tā*:] *martahy*[*ā*] = [:*hya*] : *Auramazdām* : *yadāta*(line 35)*iy* : [*avahyā* : *yāna*]*m* [: *ahatiy*:] *utā*:*jīvahyā*:*utā* (line 36): [*martahyā*] 'whoso shall worship Ahuramazda, divine blessing will be upon him both living and dead'. The crucial traces are described as follows by Cameron, *loc. cit.* 53 *sq.*: (line 19) *ya-da-*[*a*]-o-[*ta-i*]-*ya* : *ya-a*[+ .... +] : *a-ha-ta-i-ya* : *u-ta-a*, with a note that the restoration *yā*[*tā*:*taumā*] suits the space (which means there is space for seven letters and one word divider); (line 34) *ya-da-a-ta*(line 35)*i-ya*: [+ .... +]-*ma* [: *a-ha-ta-i-ya*:] *u-ta-a*, with a note that the gap preceding *-ma* is ample for at least eleven signs.

Kent's restoration is excluded by the necessity of inverting *yānam avahyā*, and the insufficient number of signs he allotted to the first gap in line 35. But Kent correctly saw that the gap must contain a noun in the nominative, subject of *ahati*, and a noun or pronoun in the genitive-dative, with which *jīvahyā* and *martahyā* agree. The ending *-ma* is either that of an unthematic gen. (*-maʰ*), or of a nom., thematic neuter (*-(a)m*), or masc. (*-maʰ*), or unthematic neuter (*-maʰ* or *-maⁿ*). If, as is more likely, the final word is a nom., the word in the gen. must be either the one beginning with *yā-* or a third word lost without trace in the middle of the gap. The second alternative is to be preferred in the absence of any suitable Indo-Iranian word beginning with *yā-* that could serve as term of reference for *jīvahyā* and *martahyā*. There is, however, so little space left for the third word in the gen. that this can only be the enclitic pronoun -šaiy. The word beginning with yā- is then almost certainly yāvai(y) 'always', which eminently suits the following combination of 'alive and dead'. By now the number of signs of the noun ending in -(a)m(a) can be exactly defined. Not less than eleven signs, including one word-divider, must be allotted to the gap in line 35. Even this minimum would mean that the gap of line 19 contains in addition to the word-divider nine signs, instead of the seven foreseen by Cameron. Clearly, therefore, the minimum number of eleven signs in the gap of line 35 is also the maximum, and the noun we are looking for has neither more nor less than three letters inside the gap.

We have already seen (note 2[3]) that according to Xerxes the faithful will be happy (šiyāta-) while alive, and Truth-owning (artāvan-) when dead. Before Cameron published his revised reading, Eilers had suggested, on the strength of Xerxes' statement, that our passage should be restored as šiyātiš avahyā ahatiy utā jīvahyā utā martahyā artam 'happiness will be his and(?) while he is alive, and Truth when he is dead'. We now find that the reward which Darius promises to the man who worships Ahura Mazdāh is the same both in life and after death. Since the reward must consist in what Darius is likely to have rated highest, the restoration a-ra-ta]-ma = artam offers itself as obvious. Admittedly one could think of other thematic neuter nouns that would fulfil the requirements of space and trace, and provide a tolerable meaning, e.g. pa-a-θ^{r}a]-ma = pāθ^{r}am 'protection', and there is the possibility of -ma standing for -ma^h or -ma^n: but no 'reward' occurs to me that would be esteemed by an ancient Iranian as highly as artam. We have tried to show in note 2[3] above that this restoration does not involve the assumption that Darius held different views from Xerxes on the highest reward to be attained in this life.

The proposed restoration of the two gaps is therefore ya-a[-va-i-ša-i-ya:a-ra-ta-ma] in line 19, and [ya-a-va-i-ša-i-ya:a-ra-ta]-ma in line 35, to be normalized as yāvaišaiy artam; the whole sentence should be translated: 'whoso worships Ahuramazdāh, Truth will forever be his, both (while he is) alive, and (after he is) dead'.

104[1]. Here is Bth.-Wo.'s rendering of the difficult stanzas 104-5: '(a) des gar lange Arme die ergreifen, (die) den Miθra betrügen; (b) auch wenn (er) im östlichen Indien (ist), er packt (ihn), (c) auch wenn (er) im westlichen (ist), er schlägt (ihn) nieder, (d) auch wenn (er) an der Mündung der Raṅhā, (e) auch wenn (er) im Mittelpunkt der Erde hier (wäre): (f) auch den erhascht Miθra, mit den Armen (ihn) umfassend. (g) Der Übelberüchtigte, (der) vom geradesten (Weg) abgegangen (ist), (h) ist unfroh in (seinem) Gemüt; (i) so denkt sich der Übelberüchtigte: (k) "nicht all das, (was) übel getan (ist), nicht alles, (was) zum Trug (getan ist), (l) sieht der blöde Miθra"'.

The following are interesting differences in the translations offered by other authors:

(a) den ergreifen, der dem Vertrag (Mithra) Gewalt antut (Lo.); den treubrüchigen packen (Gdn.); dessen lange Arme hervorgreifen, die mithramächtigen (Wi.); mit Mithrakraft (Spi.); whose arms, however far, seize the miθra-slayer (Hzf., 479, emending to *miθra.janō*); dessen lange Arme vorwärts greifen, seine, des Miθrakräftigen (Htl.); de qui les bras, forts de la force de Mithra, embrassent [le monde] (Da.);

(b) wenn einer im Osten, in Indien ist, wird er ergriffen (Lo.); wenn er von Indien den anfang nimmt (Gdn.); was im östlichen Indien ist das ergreift er (Spi.); whether in the east on the Indus (Hzf., omitting the verb); ob an dem am nächsten bei der Morgenröte Gelegenen (= im fernsten Osten) [in Indien], er holt einen mit seinem Griff heran (Htl.); soit qu'il les lève à la Rivière du Levant (Da.);

(c) wenn im Westen wird er niedergeschlagen (Lo.); und wenn er im Westen sich niedersenkt (Gdn.); und was im westlichen das schlägt er (Spi.); or in the west on the Tigris (Hzf., emending to *tigrē*); ob an dem am nächsten beim Abend Gelegenen (= im fernsten Westen), er schlägt ihn nieder (Htl.); ou les abatte à la Rivière du Couchant (Da.);

(d) ob er am strande des oceans (Gdn.); was in der Tiefe der Ranhâ (des Oceans) (Wi.); und was an den Steppen der R. (Spi.); or at the *sanakē* of the Iaxartes (Hzf.); ob am Gestade der R. (Htl.);

(e) was an den Enden dieser Erde ist (Spi.); ou à la frontière de la terre (Da.);

(f) auch den packend, o Mithra, erreiche mit den Armen (Lo., following Gdn.); du o Mithra weitergreifend umstrecke die Hände (Wi., Spi.); diesen tötet Miθra, indem er ihn mit beiden Armen umfängt (Htl.); et tout cela Mithra l'enveloppe et l'embrasse dans ses bras (Da.);

(g) der heillose (Lo.); der ehrlose (Gdn.); der Ruchlose durch die Gerechtigkeit erreicht (vernichtet) (Wi.); durch den Gerechten vernichtet (Spi.); der Besitzer eines schlechten Xvarənah (Htl.); l'homme sans Gloire (Da.);

(h) obwohl er unruhig in seinem gewissen ist (Gdn.);

(i) so thinks the man fit for bad fortune (Bailey, *Zor. Prob.* 21);

(k) nicht all diesen frevel, nicht alles sieht um es zu bestrafen (Gdn.); nicht alle diese Unthaten, nicht all diesen Betrug (Wi., Spi., Lo., Da.); nicht alle diese böse Tat, nicht jede, die zur Schädigung dient (Htl.); not all this ill-doing, all this tending to trickery (Bailey, *loc. cit.*); toutes ces mauvaises actions, tout (ce qui est fait) pour la tromperie (Benv., *Inf.* 55);

(l) Mithra der blinde (Lo., Bailey); der keine Augen hat (Gdn.); sieht Mithra auf der Erde (Wi.); M. der kunstlose (Spi.); der Nichtsehende (= wenn er nicht hersieht) (Htl.); M., l'indifférent (Da.).

The two serious flaws common to all versions of st. 105 are obvious.

(1) The situation is contradictory. Having just been caught by Mithra,

the *duš.x^varənah-* cannot very well have a low opinion of the god's vigilance. He may have thought little of Mithra *before* he was found out, but to bring this anteriority into the text one would have to assume that the present *mainyete* stands for a past tense. This, however, would be in conflict with the preceding statement: at the time when the *duš.x^varənah-* underestimated Mithra, he had no reason for being 'unhappy in mind'.

(2) The dative *aiwi.druxtəe* as generally interpreted implies a construction which is not only 'un peu dure' (Benv., *Inf.* 55), but impossible by the most elementary standards of Avestan syntax. Those who treat it as if it were an accusative have, of course, the text against them.

104[2]. The interpretation of *miθrō.aojaŋhō* is most uncertain. Gdn. (in his edition) suggested the emendation *miθō.aojaŋhō* 'speaking falsehood', on the strength of the nom. pl. *miθō.aoj\aa{}ŋhō* in *Yt* 19.95, where, however, most MSS. have *miθrō-*, and several *-vaojåŋhō*; in that passage 'violating contracts' or 'speaking falsehood' would make good sense, while 'having the strength of Mithra' is out of the question. Bth. hesitated between accepting Gdn.'s emendation and assuming a corruption of *miθraojaŋhō, derived by haplology from *miθra-δraojaŋhō 'deceiving Mithra (or the contract)'. In either case he postulated a stem in °*jah-*, i.e. a synthetic compound enlarged by the suffix *-ah-* (cf. Duch., *Comp.* 102 sq., who for his part, p. 56, wanted the stem of our compound to be **miθō-auja-*, leaving the acc. pl. ending of °*aojaŋhō* unexplained). Hzf., *Ap. I.* 185, kept the reading *miθrō.aojaŋhō*, and translated 'verkehrte Rede führend', taking *miθra-* here as a derivative of the base *miθ-* (on which see note 39[2] above). My own preference goes to Lo.'s and Gdn.'s (in his translation) understanding of *miθrō.aojaŋhō* as a compound with *aojah-* 'strength', in the sense of 'applying force *against* (=violating) the contract'.

104[3]. On the function of the suffix *-tara-* in *ušastara-* and *daošatara-* cf. Wn.-Debr., II, 2, 603, *f*, *n*, and Benv., *Noms d'agent*, 118 sq.

104[4]. On *hindu-* 'river' cf. Markwart, *Wehrot und Arang*, 132 sq.

104[5]. Lo., *ZII*, I, 202 sq., attractively replaces the reading *āgəurvayeite* by the variant *āgəurvyete* (passive present stem **gr̥bhya-*), which is carried by F_1 and Pt_1 in the parallel passage *Y* 57.29.

104[6]. Here again I follow Lo., *loc. cit.*, who interprets *niyne* as a passive, comparing *vaoce* 'he is called'.

104[7]. *sanaka-* occurs once more in *Yt* 12.19, where *upa sanake raŋhayå* is contrasted with *upa aoδaēšu raŋhayå* in *Yt* 12.18. The latter is also the name of the sixteenth country created by Ahura Mazdāh (*Vend.* 1.19).† There is no compelling reason for Bth.'s translation of *aoδa-* by

† In opposition to it Aŋrō Mainyuš produced *zyąmča daēvō.dātəm taožyāča daiṅhuš.aiwištāra* 'the daēva-created winter and the Taožyas (thus Bth.), ravagers of countries'. 'Non-Iranian' *daiṅhuš.aiwištāra* are mentioned in *Vend.* 1.17, and 'evil' *aiwištāra* in *Vend.* 1.10. The word is scarcely a noun of

'source', cf. Markwart, *op. cit.* 135, n. 2. The etymological connection with OInd. *udán* 'water', etc., by which Bth. was guided, is maintained also in W. Geiger's translation of *upa aoδaēšu r.* by 'an den Gewässern der R.' (thus also Htl., p. 161). We are thus free to derive *sanaka-* from the base *san-* 'to rise', and interpret *it* as 'source, spring'. *san-* was recognized in Av. *sanaṯ* G 5.5, Yt 14.7, 9 (which Bth. wrongly assigned to the base ¹*sand-*, cf. above, note 25¹), by Tedesco, *ZII*, 2, 39, who compared Sogd. *sn-*. Since then Wx., Yaγn., *san-*, and Parth. *sn-* have been added, cf. Bailey, *BSOS*, IX, 77. The past stem Khot. *sata-*, Wx. *sat-*, Parth. *sd-* Sogd. *st-* (*GMS*, p. 248 on 864),† excludes Tedesco's connection with Latin *scando*, New Indian *čaḍh-*, and makes it clear that OIr. *san-* (as posited by Ghilain, *Essai*, 55, 92) represents the base, IE **ken-* or **sk̑(h)en-*:(*s*)*kṇto-*. An Av. verbal noun *sāna-* has been surmised to occur in Yt 19.44, see Introduction, p. 36, n. †. [See Addenda.]

105¹. Lit. 'with a reaching round, an embrace'.

105². Nowhere, except allegedly in this uncertain passage, does *iθā̆* replace *uiti*, the adverb normally used for introducing direct speech (or thought). On the other hand, *iθā̆* frequently appears *in* direct speech, as an enlivening adverb. In the interpretation here proposed such an adverb is well in place. The wretch who having been caught by Mithra has forfeited the straightest path realizes that he had relied on the following false assumption, current among contract-breakers: Mithra sees a good deal of all the ill-doing perpetrated by us tricksters, but he does not see it *all*,‡ because sometimes he looks away; we tricksters therefore have a chance of getting away with our misdeeds.

Caught red-handed the wretch feels disabused and bitter. The initial *iθa* 'so!' conveys the belatedness of his understanding of the true situation, and his grief over it. The tortuous wording of the admission introduced by *iθa*, and the negative terms in which it is couched, effectively reproduce the reluctance with which the *duš.xᵛarənah-* pays tribute to Mithra's omniscience. Contrast with it the eager, positive, almost triumphant tone in which the poet retorts in the next two stanzas: the conditions prevailing *among mortals* as regards thinking, speaking, acting and perceiving, relying on which the *duš.xᵛarənah-* had misjudged Mithra, provide no measure for any limitations Mithra may have, since the god's powers are much superior to even the highest human attainments.

agent in *-tar*, as Bth. and Wn. maintained (cf. Duch., *Comp.* 101), but a root-noun *aiwi-štar-* from the base *star-* 'to rob, grab, plunder', cf. Greek στερέω. This base has hitherto passed unnoticed in Iranian, although it is also attested in Ossetic: Iron *stærī cæun*, Dig. *æstæri cæun* 'to go out on a plundering expedition', Iron *stær* 'raid, robbery', *stæron* = Dig. *æstæiron* 'robber, bandit'.

† The Yaγn. past stem is quoted as *santa* by Klimchitski (see above, p. 191, n. ‡), p. 102.

‡ Hence the emphasis on *all* in st. 107.

105³. *aiwi.druxtōe...apišma*, lit. '(when he is) not turning (his) face to the *aiwi.druxti-* ("act of deceit")'. It is agreed that *apišman-* is a compound of privative *a-* and **pišman-* 'Gesicht (=(1) face; (2) sight)', as Gdn. and Bth. assumed. From this it does not follow that the compound can only mean 'blind', the more so as the translation of *apišma.x^var-* by 'blindlings essend' is manifestly absurd. This epithet occurs in *Vend.* 13.47, where we read that the dog, like the thief and the *disu-*, is *tąθrō.činō* 'fond of darkness', *xšapāyaonō* 'going about at night' (see Benv., *Vrtra*, 52), *apišma.x^varō*, and *dužniδātō* 'ill-tempered, ill-disposed'.† The *disu-*, of evidently nocturnal, and possibly thievish habits, is according to tradition a wild animal.‡ On the assumption that *disu-* is a descriptive name (such as *sīždra-*, *dāδmainya-*), one may derive it from IIr. **didh-su-* (base *dhā-*) 'he who wants to lay in, store, viz. a hoarder', a desiderative formation like Ved. *ditsú* (base *dā*) 'wishing to give'.§ The animal in question may then be the hamster, who (1) is a proverbial hoarder, (2) goes about at night,‖ (3) is notoriously ill-tempered and aggressive, and (4) is considered a thief by farmers whose crops he despoils.

The thief and the hamster take care to consume their stolen goods 'out of sight, unobserved', and the dog who is thrown a bone usually takes it aside to gnaw it in peace. Thus interpreted, *a-pišman-* 'not subject to view, removed from sight' is comparable to *a-iθyajah-*, *a-mahrka-*, *a-yaska-*, 'not subject to (=out of, removed from) danger, death, illness'. Theoretically such compounds can also be used in the sense of 'not bringing danger, death, illness', cf. OInd. *a-yakṣmá* '(1) not subject to illness; (2) not causing illness'. *apišman-* may accordingly also mean 'not bringing one's sight towards, not facing, not turning the face (to)', and thereby account for the puzzling *aiwi.druxtōe*, which, as Benv. pointed out (*Inf.* 55), cannot be an infinitive, but must be the dative of the noun of action *aiwi.druxti-*. [**106¹**, see Addenda.]

107¹. In interpreting this period I have followed Wi.'s lead, against all other translators. Wo., for instance, has 'es gibt keinen Menschen auf

---

† Thus in accordance with Bth.'s interpretation of the Pahl. transl. as *duš*(?)-*hēm*, which is, in fact, the reading of the manuscript IM used by Hoshang Jamasp, *Vendidâd*, I, p. 485, n. 3.

‡ Cf. Da., *ZA*, II, 207, n. 66; Spi., *Commentar*, I, 316.

§ *disu-* instead of expected **dizu-*, like *dasva* instead of expected **dazva*, cf. Bth., *GIP*, I, p. 22, §53; II, 8. The analogy probably operated in the first place on the desiderative stem **didh-s(a)-*, to which *-u-* was merely added in secondary derivation, cf. Wn.–Debr., II, 2, 468. **didh-s-* should have resulted in Iranian **diz-*, but like *dasva* was affected partly by the corresponding formation from the base *dā-* (in this case **dis(a)- < *did-s(a)-*), partly by the characteristic *s* (in this case of the desiderative), which in parallel formations remained unchanged after dentals that were not voiced aspirates.

‖ Alfred Brehm, *Die Säugetiere*, II (1914), p. 324: 'gewöhnlich ist die erste Hälfte der Nacht und der Morgen vor Sonnenaufgang seine Arbeitszeit'. The hamster is said to have reached Europe from Central Asia.

der Welt, (der) in solchem Maße mit (seinen) Ohren hört, wie der geistige M., der scharfhörige, mit tausend Fertigkeiten begabte; jeden, (der) trügt, erblickt er'. Spi., somewhat unfairly, translated as if *vīspəm* were preceded by the rel. pron. *yō*. I do not see that the statement as it appears in our translation is illogical. To catch a contract-breaker Mithra must see him, but to detect him his 'thousand ears' and 'thousand perceptions' serve him at least as well as his 'ten thousand eyes'. One may even argue that if Mithra has only 1000 ears (*hazanrā.gaoša-*) as against 10,000 eyes (*baēvarə.čašman-*), this is because the perceptive strength as between ears and eyes was held to be in the ratio of ten to one. The poet will then be saying that even man's most sensitive organ at its best cannot compare with Mithra's vision.

Whether or not we choose to link *vaēnaiti* with *surunaoiti*, it is likely that ancient listeners to the Hymn were prepared for the reference to Mithra's vision by the foregoing praise of his *āsnō xratuš* (cf. st. 141, where Mithra is *bayanąm...aš.xraθwastəmō*). For, as Bailey pointed out in *Zor. Prob.* 98, n. 2, a *Dēnkart* passage states that 'vision in men derives from the possession of the *āsna xratu*; this *āsna xratu* is itself the faculty of vision of the *jān*, whence the eye is ordered'. Accordingly, 'insight', Germ. 'Einsicht', seems an apt translation of *āsna- xratu-*, and may be applicable even to *xratu-* alone. A return to the traditional interpretation of *xratu-* as a thinking activity is seen in Lentz, *AAWL*, 1954, pp. 967, 1006; thus also Bailey, *op. cit.* 85, translates MPers. *xrat* by 'reason', Henning, *Mir. Man.* II, 350, by 'wisdom'. For a review of various interpretations of *xratu-* cf. Hzf., *Ap. I.* 235 *sqq*. Notice the divergent meaning of Khot. *grata, grra* 'commandment, injunction, admonition', see Bailey, *BSOAS*, x, 901.

107[2]. *uγra* (*vazaite*) *xšaθrahe* 'der starke des Reichs' (Wo.); 'der mächtige...der über die Herrschermacht verfügt' (Lo.); 'in seinem reiche fährt der gewaltige' (Gdn.); 'alles Gewaltige der Herrschaft (= alle Herrschergewalten) fährt er (nämlich in seinem Kriegswagen)' (Htl.); 'dans le terreur de sa royauté' (Da.); 'gewaltig an Herrschaft' (Wi., Spi.). Is *uγra* a blunder for *uγrō* under the influence of the preceding *amava*?

108[1]. On the implications of 'bad sacrifice' see above, Introduction, p. 63.

109[1]. *x^{v}ainisaxta-* 'mit schönem (Waffen)gerät, Zeug' (Bth.); 'glänzend gerüstet(?)' (Spi., Lo.); 'arrangée belle' (Da.); 'mit... versehen (Bedeutung von *x^{v}aini-* unbekannt)' (Htl.); 'in sich mächtig' (Gdn.); 'selbstbefestigt' (Wi.); 'engines of war (*lit.* manufactured with skill)' (Hzf., 474 *sq*.). In my opinion *saxta-* is the Av. equivalent of NPers. *saxt* 'hard, strong, firm (etc.)'. The latter word was connected by Horn, *NPers. Et.* 160, with OInd. *śaknóti* 'to be strong; to be able'; in support of this etymology Hübschmann, *Pers. St.* 74, quoted the semantics of

COMMENTARY   [109.1

Gr. κρατύς 'strong' beside Goth. *hardus* 'hard'. To the same base, which in Avestan (¹*sak*-) means 'to understand, learn', belong Khot. *sāj-/sīya-* 'to learn, study', Sogd. *fs'c-/fsyt-* 'to teach, tame', Buddhist Sogd. *βsys-* 'to be tamed, get used to', and Parth. *hw'bs'gyft* 'submission, docility'.

To this base Ghilain, *Essai*, 68, also assigned Parth. *s'c-*, *ps'c-/psxt-*, and *nys'ž-*, 'to prepare, arrange', adding on p. 50 under the same heading Parth. *scyd* (see Mary Boyce, *MHCP*, p. 164, 1*b*) 'it is fitting', which corresponds to MPers. *szyd* (*Mir. Man.* II, 299.2), NPers. *sazad*, Sogd. *s'ct*, *s'št*. It is, however, difficult to see how the divergent meanings can be reconciled, and accommodated under a single base. Henning treated MPers. *ps'z-/ps'xt-*, *hs'z-/hs'xt-*, 'to prepare' (to which belongs NPers. *sāxtan* 'to prepare, manufacture'), as causatives of MPers. *sz-* (in *szyd* 'it is fitting' and *pscg*, *pszg* 'fitting, suitable'), without pronouncing on the ultimate identity of the base *sak-* in question (see *ZII*, IX, 170 and 186). Bth. confined himself to the remark that neither NPers. *sazad* 'it is fitting' nor *sāxtan* 'to prepare' should be derived from Av. ²*sak-* 'to pass (of time)' (see *Wb.* 1554, n. 5). He was, however, mistaken.

It being agreed that MIr. *sāč-/sāž-/sāz-* 'to prepare' is causative of *sač-/saz-* (in Sogd. *sāč-*) 'to fit, be suitable', the latter cannot be derived from Av. ¹*sak-*, but clearly belongs to ²*sak-*. The Avesta itself shows how the development came about. In *Vend.* 18.16 Būšyastā tells man in the morning: *x^vafsa darəyō mašyāka, nōiṯ tē sačaite*,† which Bth. correctly translated 'sleep long, man, your time is not (yet) up'. From this idiom we gather that the impersonal expression *tē sačaite meant 'your time is up', which by a short step would lead to 'you must, ought to, it behoves you'. The impersonal construction of Pers. (etc.) *sazad* continues the OIr. usage as attested in Avestan. Such direct evidence of the descent of *sazad*, hence also of *sāxtan*, from ²*sak* is welcome, but the derivation from this base could have been, and has been, taken for granted even without it. Already Lagarde, *Gesammelte Abhandlungen*, 300, illustrated the transition from 'passing' to 'being suitable, fitting' by referring to Gr. καθήκει and προσήκει. Equally to the point is the German adaptation of French *passer* to convey the sense of 'being suitable' in *passen*, *passend*.

Thus the Iranian family of ¹*sak-* (OInd. *śaknóti*) is reduced to the Av. verbal forms of this base, and Av. (*x^vaini-*)*saxta-*, NPers. *saxt*, Khot. *sāj-*, Sogd. *fs'c-*, Parth. (*hw*)'*bs'g-*.

The group of forms belonging to Av. ²*sak-* is correspondingly increased. To the above-mentioned verbs for 'preparing' and 'being suitable' one may add Sogd. *s'cyy* 'duty', *s'ckw* 'suitable', *nws'cy* 'unsuitable', *pts'c-/ptsyt-* 'to adorn, arrange', *pts'k* 'order, arrangement', *'ns'c-/'ns'yt-* 'to join, fix', *'ns'k* 'preparation', NPers. *sēcīdan*, *sazā*, *sazāyīdan* (see, *GIP*, I, pp. 136, n. 5, 297), and Arm. *patšač* 'suitable' (see

† The whole passage is quoted below, p. 293.

Hübschmann, *Arm. Gr.* 225). The meaning 'to pass' is preserved in Pahl. *sač-* (etc.) 'to pass away' (cf. Zaehner, *Zurvan*, 473; Nyberg, *Hilfsbuch*, II, 199), Sanglechi (etc.) *šəxs-* 'to pass over' (see Morg., *IIFL*, II, 414), and Parth. *'wsxt-* 'to descend'; corresponding to OPers. *θakata-* we have Parth. *sxt*, Sogd. *syty'*, Khot. *skyätä*; in addition cf. Khowar *ju-saxa* 'a period of two years' (Morg., *BSOS*, VIII, 664), Pš. *sa*ž, *sa*ṣ̌*kāl* 'this year' (Morg., *EVP*, 72), NPers. *saxš* 'old (clothes)' (Morg., *NTS*, XII, 265).

109[2]. On the hendiadys $x^v ainisaxtəm\ pouru\,.\,sp\bar{a}\delta\partial m$ see note 47[1].

109[3]. Two ways of construing this stanza, and the parallel st. 111, have been suggested. One consists in treating lines 5–10 as an amplification of *vahištəm*, and taking the last three lines in isolation. As an example Wo.'s translation may be quoted: 'Wem soll Ich, (ohne daß) er in (seinem) Sinn daran denkt, eine starke Herrschaft zuweisen, mit schönem Gerät, mit zahlreichem Heer, die trefflichste eines allherrschenden Machthabers, (der) aufs Haupt schlägt, eines tapfern siegreichen unbesieglichen, der die Strafe zu vollstrecken befiehlt?—Sofort bei der Bestimmung wird die vollstreckt, sobald er ergrimmt (sie zu vollstrecken) befiehlt. (Aber) auch des gekränkten, nicht befriedigten Sinn besänftigt er durch Miθra bei guter Befriedigung des Miθra'. The other construction is the one Widengren applied in *Hoc.* 101 *sq.* It consists in making a break after *vahištəm*, and co-ordinating *ṭ̌bištahe axšnuštahe* with the genitives in lines 5–7: 'Wem soll ich gewaltige Herrschaft, ..., alles Vorstellungsvermögen übertreffende, verleihen als das Beste? Auch eines allherrschenden Sātars, welcher die Schädel schlägt, ... auch eines angefeindeten, nichtbefriedigten Denken beruhigt Mithra, wenn Mithra wohlbefriedigt ist.' In Widengren's own translation, modelled on that of his immediate predecessors, and attended by a wilful interpretation of his own (see above, Introduction, p. 54, n. †), this interpunction did not produce a sensible contrast between sts. 109 and 111. However, as soon as we combine it with Wi. and Spi.'s interpretation of *sāθrasčiṭ hamō.xšaθrahe kamərəδō.janō* (see note 109[6]), the contrast becomes clear: the same 'conqueror unconquerable', killer of tyrants, will respect a treaty-abiding rival installed by Mithra, but will be incensed by a breach of contract even on the part of an otherwise friendly ruler. We may take it that by being 'incensed' (*scil.* into making war), he becomes instrumental—thanks to his invincibility—in overthrowing the treaty-infringing ruler, whom Mithra thus deprives of his kingdom, however impregnable it may be.

109[4]. The construction of the last three lines was correctly understood only by Bth., who, however, took *miθrahe* for the name of the god. Lo. saw that by *miθrahe* the contract was meant, but spoiled his version by using the reading *miθrō* in place of *miθra*. The subject of *rāmayeiti* is, of course, the ruler on whom Mithra bestows the kingdom.

109[5]. *axšnušta-*, with *š* transferred from the present to the past stem, cf. *BSOAS*, XVII, 481.

COMMENTARY [109.6

109⁶. The words *vahištəm...aurvahe* are rendered by Wo.: 'die trefflichste (Herrschaft) eines allherrschenden Machthabers, (der) aufs Haupt schlägt, eines tapferen'. Similarly Htl., Hzf. (pp. 473, 475), and Lo. (who for *hamō.xšaθrahe* has 'alleinherrschend', but in *Yt* 13.18 'allbeherrschend', and elsewhere 'selbstherrschend'). Da. translates 'ce bien suprême...la souveraineté du brave tyran, souverain absolu, qui abat les têtes'; Gdn. has 'das reich eines oberkönigs, der zu roß die köpfe (der feinde) erschlägt'. All these renderings are obviously weak. Nowhere else does *kamərəδō.jan-* appear without the owner of the evil head being mentioned in the accusative (*Vend.* 4.49) or genitive (see above, p. 160, n. †), and it is awkward to have to make of the 'daēvic' *sātar-* (cf. note 34⁴) a 'brave tyran', or, with Bth., a neutral word for 'ruler', for which elsewhere *sāstar-* is used. We may then try to revert to the view of Wi. and Spi., who assumed that *hamō.xšaθra-* could have two meanings, according as the first member of the compound was ¹*hama-* 'same, *idem*' or ²*hama-* '*omnis, totus*'.

The versions adopted by these two scholars are immediately convincing: 'der des Feindes, des Nebenbuhlers Schädel schlägt', 'der dem gleiche Herrschaft besitzenden Feinde den Schädel zerschlägt'. In this way also °*čit*, 'even', comes into its own, while Wo.'s translation treats it as pointless. It was Gdn. who discarded Spi.'s interpretation, on the ground that it did not suit st. 111. The objection no longer applies once we adopt Widengren's interpunction of the stanza, cf. note 109³.

Wi.'s and Spi.'s interpretation of *hamō.xšaθra-* suits *Yt* 15.54 at least as well as the usual one, which in this case is also Spi.'s: ...*yaθa anyåscit̰ xšaθrāt̰ xšayamnå hamō.xšaθrō.xšayamnå* 'wie auch die andern über das Reich herrschenden, als Allherrscher herrschenden'. The context is unfortunately obscure, but this much can be said: if there are 'others ruling', one of them cannot very well have *all* the power; hence the translation 'like also others ruling over the realm with equal power' has a better chance of being correct.

On the other hand ²*hama-* prevails when *hamō.xšaθra-*, is used with *sāstar-*. This happens in the line *sāsta daiṅhəuš hamō.xšaθrō*, which occurs four times in different contexts: *Yt* 13.69, 14.13, 15.50, and 13.18. In the first three passages Bth.'s 'allherrschend' undoubtedly makes good sense (while 'gleichherrschend' does not), and must be accepted. *Yt* 13.18, however, stands to gain considerably from an interpretation of *hamō.xšaθrō* as containing ¹*hama-*. The usual understanding of this stanza was most neatly put by Lo. as follows: *āat̰ yō nā hīš hubərətå barāt̰* | *jva ašaonąm fravašayō* | *sāsta daiṅhəuš hamō.xšaθrō* | *hō aṅhāiti zazuštəmō* | *xšayō kascit̰ mašyānąm* | *yō vohu.bərətąm† baraite* | *miθrəm yim vouru.gaoyaoitīm* | *arštātəmča frādat̰.gaēθąm varədat̰.gaēθąm*

† Read (with several MSS.) *vō hub°*, cf. Tedesco, *ZII*, 2, 45, Altheim, *ZII*, 3, 35. Note the variant °*bərətəm*.

'und welcher Mann (sie) die Schutzgeister der Frommen bei Lebzeiten gut gepflegt hält, ein allbeherrschender Gebieter des Landes, der wird, wer er auch sei, der siegreichste Herrscher der Menschen;—der euch gut gepflegt hält, den Miθra mit den weiten Triften und die äckerfördernde, äckermehrende Geradheit'. Here Spi. did contemplate translating *hamō.xšaθrō* by 'gleich am Reiche', but as the construction of the stanza eluded him, he finally adopted Justi's 'Allherrscher', cf. *Commentar*, II, p. 597.

It is not quite clear to me whether Lo. referred 'euch' (*vō*) to the Fravašis, or to Mithra and Arštāt. The latter alternative is preferable, since the change from speaking of the Fravašis in the third person to addressing them in the second person in association with two other gods would be harsh; but if *vō* does not refer to the Fravašis there is no connection between the two parts of the stanza, unless *yō* takes up an implied demonstrative pronoun governed by *hamō.xšaθrō*. Just as OPers. *brdiya^h...hamapitā ka^mbujiyahyā*, Beh. I, 30, means 'Smerdis...who had the same father as Kambyses (gen. or dat.)', so *hamō.xšaθrō hō aŋhāiti...(aētahe* or *aētahmāi) yō...baraite* should mean 'he will have the same power as he who holds...'. For the ellipse of the dem. pron. see above, p. 161, n. †; cf. also our interpretation of the gen. governed by *haθrākō*, note 66⁴.

Accordingly, the following literal translation of *Yt* 13.18 can be suggested: 'The man who in his lifetime treats them well, the Fravašis of the owners of Truth, he—(being) a ruler of a country, ruling most absolutely (*zazuštəmō*) over men, whoever he (be)—will be having the same power as (he) who treats well you, (namely) grass-land magnate Mithra and world-furthering, world-promoting Justice'. On 'Justice' in this connection cf. below, note 139².

110¹. Illness and death are bestowed on behalf of Mithra by Vərəθrayna in *Yt* 14.47 quoted above, p. 193.

110². The parallelism with *ištīm pouruš.x^vāθrąm* in st. 108 would lead one to expect *dučiθrəm* to be an adj. agreeing with *ainištīm*. Hence Bth. preferred the variants with °*θrĭm*. Wo. has 'quälende Armut', Lo. 'Unvermögen und schlechten Samen', Gdn. 'mangel und noth'. Earlier commentators relied on Westergaard's emendation *dužāθrəm*, cf. Gdn., *Studien*, 26. I would tentatively regard *dučiθra-* as issuing from an original compound **duž-jīθra-*, containing **jīθra-* 'life'. For the *ablaut* degree of the base cf. Wn.–Debr., *Ai. Gr.* II, 2, 708. Beside **jīθra-*, and probably replacing it at an early stage, OIr. had **jīvaθra-*, which survives in Parth. *jywhr*, MPers. *zyhr*, see Bth., *ZWb.* 52, Tedesco, *MO*, xv, 198. After the group -*žj*- had been simplified to *j* and *ī* had become *i* as in *jiṭ.aša-*, **dujiθra-*, no longer supported by the simplex **jīθra-*, was adapted to recall the familiar *čiθra-*. A parallel can be quoted from *Vend.* 5.4, where some scribes, failing to understand *išasəm jiṭ.ašəm*, wrote the more familiar-looking, but meaningless *išasəmčiṯ ašəm*. Such confusion is in line with the occasional interchange of *č* and *j* in the MSS., cf. *tanč/jištəm* Yt 10.141, *saočintąm*:°*jəntąm* Vend. 2.8, *vīčirå:vījarå* Yt 13.40; J₂ has *yavaēčyō* for °*jyō* Y 39.3, and *jayəmā* for *čay*° Y 38.3; cf. Bailey, *Zor. Prob.* 193. If we follow Bth. in his reading *dučiθrīm* we shall analyse the form as an -*ī*- fem. of an -*a*- adj., analogous to *zaranaēnī*- and *hu-puθrī*-.

111¹. On the consequences of this 'incensement' see the end of note 109³.

112¹. Bth. recognized, in *ZWb.* 144, that *frašna*- belongs to Pahl. *plš* 'spear' or sim. For subsequent discussions see Bailey, *TPS*, 1953, 31.

112². It would be strange if Mithra, who is referred to as *daiṅhupati*- in sts. 78 and 99, and as *vīspanąm dahyunąm daiṅhupati*- in st. 145 and Ny 1.7 (quoted note 113¹), here bore the modest title of a *vīspati*-, as all interpreters assume. A glance at the Apparatus shows that *vīspaitīm* is an emendation, the correct reading being *vīsūptīm*, an Avestan compound which hitherto has passed unnoticed; its meaning is obviously 'broad-shouldered', cf. Ved. *vyàmsa*.

112³. Earlier translations of lines 5 *sqq.* are as follows: 'Klar (sind) Miθra's Wege—wenn er dies Land besucht, wo er in guter Pflege gehalten wird,—: weit (und) tief zur Weide' (Wo.); 'Leuchtend sind M.'s Wege (Htl. "Auszüge"), der dás Land besucht, wo (etc.); breit und tief sind seine Triften' (Lo.); 'The signs of M. are friendly, who visits that country (etc.); the paths (*sic*) of its pastures are deep' (Hzf., 427); 'sichtbar ist des M. kommen' (Gdn.); 'mannichfach sind des M. Wege' (Wi.); 'wo er wohl verehrt die tiefen Ebenen zu Triften macht' (Spi.).

On *čiθra*- see note 64⁶. For *frayana*- Hzf.'s explanation deserves preference over the usual derivation from *i*- 'to go'. The word is the middle present participle *fryāna*- of *frāy*- 'to love', with *ā* shortened as in the cases quoted above, p. 167, and -*ay*- replacing -(*i*)*y*- as in *aspaya*- ∼ OInd. *áśvya* (cf. Bth., *GIP*, I, p. 155.11) or the loc. sg. masc. *bərəzantaya ašavanaya* (see Bth., *Wb.* 960, n. 6). [See Addenda.]

112⁴. *hubərətō* was analysed as acc. pl. of *hubərət*- in note 48⁵. Here we may compare it with the forms *hu-bərətå* and °*bərətąm*, which Bth. quoted under one entry with *hu-bərətō*, describing all three as 'absolutives'. The passages are (apart from the present one):

(*a*) Yt 13.18 (quoted note 109⁶) *yō nā hīš* (viz. the Fravašis) *hubərətå barāṯ*; here °*tå* is evidently acc. pl. fem. of *hubərəta*-;

(*b*) Yt 13.18 (quoted note 109⁶) *yō vō hubərətąm* (var. °*bərətəm*) *baraite miθrəm...arštātəmča*;

(*c*) Yt 15.40 *yaṯ nmānō.paitīm vindāma...yō nō hubərətąm barāṯ* 'that we may find a husband who will treat us well'.

In (*b*) *hubərətąm* stands for *hubərəta*, acc. pl. masc. covering both the masc. *miθrəm* and the fem. *arštātəm*. -*ąm* not infrequently interchanges with -ą, cf. Bth., *GIP*, I, p. 158, sect. 54*e*, and above, note 71⁴. In the present case the intrusion of *m* was favoured by the immediately following labial of *baraite*, cf. *duždạfəδrō*: *duždạm.fəδrō* in *Vend*. 19.43 (see Bailey, *BSOS*, VI, 597, *Zor. Prob.* 224). The variant °*bərətəm* inspires less confidence, although it, too, is understandable as aiming only at *miθrəm*, to which *arštātəm* was loosely added. In (*c*) one would expect *hubərətå*, since girls are speaking. However, seeing that in the preceding stanza we read *kainina* yōi *anupaēta mašyānąm*, it is evident that *hubərətąm* is merely the acc. pl. masc. used instead of the fem.

There is thus no need to postulate 'absolutives'. Avestan has an idiom consisting of *hubərəta-* in the acc. governed by *bar-*, which corresponds to OPers. *hubrtam bar-* and Ved. *súbhṛtam bhar-*, cf. Wn., *BSOS*, VIII, 823 *sq*. This is evidently an Indo-Iranian 'figure étymologique', cf. Benv., *Mages*, 22, which in no way precluded the use of *hubṛta-* 'well treated' in other contexts. We have met *hubərəti-*, the noun of action to *hubərətəm bar-*, in st. 78. The existence of *hubərət-* 'treating well' beside *hubərəta-* is no more surprising than that of OIr. **duš-kṛt-* (Av. *duš.kərət-*) 'doing ill' beside **duš-kṛta-* (OPers. *duškrta-*, Aram. *dwškrt*') 'ill-done'. What is daring is the use of the acc. of *hubərət-* as object of *bar-*; such wording was almost certainly intended as a playful allusion to the common expression *hubərətəm bar-*, of which the god is the object (cf. (*b*) above, where one of the two gods is Mithra) and the worshipper(s) the subject. There is no reason why we should not credit our poet with such flashes of originality.

As shown by *yaθa*, *hubərətō* is not an adjective belonging to *čiθrå*, but is substantivized in the sense of 'persons who treat well'. *baraiti*, accordingly, fulfils a double task: it is the verb of the main clause, *and* of the comparative clause which *yaθa* introduces. In pedestrian Avestan we should have *čiθrå*... *yaθa hubərətō baraiti baraiti* 'he treats the clans as he treats those who treat (him) well'. The parallel passage in st. 48 displays the same syntactic economy.

112⁵. *paθanå* and *jafrå* are neuter plurals with fem. endings. Spi.'s 'tiefe Ebenen' finds support in the use of Oss. *fætæn* (< *paθana-*) both as adj. 'broad' and subst. 'surface, square'. But what can be envisaged for *paθanå* is also admissible for *jafrå*, and 'wide depths (= valleys, cf. *jạfnu-* "valley")' seems to me to make better sense than 'deep plains'.

112⁶. *vasō.xšaθrō* 'at will', that is, unimpeded by raids of *miθrō.-drug*-s, cf. sts. 38, 86.

113¹. The first two lines have been rendered: 'dann komme er uns zu Hülfe, o Mithra, hoher Herr' (Wi.); 'deswegen möge uns zum Schutze herbeikommen Mithra und Ahura die großen, ja Mithra und Ahura die großen' (Spi.; twice because dual??); 'dann sollen uns zu hilfe kommen M. and A. die hehren' (Gdn., followed by all later translators except Hzf.); 'Then he shall come to our help! Mithra-Ahura you high ones!' (Hzf., 426, 428).

In the difficult lines that follow, there is almost unanimity in subordinating the *yat̯* clause to the preceding *taδa* clause: 'then may he (*or* the two) come...when the whip raises...; then the sons...will fall....'. Only Hzf. differs: by suppressing *yat̯* he eliminates subordination altogether.

It seems that the bearing *Ny* 1.7 may have on the interpretation of this stanza has not been considered:

Ny 1.7   *miθrəm vīspanąm daḧyunąm*
           daiṅhupaitīm yazamaide
           yim fradaθat̯ ahurō
           mazdā̊ xᵛarənaŋuhastəməm
           mainyavanąm yazatanąm;†
           tat̯ (var. *taδa*) *nō jamyāt̯ avaŋhe*
           miθra ahura bərəzanta;

'Mithra we worship, who in (*lit.* of) all countries is the head of the country (see below, note 145[1]), whom Ahura Mazdāh created as the most Fortune-endowed of supernatural gods! May he (Mithra) *therefore* come to our help, O Mithra and Ahura the exalted!' Here the words *tat̯* (var. *taδa*) *nō jamyāt̯ avaŋhe* follow upon a statement of Mithra's exalted position; Wo. translated '*alsdann* mögen kommen', Da., more convincingly, 'que viennent *donc*...'. *tat̯* (*taδa*) in this context seems to have motivating, not temporal value. Another example of this usage of *tat̯* (*taδa*), which has a parallel in the English usage of *then* in the sense of 'accordingly', is found in *Yt* 7.3: *måŋhəm...yazamaide:tat̯* (var. *taδa*) *måŋhəm paiti.vaēnəm, tat̯* (var. *taδa*) *måŋhəm paiti.vīsəm* 'we worship the moon: therefore I look (*or* looked) out for the moon, accordingly I seek‡ (*or* sought) the moon'; here for *tat̯* (*taδa*) Bth. has 'nunmehr, jetzt(?)', Lo. 'nun', while Da. ignores the word. In *Ny* 1.7 the words *tat̯* (*taδa*)... *bərəzanta* are a logical sequel to what is said in the preceding five lines. Mithra's supremacy, subject to his being a creature of Ahura Mazdāh's, having been stated, Mithra alone is asked to attend, but *both* gods are invoked to bring about the attendance. Since in *Ny* 1.7 the dvandva *miθra ahura* is not yet inverted (see Introduction, p. 44), it is fair to assume that the whole of that stanza up to *bərəzanta* § is not less ancient than the Mithra Yašt; it is, in fact, a condensed hymn to Mithra, in which the essentials about the god are stated, so as to induce him to lend his assistance.

† Up to here the text is identical with the closing formula of *Yt* 19.35.

‡ *paiti.vīsəm* and *aiwi.vīsəm* (which follows in the text), do not belong, in my opinion, to ¹*vaēd*- 'to know', as Bth. thought, but to ²*vaēd*- 'to find'; cf. Sogd. *parwēδ*- and *frawēδ*-, Henning, *BSOAS*, xi, 484.

§ The final sentence of *Ny* 1.7 ('we worship the swift-horsed (etc.) sun') belongs more to what follows in *Ny* 1.8 ('we worship...Tištrya,...we worship the Tištryaēnī-s,...we worship Vanant..., Θwāša..., Zrvan..., etc.') than to the preceding invocation of Mithra. [See Addenda.]

It thus becomes likely that the first two lines of st. 113 also serve as conclusion to what precedes: we worship Mithra who treats well the clans dear to him, as they treat him well; therefore, (considering that we, too, treat him well,) may he come to our assistance. The pattern of the closing formula of the 'condensed hymn', which, we may suppose, was familiar to the author of our Yašt, would inevitably prompt the addition of the invocation *miθra ahura bərəzanta* after the words *taδa nō jamyāṯ avaiṅhe.*

We are now free to interpret the remainder of st. 113 as consisting of a subordinate clause introduced by *yaṯ*, followed by a main clause introduced by the second *taδa.*

113². Bth.: 'wenn laut die Peitsche knallt und (wenn) die Nüstern der Pferde in Aufregung geraten, (wenn) die Peitschen sausen, die Sehnen schwirren (und) die spitzen Pfeile'; Hzf., 426: 'the whip shall call with high voice, and the horses' nostrils shall snort, the whips shall crack, the bow-strings whiz, the arrows [verb corrupt]'.

The impossible notion, introduced by Wi. and still maintained by Duch. (*BSOS*, IX, 863), that *srifa-* is a metathesis of Ved. *śíprā* 'hair' (previously thought to mean 'nostrils', cf. M. Leumann, *IF*, 39, 209, Charpentier, *KZ*, 46, 26 sq.), has had the effect that all translators saw in it the subject of *xšufsąn*. This even applies to Scheftelowitz who, disbelieving this etymology, translated *srifa* by 'snorting', rightly, I think, but on the strength of an equally impossible etymology (OInd. *riphati, ZII,* 2, 273, n.). If we work our way backwards from *tiγrånhō aštayō,* we find that the verb three times precedes its subject, which in the case of *xšufsąn* will be *aštrå.* This consideration, combined with the single occurrence of °*ča,* and precisely in a position where it suggests that *srifa* is co-ordinated to *aštra* as an additional subject of *barāṯ,* settles both the construction of the period and the approximate meaning of *srifa*: it is a noise produced by horses, hence, presumably, 'snorting' or 'neighing'. Conceivably NPers. *sarfāk* 'sound, voice, clamour' is connected.

113³. The present stems *xšufsa-, kahva-,* and *naviθya-* only occur here. For the first OInd. *kṣobhate* 'agitatur', NPers. *āšuftan* 'agitare', etc., have long been compared. This meaning makes sense also with *aštrå* as subject. For *kahva-* we are entirely reduced to guessing from the context. *naviθya-* or *nava(i)θya-* can scarcely be a side-form of **nuiθya-,* as Bth. wanted. It is more likely a denominative of an unknown **naviθ(a)-* or **navit(a)-.* From the former, or a derivative **navit-ya-* (becoming **naviθya-*) of the latter, one might at a guess derive Khot. *nūha, nauhä,* 'point' (cf. Konow, *NTS,* XI, 58), for which Leumann, *AKM,* XX, 452, postulated a word **nautha-.* To the Khot. word, with a different suffix, belong Sogd. *nwk,* NPers. *nōk,* 'point', cf. Henning, *BBB,* p. 61. One might then assume that **naviθ/t(a)-* was a pointed missile, say a dart, and attribute to the denominative verb *naviθya-* the meaning 'to dart'.

COMMENTARY   [113.4–115.1

113⁴. ²ašti- 'arrow' (or sim.) is etymologically connected with Russ. *ость* 'awn', OInd. *apāṣṭhá* 'barb of an arrow', cf. Pokorny, *IEW*, 22. On ³ašti- 'palm' see Henning, *TPS*, 1948, 69. On yet another ašti- in Gāthic see below, p. 285.

113⁵. In the last two lines I have followed Bth.'s interpretation, except as regards *frā.vərəsa-*. For this word he has 'des Haars beraubt, skalpiert', Wi. 'die Haare voraus = kopfüber', Spi. 'aufgeschichtet an den Haaren', Htl. 'mit vorüberhängenden Haaren', Hzf. (p. 426) 'their hair up'. The assumption of an OIr. form *u̯r̥sa-* of the word for 'hair', beside *u̯arsa-*, is unsafe, since even NPers. *gurs* has been rightly held to represent *u̯arsa-*, see Horn, *GIP*, I, 2, p. 64. The consonantal nature of *r* in this stem is confirmed by Sogd. *wrs* being a heavy stem, cf. *GMS*, §§ 485 sqq.; Pš. *weṣṭə* is derived from *varəsa-* by Morg., *TPS*, 1948, 70; the meaning of Khot. *bilsahai*, occurring only once, requires confirmation, cf. Bailey, *Donum Natalicium H. S. Nyberg Oblatum*, 3. In any case an alternative and, to my mind, more satisfactory explanation of °*vərəsa-* is at hand: the word may go back to *u̯r̥t-sa-*, an -*s*- extension of *u̯r̥t-* 'to turn, roll'. *frā.vərəsa-* is then a synthetic compound, comparable in formation to *arənaṭ.čaēša-* (see note 35¹) < *(*arna-*)*čai-ša-*. This interpretation is supported by Khot. evidence, cf. *ābei'sa* 'whirlpool' < *ā-vartsa-*, *ggei's-* 'to turn round' < *vart-s-*, etc. (see Konow, *NTS*, XI, 39, 48, Bailey, *JRAS*, 1954, 32).

115¹. In *Y* 19.18 we read: *kaya ratavō? nmānyō vīsyō zantumō dāhyumō zaraθuštrō puxδō åŋhąm dahyunąm yā anyā rajōiṭ zaraθuštrōiṭ; čaθru . ratuš raya zaraθuštriš. kaya aiṅhå ratavō? nmānyasča vīsyasča zantumasča zaraθuštrō tūiryō*. 'Which are the *ratu*-s? The *nmānya* (= the one of the house), the *vīsya*, the *zantuma*, the *dāhyuma*, and, fifthly, Zarathuštra; (thus) with (all) the(se) countries other than the Zarathuštrian Raγa; the Zarathuštrian Raγa has (only) four *ratu*-s. Which are its *ratu*-s? The *nmānya*, the *vīsya*-, the *zantuma*, and, fourthly, Zarathuštra.' Evidently Zarathuštra was the Ratu General, to whom even the *dāhyuma*-s of the various countries were subordinated; in 'Zarathuštrian Raγa', however, he was himself the *dāhyuma*.† The fivefold religious hierarchy here outlined corresponds to a fivefold division of temporal authority, see below, note 145¹.

The *nmānya*, *vīsya*, etc., of *Y* 19.18 are evidently persons, but in the other passages quoted by Bth., *s. vv.* ¹*nmānya-*, ¹*vīsya-*, etc., Nmānya is the name of the genius of the house, *Vīsya* of the genius of the clan, and so forth, cf. Da., *ZA*, I, lv, 30 *sq*. In the series of genii Zarathuštra is replaced by the *zaraθuštrō.təma*. The genii, then, took their names from the corresponding human *ratu*-s.

---

† Cf. Marquart, *Eranšahr*, 122 *sq*., who, however, took *zaraθuštrō* in this passage to mean a person holding the title 'Zarathuštra', not the prophet himself. This is not the view taken by Bth., *Wb*. 1671.

The situation was thought by Bth. to differ again in Y 17.18, where the Fravašis, in the plural, are said to be nmānyå vīsyå zantumå dāhyumå zaraθuštrōtəmå. These words Bth. interpreted as the fem. pl. of adjectives ²nmānya-, ²vīsya-, etc., meaning 'connected with the genius Nmānya, the genius Vīsya, etc.' This interpretation, unnecessarily complicated in the case of Y 17.18, becomes almost grotesque when extended to our stanza: in describing Mithra as a ratu- that is nmānya-, our author is thought to mean a divine ratu- who is connected with a genius that bears the name of the human ratu- of the house! In my opinion the presence of the word ratvō is a clear indication that nmānya, vīsya, etc., are here to be taken as human ratu-s, as in Y 19.18, the successive Ratu Generals who replaced Zarathuštra after his death having assumed the title of zaraθuštrō.təma-. The same explanation can be applied to Y 17.18: because the Fravašis have protective functions which partly coincide with those of the ratu-s called nmānya, vīsya, etc., they are, as a figure of speech, addressed by the titles of these ratu-s; since the titles happen to be substantivized adjectives, they are capable of taking fem. endings when applied to the Fravašis.

There is nothing bewildering in Mithra or the Fravašis being thought of as holding functions which are normally the prerogative of men; the metaphor is the same as when Mithra is addressed as daiṅhupati- (cf. sts. 78, 99), or Haoma as pati- of the house, the clan, the tribe, and the country (see Y 9.27). I would therefore agree with Lo.'s 'o Mithra, ..., o Meister im Haus, im Dorf, im Gau, im Land, o du höchst zarathuštrahafter', except in so far as this rendering obliterates the technical character of these terms, more especially of zaraθuštrō.təma-, for which Spi. and Gdn. correctly have 'Oberpriester'.

116[1]. Wi.'s translation of suptiδarənga- by 'Schultermagen = Geschwisterkinder' has been accepted by most interpreters. Da. defined the word more closely as 'qui s'appuie contre l'épaule l'un de l'autre', and, in his Addenda, ZA, III, 197, as 'cousins'. Bth. has 'Gaugenosse' without explanation. Nearest to what seems to be the right meaning is Hzf.'s rendering of haša suptiδarənga by 'business friends' (pp. 491 sq.), although his connection of supti- with Arab.-Pers. suftaja 'money-order, bill of exchange' is very unreliable, and that of -δarənga- with OPers. haⁿdugā- most improbable. More promising is B. Geiger's comparison of °δarənga- with Aram. 'drng, which occurs in the company of words meaning 'partner' (one of which is hngyt, cf. note 116[2]), see E. G. Kraeling, The Brooklyn Museum Aramaic Papyri, Nos. 9.18, 10.12, 11.9. Geiger, however, did not attempt to account for the form of °δarənga-, beyond attaching it to the base drang-. That the word represents *drənga- (= *dranga-) with svarabhakti a is a possibility on which one hesitates to rely very much. A form *°δarənga- (= *δr̥nga-) would be acceptable as a parallel to Sogd. °trnng, tryt-, from *°tr̥ng-, tr̥nxt- (see GMS, §§ 152 a sq.), but no convincing reason is available why -ərə- should here have been

COMMENTARY [116.1–116.3]

replaced by -*arə*-. Perhaps the best solution lies in combining Geiger's connection of °δ*arənga*- and *drang*- with Benv.'s analysis of °δ*arənga*- as representing δ*ar*- followed by two determinatives (*Origines*, 28). One may then postulate an IE alternation *dher-n-gh- (Av. δ*arəng*- < *δ*arng*-): *dhr-en-gh- (Av. *drang*-).

Geiger's affiliation of our word with *drang*- 'to hold, make firm', cf. Av. *ā-drənja*- 'to fix', introduces the possibility of attaching to it a business connotation, namely by comparing *avadranga* 'earnest money', an Iranian LW in Buddh. Sanskrit, on which see Bailey, *JRAS*, 1955, 14 *sq*. There is no need to dissociate *supti*° from the Av. word for 'shoulder' (cf. also *vīsupti*-, above, note 112²); we seem to be dealing with a sphere of activity in which the shoulder comes in metaphorically also in English, cf. *to shoulder an expense, responsibility*, etc. Accordingly, δ*arənga*- may be interpreted as 'liability, commitment' (a development of 'hold'), and *suptiδarənga*- as a bahuvrīhi meaning 'who has a liability on his shoulders, shouldering a liability'. The contract between two friends will then concern the obligation of mutual friendship and assistance they have undertaken.

116². *haδō.gaēθā*- is of almost identical formation with Aram. *hngyt* (which occurs in a series together with Aram. '*drng*, see prec. note), Khwarezmian *angēθ*, and Parth. *h'mgyh̲*, see Mary Boyce, *MHCP*, 188.

116³. *huyāyna*. On this word Wi., who translated it by 'husband and wife', wrote: 'ich identificire den ersten Theil mit dem N. Pers. *šui* (read *šōi*) Mann; *huya* kommt von *su* erzeugen'. This Spi. curtly rejected, no doubt because of the impossible connection with NPers. *šōi* 'husband', and replaced by an equally impossible interpretation as 'die gut Opfernden' (base *yaz*-!). Gdn. has 'zwischen eigenen weibern (d. h. *eines mannes*)'. Bth. suggested 'room-mate' without explanation, Lo. doubtfully followed suit. Da. and Htl. left the word untranslated. Finally W. Krause, *KZ*, 56, 305, suggested that the word stands for *ha-yāyana*- and meant 'belonging to the same liver'. This was accepted by Duch., *Comp.* 187, and Hzf., 491 *sq*., but is manifestly impossible: neither can OIr. *ha*- be represented by Av. *hu*-, nor does OIr. *k*, intervocalic or in direct contact with *n*, become *γ* in Avestan. It is obvious at a glance that Wi.'s explanation offers everything one can possibly hope for in the case of an unknown word occurring only once: excellent meaning, excellent etymology, and strict adherence to the text. *huya*- as 'begetter', to OInd. *sū́te*, *sūyate*, must be the descendant of an IIr. word *suya- for 'husband', which escaped being stigmatized as 'daēvic' because its etymological connection with *hunā*- 'to bring forth' and *hunu*- '(daēvic) son' had been forgotten, possibly, too, because it had ceased to be used except in the present dvandva with *γnā*-. From the semantic point of view there is no objection to the IE base for 'begetting' yielding verbal nouns not only with the meaning 'son' (Gr. *υἱύς*, Agnean *se*, Kuchean *soy(ä)*, etc.), but

267

also with the meaning 'father, husband' (*suio-). The long $\bar{a}$ of $huy\bar{a}°$ is probably the dual ending, cf. zastāišta-, zastā.maršta-, Duch., Comp. 11.

116⁴. Bailey, *JRAS*, 1953, 97, attractively derives Khot. *bišta*- 'disciple' from °višta- in hāvišta-. The long $\bar{a}$ of hā- may be due to the position before u, see note 2⁵ above.

117¹. There are significant differences in the interpretations proposed of the scale of contracts listed in sts. 116 sq. Wi.: 'Zwanzigfach ist der Mithra zwischen..., hundertalterig zwischen..., zehntausendalterig ist der Mithra der mazd. Lehre'; Spi.: 'Zwanzigfach ist Mithra unter..., hundertfach..., zehntausendfach ist Mithra bei dem, welcher am mazd. Gesetz festhält'; Gdn.: 'zwanzigfältig ist Mithra (die treue)..., zehntausendfältig ist das verhältniß innerhalb des Mazdaglaubens (d. h. konkret innerhalb... der gesammtheit der gläubigen)'; Da.: 'Mithra est vingt fois entre... mille fois entre... dix mille fois vaut le Mithra de la Religion mazdéenne (= le contrat entre le fidèle et la Religion)'; Bth.– Wo.: 'Zwanzigfach (bindend) ist Mithra zwischen..., hundertfachen Halt (bietet er) zwischen... zehntausendfachen Halt bietet Mithra (dem), der zur mazd. Religion (gehört)'; Lo.: 'Zwanzigfach ist der Treubund (Mithra)..., von hundertfältiger Dauer..., von zehntausendfältiger Dauer ist der Treubund (Mithra) des mazdayasnischen Glaubens'; Htl.: 'Wie 20 ist M. in..., zu 100 werdend in..., zu 10000 werdend ist der M. des mazd. Herzenslichtes'; Hzf., 491: 'The *miθrā, societates* are valid 20-fold between..., 100-strong between..., 10,000-strong between the māzd. religion.'

Opinion thus differs on the meaning of *miθrō*, and the value of the second component of *satāyuš*, etc. On the first point, seeing that up to *hazaŋrāiš* the parties concerned are always two, it seems obvious that *miθrō* here stands for 'contract'; those who deny this have the obligation, as yet unfulfilled, of producing strong proof to the contrary. In the case of the ten-thousandfold contract of the Mazdayasnian Religion the contracting parties can safely be held to be the Religion on the one hand, and each of the faithful on the other. As regards the second point it is unlikely that numeral adjectives of different function should have been used in passing from ninety to higher figures. I see no objection to assuming that *satāyu-*, etc., contain *āyu-* 'duration' (cf. Lo.'s translation, and Duch., *Comp.* 177), but would conclude from the present passage that 'lasting one hundred' had come to be used in the sense of 'hundredfold'.

117². Benv., after stating that the last two lines came about as a result of two successive interpolations, and that *amahe* is a 'corruption' of *hamahe*, concludes: 'c'est perdre son temps que d'y chercher une suite cohérente' (*Vṛtra*, p. 24). In fact, provided one refrains from emending and takes the words at their face value, the two lines disclose a sensible and grammatically correct statement, which has a logical link with what precedes. *ava* is the adverbial instr. of the dem. pron. *ava-*, as *tā* 'therefore' of *ta-*, *yā* 'because' of *ya-*, *kā* 'how' of *ka-*, *anya* 'otherwise' of *anya-*. *ava* 'thereby' (that is, by respecting the various contracts in accordance with the marks assigned to them in the preceding scale) is co-ordinated to *aθa* 'thus'. *ayąn*, nom. pl. n. of *ayan-* (cf. *ayąnča* Y 57.17) governs both *amahe* and *vərəθraynahe*, taking a different verb (in the singular, because the subject is neuter) for each dependent genitive. Of all translators only Htl. ventured to offer a literal translation of the un-emended text, which, though grammatically possible, arouses misgivings on account of its artificial syntax: 'mit so viel Kraft (meaning presumably "with so much *of* strength ") ist er (= Mithra) verbunden und wird es sein am Tage der Feindestötung'.

118¹. Lo., *ZII*, 2, 216, was the first to realize that the speaker in this stanza is Mithra, and not the worshipper as was previously thought. This discovery threw new light on *aδara dāta* and *upara dāta*, which Bth. in the circumstances could not but misunderstand ('mit untenhin (obenhin) geweihter Verehrung'). Yet neither Lo.'s nor Htl.'s rendering of these words carries conviction: 'Auf gegen (im?) Norden vollzogene Verehrung hin will ich nahen, auf gegen (im?) Süden vollzogene' (Lo.); 'Durch die untere (= auf Erden) gespendete Verehrung will ich kommen, durch die obere gespendete (also infolge der durch die geistigen Opferwürdigen gespendeten)' (Htl.). We are obviously in the presence of two com-pounds, *aδara.dāta* and *upara.dāta*, which serve as adjectives to *nəmaŋha*. Since the function of °*dāta-* here is most likely the same as in *ātərəδāta-* 'given by fire', *nəmō aδara.dātəm* will mean 'homage paid by an *aδara-*, a lowly person', cf. Lat. *supplices inferioresque*. For the idiom *nəmō dā-* one may compare *stūtō garō vahmǝ̄ng...dadəmahičā* 'praise-songs and prayers we offer' Y 41.1. *upara°* will correspondingly refer to highly placed persons. One is reminded of the contrast between the lowly (*ādrəng*) and the exalted (*ərəšvånhō*) in the Gāthic stanza Y 29.3.

118². On the use of the pronoun *ava-* in referring to the region of the sky see below, p. 293.

118³. It is not surprising to find Mithra addressing Zarathuštra, since apart from Ahura Mazdāh other divinities do so also, e.g. Anāhitā (cf. note 45³), Haoma (Y 9.1 *sqq.*), Aši (*Yt* 17.21 *sq.*), and even Aŋrō Mainyuš (*Vend.* 19.6, 8). A meeting of Mithra and Zaruthuštra seems to be referred to in the unclear gloss to Y 9.1 *miθrō zyāt̰ zaraθuštrəm*, according to Bth., *Wb.* 1659 'Mithra knew Zarathuštra'; differently J. M. Unvala, *Hōm Yašt*, p. 4, cf. Henning, *Sogdica, Errata* sheet, and *BSOAS*, XI, 722 on 671. Zarathuštra is seen invoking Mithra in *Vend.* 19.15.

119¹. That Ahura Mazdāh is the speaker of sts. 119 *sq.* becomes clear in st. 121. The chief god first addresses Zarathuštra, then turns to Mithra to assure him that he shall be worshipped by the whole animate creation, and finally (in st. 120) announces to men in general that, since Mithra is their protector, they are to drink libations in his honour.

119². It is the general opinion that *vayaēibya patarətaēibya* occurs again in the Yašt dedicated to Sraoša, *Y* 57.28:

āsyaṇha aspaēibya āsyaṇha vātaēibya
āsyaṇha vāraēibya āsyaṇha maēγaēibya
āsyaṇha vayaēibya patarətaēibya
āsyaṇha hvastayā̊ aiṅhimanayā̊

'(the two [actually four] runners are) faster than two horses, faster than two winds, faster than two cloudbursts, faster than two clouds, faster than two...birds, faster than two (arrows) which are being shot well-shot'. Here a considerable number of MSS. have the variant *hupatarətaēibya*. While it is hard to see how such a variant could have arisen by mistake, an original *vayaēibya hupat°* would be liable to be changed to *vay° pat°* by imitation of the present passage in our Hymn; for the Mithra Yašt altogether greatly influenced the wording of the Hymn to Sraoša, as B. Geiger has pointed out (cf. note 41¹, end of first para.). If in *Y* 57.28 we restore the reading *vayaēibya hupatarətaēibya*, as I think we must, then Bth.'s translation of the simple *patarəta-* by 'im Flug begriffen', and his analysis of the adj. as a *-ta-* extension of an adverb **patarə* 'im Flug', will be found inadequate. This highly artificial explanation should give way to a straightforward interpretation of the word as representing a *-ta-* extension of **ptar-* 'wing', cf. Gr. πτερ-όν, πτέρ-υξ, etc., see Benv., *Origines*, 28. The first *a* is then a svarabhakti vowel, as in Av. *patar-* 'father', <*ptar-*. Accordingly, *p(a)tar(ə)ta-* will mean 'winged',† as Ved. *párvata* (to *párvan* 'knot') means 'knotty', and *hu-p(a)tar(ə)ta-* 'having good wings, well-winged', as *hu-kərəp-ta-* means (in Lo.'s interpretation) 'having a beautiful body', cf. Wn.–Debr., II, 2, 588.

Having dissociated our passage from *Y* 57.28, and reinterpreted the meaning of *patarəta-*, we are in a better position to understand the stanza than Bth. was. In his opinion *pasubya staoraēibya* and *vayaēibya patarətaēibya* together formed one dvandva, of which the first component was *p° st°*, the second *vay° pat°*. The assumption of such a ponderous structure is hardly compatible with the following relative clause, which, if Bth. were right, should either apply to both parts of his complex dvandva (which clearly is not the case), or *form part* of the second component of the dvandva, thus rendering the whole structure even more ponderous and unlike anything known as dvandva. On the other hand, the two words *pasubya staoraēibya* have every appearance of an ordinary dvandva, and were so taken by Spi., Da., Htl., and Hzf. (p. 477). Their meaning is accordingly not 'with two heads of small and large cattle', as all other translators have it, but simply 'with small and large cattle'. If this is granted, then *vay° pat°* almost certainly also form a dvandva that com-

† This was anticipated by Hzf., *Zor.* II, 477, when, without explanation, he printed *ftartā* and translated 'winged creatures'.

prises two kinds or sizes of birds. Bearing in mind the formation of Germ. *Geflügel* 'domestic fowls, poultry', a derivative of *Flügel* 'wing', one may venture to guess that Av. *patarata-*, as a derivative of **ptar-* 'wing', had the meaning of *Geflügel*. Here, then, we must disagree even with Spi., who translated 'mit großen und kleinen Vieh, mit zwei Vögeln'.

We now come to the main point. Only Hzf. paid attention to the fact that when *yaz-* means 'to immolate' it takes the double accusative: *tąm yazata haošyaŋhō...satəm aspanąm...hazaŋrəm gavąm...* 'to her H. immolated 100 horses, 1000 oxen' *Yt* 5.21 (etc.), whereas *yasna-*, *zaoθra-*, etc., when used with *yaz-* are always in the instrumental. Hzf.'s reaction was to replace the instrumentals *pasubya*, etc., by accusatives. However, the obvious lesson to draw from Hzf.'s observation is that *pasubya*, etc., are comitative instrumentals, and the widely held theory based on this passage, that the Avestan Mithra exacted blood-sacrifices from his Zoroastrian worshippers,† is nothing but a myth invented by modern interpreters.

119³. *fravaz-* means 'to fly' also in Sogdian (B. *βrwz-*, Man. *frwz-*, cf. *GMS*, §§329, 617, 1638) and Waxi, etc. (*rəwəz-*, see Morg., *IIFL*, II, 538, *NTS*, I, 66). Both with Av. *vaz-* and *fravaz-* it is often difficult or impossible to decide whether 'flying' or 'driving' is meant, cf. sts. 16, 48, 67, 70, 86, 99, 100, 101, 107, 126, 133 of the present Yašt. Clear cases of 'flying' expressed by *vaz-* occur in sts. 20, 39, 127, 129. On the other hand *fravaz-* is clearly used in the sense of 'driving' in st. 124.

120¹. No attempt has been made so far to interpret the second and third lines by respecting the inflectional endings. Bth.–Wo. have: 'Zugewiesen (und) geweiht (ist) der Haoma, (sind die Zaoθra's,) die der Zaotar weihen und opfern soll', while Lo. translates: 'Haoma, der Priester ist (gewidmet und geweiht. Sie sollen ihn widmen und opfern [verehren?]?)'. Lo.'s interpretation breaks down over *yå*, which, as Bth. saw, can only refer to an implied *zaoθrå*. On the other hand Bth. was misled by the nom. sg. *zaota* into disregarding the endings of the two verbs. As these are of the 3rd plur., the subject is necessarily the worshippers. Consequently *haomō...zaota* can only mean 'Haoma (is) the *zaotar-*'. We must accordingly scrutinize *zaotar-* more closely, to see if the word is capable of governing the acc. *yå*, which can then be taken as anticipating an implied accusative *tå* (*zaoθrå*), object of *vaēδayånte* and *yazånte*: '(the libations of) which consecrated and dedicated Haoma is the pourer, (these libations) they shall dedicate and sacrifice'. On the omission of the dem. pronoun see above, p. 161, n. †.

† This theory is not only implicit in every translation known to me of *Yt* 10.119, but is expressly stated by Cumont, *TMMM*, I, 227, and Benv., *Rel.* 26; Duch., *Orm.* 63, speaks of a 'culte non sanglant' of Mithra in the Avesta, without mentioning our stanza or his predecessors' interpretation of it.

The word for 'priest', *zaotar-*, Bth. stated in *Wb.* 1653, goes back to Indo-Iranian times (cf. Ved. *hótṛ*), when two meanings coalesced in **ž́hautar-*: (1) 'he who performs the libation' (Ved. *juhóti* 'to pour'), and (2) 'he who calls the gods' (Ved. *hávate* 'to call'). Gdn., *Indo-Iranian Studies...Sanjânâ* (1925), 277 *sqq.*, argued that in one Avestan passage, *Y* 11.1, *zaotārəm* is still used as a verbal noun of agent, not merely as the ordinary word for 'priest'. His argument is unconvincing, because he has to attribute to *zaotārəm* the meaning 'he who sacrifices (the cow)', whereas *zaotar-* as a noun of agent to Ved. *juhóti* should mean 'the pourer'. Moreover, there is no compelling reason why *zaotārəm* in *Y* 11.1 should not simply mean 'priest'. The present context, on the other hand, seems to be a case where one may profitably walk in Gdn.'s footsteps. As a verb, **zu-* 'to pour' does not occur in Old Iranian, but the base is attested on the one hand in Av. *zaoθra-* 'libation, poured offering', on the other hand, as noticed by H. W. Bailey, in Khot. *ysuma-*, Pš. *zwamna*, 'broth' (cf. *BSOS*, VIII, 920), and *ṇiysuna-* 'flow out, discharge, menstruate' (cf. S. Konow, *A Medical Text in Khotanese*, 93).

It is more than likely that Avestan priests should from time to time have given a thought to the possible original meaning of the Av. word for 'priest', *zaotar-*. When doing so, they could hardly have avoided suspecting a connection with their word for 'poured offering, libation', *zaoθra-*. The pattern of *pātar-* 'protector':*pāθra-* 'protection' (cf. Benv., *Inf.* 43), or *dātar-* 'giver':*dāθra-* 'gift', would, one would think, forcefully suggest to them that the real meaning of *zaotar-* was 'pourer'. Naturally it would not escape them, either, that *zaotar-* could also be regarded as the noun of agent to *zu-* 'to invoke'. Hence their definition of *zaotar-* was probably much the same as Bth.'s. It is only to be expected that a reflex of their views on the subject should occasionally turn up in their writings. The present passage has a good chance of containing such a reflex. The author had already defined Haoma as a *zaotar-*, 'priest', in st. 89. In returning to Haoma in the present stanza he could count on his listeners appreciating his etymologizing use of the word *zaotar-* to describe precisely that priestly function of Haoma which makes the etymologically related *zaoθrā-*s (represented by *yå*) the object of *zaota*.

The verbal construction of *zaotar-* with the acc. is linked with its stress as deduceable from Ved. *hótṛ*. This type of *-tar-* noun frequently takes its object in the acc. like an active participle, cf. Benv., *Noms d'agent*, 19 *sq.*, while the type which in Vedic has the stress on the suffix behaves like a noun and takes the gen. Our stanza accordingly indicates that *zaotar-* should not be regarded as corresponding to a hypothetical Vedic **hotṛ́*, as Benv. would allow (*op. cit.* 25). By yielding sense only when analysed as governing the acc. *yå*, *zaota* discloses its adherence to the accentual and functional pattern of Ved. *hótṛ*.

As to the substance of the statement, the poet evidently considered that the god of the plant Haoma, which 'consecrated and dedicated' forms the most precious ingredient of the libation, is himself 'consecrated and dedicated' as he pours out the libation in his capacity of priest. Haoma, god and priest, thus graciously offers himself in the shape of pressed Haoma-juice as a sacrifice to Mithra. In so doing he, the priest of Ahura Mazdāh (st. 89), sets a shining example to the Mazdāh-worshippers, who are being told by none less than Mazdāh to worship Mithra. [On Haoma as god, priest, and victim, cf. now also R. C. Zaehner, *The Teachings of the Magi*, 126, 129.]

120². On $x^v ar$- meaning 'to drink' see Tedesco, *ZII*, 2, 44.

122¹. Such may have been the meaning of the last line, as Da. noted, who, however, thought that 'the Staota Yesnya *and* the Vīspe Ratavō' was intended. In this he was followed by all later translators. But the structure of the Avesta as handed down to us need not be as ancient as this stanza. At an early date the Staota Yesnya may well have *included* a section called Vīspe Ratavō.

124¹. All translators except Htl. ('dem geopfert hat Ahura Mazdāh... mit für die Nichtvernichtung erhobenen Armen') connect *uzbāzāuš* with *miθrō*, and there is general agreement that *amərəxtīm* is here an abstract, though why Mithra should lift his arms to 'indestructibility' is not, and cannot be, explained. Bth. assumed that in *Yt* 19.11, 89 *amərəxtiš* referred to the *Saošyant-* and meant 'indestructible'. Although our interpretation of *amərəxtīm* as 'the indestructible' may therefore suit the theory that the Roman Mithras is a *Saošyant-*,† it must be said that there is no compelling reason for accepting Bth.'s interpretation of *Yt* 19.11, 89. Lo. achieves in *Yt* 19.11, 89 at least as good sense as Bth., although to him *amərəxtiš* means 'indestructibility'. The Av. word is of course capable of either explanation. *amərəxti-* 'not subject to destruction' as epithet of Mithra agrees well with *aiθyejaŋha* 'not subject to danger' as epithet of *miθra ahura* in st. 145.

124². Bth. took *vavazānəm* as an 'absolutive', 'beim Fahren, beim Lenken (des Wagens)', comparing *vaŋhānəm* in *Yt* 5.126: *arədvī sūra anāhita...frazušəm aδkəm vaŋhānəm pouru.paxštəm zaranaēnəm*. Here one may prefer to assume that an original *vaŋhāna, nom. sg. fem. of the pres. participle middle of *vah-*, was changed to °*nəm* under the influence of the surrounding masc. accusatives: 'Arədvī Sūrā Anāhitā, ...wearing a long-sleeved (see above, p. 220, n.), much-embroidered, golden coat'. As to *vavazāna-*, it is best regarded as a regular shortening of **vāvazāna-* (see note 103¹, first para.), an adjectival intensive middle perfect participle of *vaz-*, comparable to OInd. *vāvaśāná, vāvasāná*, etc.,

† See Cumont, *TMMM*, I, pp. 187 *sq.*, 205. According to Zaehner (*BSOAS*, XVII, 240) the appellative *Sebesio* in Mithraic inscriptions is 'easily explained as *savišyō (Voc.?) = Av. *saošyā*'. In which Iranian language could *saušyant-* be expected to turn up as **sauišyant-*?? Cf. above, Introduction, p. 68, n. ‡.

cf. Wn.-Debr., II, 2, 271. *vāvazāna- is, as it were, the passive of Ved. vāvahi 'driving well, or fast' (said of the driver).

124³. *hāmō.taxməm.* Only Htl. took *hāmō°* in the sense of 'all' ('den allgewaltigen'), which it never has in other compounds. Wi. and Spi. have 'equally strong', Da. 'qui roule d'une force uniforme', Bth. 'gleichmäßig tüchtig'; Gdn. hesitated between 'gleichmäßig fest' and 'gleichmäßig laufend'; his latter alternative was adopted by Lo. and Hzf. ('smooth-rolling'). Obviously the last three authors were thinking of the base *tak-* 'to run', but they offered no explanation of the suffix. Hence Duch., *Comp.* 129, maintained Bth.'s unlikely translation. In my opinion *hāmo.-taxma-* is a thematization of **hāmō.taxman-* 'having (always) the same course = running evenly'; cf. the stems of *duždāma-* and *kamərəδa-*, Duch., *Comp.* 37.

125¹. *ham.ivā-*. It is a safe guess that one of the many *hapax legomena* in the last three lines of the stanza must be a word for 'yoke'. We shall see in the following notes that the other nouns in this passage can be identified with reasonable likelihood as referring to the shaft and to *parts* of the yoke. By a process of elimination the meaning 'yoke' thus falls to *ham.ivā-*, which hitherto was thought to mean the 'pole' or 'shaft' of the chariot. This identification is in keeping with the expectation that the chief item should be the first to be mentioned. Etymologically one may assume that *-miv-* stands for *-myuv-* as *-niv-* for *-nyuv-* in *mainivasah-* (cf. note 13¹) from **mainyu(v)-asah-*; hence **ham-yu(v)-ā-*, to *yu-* 'to join', is a double yoke intended for a συζυγία, a pair of draught animals.

125². Gdn. already recognized that *simā-* belongs to NPers. *sīm* and Ved. *śámyā*; but in translating it by 'jochbalken' he paid insufficient attention to the meaning of these cognates, which is 'pin of a yoke', similar to Arm. *sami-k̈* 'legni curvi uniti al giogo, che si pongono intorno al collo di bue', see Hübschmann, *Arm. Gr.* 488, and *Pers. St.* 79. Beside *simā-* we have Av. *səmī-* in the compound restored by Reichelt (*Wörter und Sachen*, XII, 288) as *yuyō.səmi* ~ OInd. *yugaśamyá* 'yoke together with the pin', cf. also Duch., *Comp.* 45 sq., and Benv., *J As.* 1938, 533. From dialects Morg. has added Munji *sām* 'yoke-peg' and Sistānī *simāk* 'yoke-key', see *IIFL*, II, 247. From the IE point of view the relation between Av. *simā-*, etc., Gr. κάμαξ on the one hand, and Av. °*səmi* (< **sami*), Ved. *śámyā*, etc., on the other, may be explained by an alternation **k̂Hm̥-* : **k̂Hém-*. Bth.'s interpretation of *simā-* as 'Halsring, d. i. ein vom Joch ausgehendes, den Hals des Pferdes umfassendes Rundholz' is not satisfactory, since what closes the 'ring' beneath the horse's neck appears to be the *simōiθrā-*.

125³. On *simōiθrā-* again there is no agreement. Gdn. has 'horse-collar', Bth. and Lo. 'yoke', Justi 'central part of the yoke' (*Handbuch*); Hzf., 460, vaguely rendered *simąmča simōiθrąmča* by 'yoke and what belongs to it'. As *simōiθrā-* obviously contains *simā-*, its relation to the latter is quite likely that of Arm. *sameti-k̔* to *sami-k̔*. *sameti-k̔* is the strap which passing underneath the neck of the animal connects the two yoke-pegs. In Hübschmann's opinion the word represents **sami-a-ti*, *-a-* being the compound vowel, and *ti* belonging to Gr. δέω 'to bind'. We may similarly analyse *simōiθrā-* as consisting of *simā* + *-it-* (reduced grade of *yat-*, which in OInd. occurs with the meaning 'to join, connect') + suffix *-rā-*.

125⁴. *dərəta* has been taken since Wi.'s days as qualifying *aka* and meaning 'split'. But the idea that the inhabitants of Aryana Vaējah should have split hooks to make brackets inspires little confidence. Lo.'s divergent view that *dərəta* means 'solid' is therefore preferable, whether the adj. refers to *aka*, as Lo. thought, or to *upairispātā*, as we think. Etymologically *dərəta* may then belong to Av. ³*dar-*, OInd. *dhar-*, Lat. *firmus*, to which Hzf., *Ap. I.* 136, assigned OPers. *daršam* 'strongly'; perhaps Av. *daršiš vātō* 'strong wind' and *darši.dru-* 'having a strong club' should be added.

125⁵. In *upairispātā* already Justi, *Handbuch*, saw a substantive. The form he considered to be loc. of °*spāiti-*, the meaning, 'thongs by which horses are connected with the shaft', lit. 'Ueberwurf', to *spā-* 'to throw'. Da. considered the stem of the substantive, which he did not translate, to be °*spāt-*. For Hzf., 460, the word meant 'with ends unbent' or, as subst. dual, the 'wither-forks'. Other translators saw in the word an epithet of *aka*, meaning 'dick' (Wi.), 'darübergeworfen' (Spi., Bth., Lo.), or 'von oben durchgesteckt' (Gdn.). We, at this stage, require a word meaning 'shaft'; the plough-shaft is called in Yidγa *åwusp*, in Wx. *wašp*, in Sanglechi *āwišp*, all of which Morg., *IIFL*, II, 194, 383, 550, has attractively connected with *upairi-spā-* (assuming a difference of preverb). To this group of words also belong NPers. *farasp* 'roof-beam', see Henning, *BSOAS*, XII, 315, n. 2, and Yidγa *frāspïy* 'rafters'. The base, distinct from that of Av. *fraspāt-* (see note 30²), may be *spā-* 'iacere', as Justi and Bth. thought; it is here used, perhaps, in an 'architectural' sense, as Lat. *iacere* is in *proiectura, obex, obiex*.

If we compare Morg.'s drawing of a Pamir plough-yoke (Appendix to *IIFL*, II) with the description in our stanza of the yoke to which Mithra's horses are harnessed, the only difference we shall find is that the metal hook by which in our stanza the shaft is fixed to the yoke has been replaced in the Pamir region by a rope.

(1) *ham̐.ivā-* 'complex yoke'.
(2) *simā-* 'yoke-pin'.
(3) *simōiθrā-* 'yoke-strap'.
(4) *upairispāt-* 'shaft, pole'.
(5) *aka-* 'hook'.

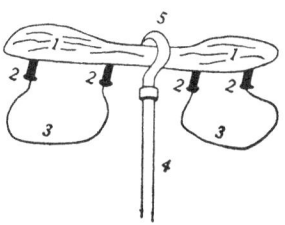

125⁶. *xšaθrəm vairīm* 'metal' is in apposition to *aka*. As Gdn. put it, 'der accusativ *xš° v°* vertritt hier die stelle eines nicht bildbaren adjektivs'.

126¹. On Rašnu flying this time on Mithra's right side (as against st. 100), see Introduction, p. 39. *razištō*, which I understand literally as 'very straight (=upright)', is taken by most interpreters in the sense of 'most just'. Lo. has 'der ganz genau abwägende Rašnu'. For *upa.-raoδištō* there is the choice between 'der am besten abwehrende' (Bth., Lo., and Htl., to ³*raod-* 'to check'; Gdn., *Lesebuch*, has 'der bedrängendste(?)'), and 'der aufgewachsenste' (Wi., and everybody else). In view of Rašnu's epithet *bərəzō* 'tall' in st. 100 we should not hesitate to follow the latter view.

126². I take *vaŋhaiti* as nom. sg. fem. of the pres. partc. of *vah-*. When through some misunderstanding the subject *razišta čista* and the following two epithets were changed into accusatives, *vaŋhaiti* and *spaēta* escaped alteration because their connection with the subject was obscured by the intervening words *spaēta vastrå*, which are governed by *vaŋhaiti*. Wo. has: 'weiße Kleider hat die weiße an', as a parenthesis.

126³. *razištam čistam...d° m° upamanəm* 'die gerechteste Unterweisung...: ein Gleichniß der mazdayasnischen Lehre' (Wi.). With Gdn. an 'and' crept in, which was a serious step backward from the recognition of the meaning and function of *upamana-* which Wi. had nearly reached: 'die...gerechte...*und* der aufseher über den Mazdaglauben', 'Cista...*et* l'Imprécation de la Religion' (Da., followed by Lo.), 'Čistā...*und* der Upamana der...Religion' (Wo.). Differently, but not better, Htl.: 'in weiß leuchtende Kleider, die der weißleuchtenden mazdayasnischen Daēnā, ist sie gekleidet—, (fährt) der Upamana'. On Razišta Čistā as Double of the Religion see note 9⁴ above.

127¹. *upa-vaz-* elsewhere means 'to drive or fly *towards*'; hence Wo. rightly translated 'herzugefahren kam', Lo. 'herbei fuhr', Spi. 'herzufährt'. Da.'s 'près de lui va' and Htl.'s 'es fuhr dabei' are imprecise. If one considers from where Dāmōiš Upamana may have come 'herzu', it seems less than likely that he travelled towards Mithra from the opposite direction, or from a direction above or below, or right or left, of Mithra; as Wi. ('nach fährt') and Gdn. ('hinterdrein fährt') saw, the Upamana of Dāmi would most probably come up to the god from behind. Such interpretation is, of course, in harmony with the statement in st. 68 that Dāmōiš Upamana 'releases Mithra's chariot well-released', and helps to prevent confusion between Dāmōiš Upamana and Vərəθraγna (see above, p. 168).

127². *yūxδa-* 'dexterous', with Bth.; Spi., followed by most other translators, has 'gerüstet, streitbar', Da. 'rapide'. On *yūxδahe pāirivāzahe* as a possible hendiadys see note 47¹.

127³. *pāirivāza-*: 'bis zum Ende fahrend' (Wi.); 'an den Seiten herumfahrend' (Spi.); 'qui s'élance de tous côtés' (Da.); 'überrennend' (Bth.). My translation is prompted by the meaning of NPers. *parvāz* 'flight, leaping'.

127⁴. Wi., reading *nixšta*, guessed 'zunächst' or 'zuniedrigst'. Justi, *Handbuch*, proposed the derivation from *ni* + *stā*, and translated 'außerhalb'. This was accepted by Spi.; Gdn. translated 'aus ihm heraus', and commented 'gehört jedenfalls zu skr. *niṣṭhā*'. Da. has 'derrière lui' as a 'pure conjecture'. Bth. relied on the reading *nixšata*, which he translated 'niederwärts von', adding the comment: 'die Bildung ist undeutlich; *nixš zu *nī* wohl wie gr. πέριξ zu περί; vgl.... *nyânk-*'. Lo. and Htl. follow Bth.

Perhaps some progress can be made with the help of Oss. *nix* 'forehead, front, finger-nail, claw', which Miller connected with Ved. *nakhá*, Gr. ὄνυξ, etc. This is one of the few cases where OIr. *a* is represented by (Iron) *i*, (Digor) *i*, before a consonant other than *n* (see Miller, *Ossetisch*, 18). It is noteworthy that Ossetic has a verb *nixin* (Dig. *nixun*) 'to scratch', which *may* be a denominative of *nix* 'finger-nail', but has a past stem *nixt-*, as if from OIr. **nixt-* (or, just possibly, **naxt-*). Within Iranian *nix* stands out as the only representative of the word for 'finger-nail' which, admitting the change of *a* to *i*, coincides with Ved. *nakhá*. Elsewhere in Iranian we find forms that go back to either OIr. **naxar-*,† or **naxaka-*,‡ or **nāxan-*.§

It will be seen that Oss. *nix* 'finger-nail' is sufficiently unusual to make it inadvisable to charge it with the further oddity of having developed the meaning 'forehead, front'. Clearly two different words have coalesced in Iron *nix*. Digor *nix* only means 'finger-nail', but both dialects preserve the 'front' word in Iron *tærnix*, Digor *tærnix*, 'forehead'.|| On the other hand, Digor is alone in having a word *nixagæ*, *nixæg*, 'curl on the forehead or temple'.¶ In Iron the external locative *nixmæ* is used adverbially in the sense of 'facing, against'.

† Sanglechi *narxax*, Wx. *dogər*, etc., see Morg., *NTS*, I, 62; cf. the OInd. adj. *nakhará*. ‡ Orm. *naxk*, Pš. *nūk*, see Morg., *NTS*, v, 25.
§ With prosodic metathesis from **naxān-*, strong stem of **naxan-*? Cf. the pattern of Ved. *pāvaká*: Parth. *pw'g*, Bth., *ZWb.* 97, n. 4. The examples are NPers. *nāxun*, Sogd. *n'yn*, and, perhaps, Yd. *anaxno* (with secondary shortening, or from **naxan-*?).
|| *tær-* is from **tāra-*, with the shortening in initial compound position which is found in *ævd-særon* 'seven (*avd*)-headed', *ærdxord* 'confederate' (cf. *BSOAS*, XVII, 484, n. 2), and a few other cases, see Abayev, *ap.* A. M. Kasayev, Осетинско-русский словарь, p. 528. **tāra-* is attested in Sogd. *t'r* 'forehead', Khot. *ttāra* 'forehead'. In the Sogd. *Dhyāna*-text, 214 *yry t'r* is 'mountain-top', hence Oss. *tær-nix* could also be taken as literally meaning 'top-front'.
¶ *-æg* in *nixæg* may go back to *aka-* 'hook', attested in st. 125 of our Hymn. *nixagæ* would then be a case of lengthening in final compound position, cf. *raidzast* 'bright, cheering' (with *cæst* 'eye'), and other examples quoted by Abayev, *loc. cit.*

nix 'front, forehead' may indirectly be connected with Av. *ainika-* 'face, front', OInd. *ánīka*, with shortened *i* as in *æxsîr* 'milk' against Ved. *kṣīrá*. It is agreed† that *ánīka* is from IE **eni + *ok^u-* 'eye', and that the IE prepositions **eni* and **ni* are related, cf. Pokorny, *IEW*, 311. Oss. *nix* might be from OIr. **anī°*, but if, as I believe, Av. *nixš(a)ta* is connected, the Oss. *n-* must have been initial from the beginning. The Oss. spirant may belong to a variant **ax-* of OIr. **ak-* 'eye', which originated in direct contact with certain consonants; thus the nom. sg. corresponding to Gr. ὤψ would be **āxš* in OIr., and the IE gen. **ok^unós* (cf. Hübschmann, *Arm. Gr.* 413) would become OIr. **axnah*.

In Av. *nixš(a)ta*, *x* is conditioned by the following *š* < IE *s*. We may start either from **nīk-* < **ni-ak-*, parallel to *anīk-* < **ani-ak-*, OInd. *prátīk(a)* < **prati-ak-*, or from **nīxš-* < **ni-axš-*, cf. Av. *aiwyāxš-*, Bth., Wb. 310 sq., which has secondary *ā* after *y*. If the correct reading is *nixšta* we may assume a compound **nīk-sta-* or **nīxš-sta-* 'standing = being in front', treated as a thematic adj. like Skt. *niṣṭha*; its instr. sg. could be used as an adverb (cf. Reichelt, *Aw. Eb.* §455) governing the ablative. If the correct reading is *nixšata*, this may similarly be an adverbial instrumental, this time of **nīk-hant-* (IIr. **nīk-sant-*) or **nīxš-hant-* 'being in front'; such a formation would be parallel to **upa-hant-* 'being (intent) on', the theoretical participial adj. to the Av. noun *upaṇha-* 'being (intent) on', < *upa- + *aṇha-*.

127⁵. Excepting Wi., Spi., and Htl., all translators insert a gratuitous 'and' between *ātarš* and *x^varənō*, as most of them had done between *čistąm* and *upamanəm* in st. 126 (see note 126³). Spi., too ingeniously, translated 'das Feuer, welches entzündet hat den starken königlichen Glanz (Majestät)'. Htl. correctly has 'der Ātar, der entflammte, welcher das mächtige königliche Xvarənah ist', but failed to realize the import of this identification; similarly Wi., who states that *yō* after *uγrəm* is 'disturbing'. In fact it is perfectly correct: by using the masc. pronoun, the author shows that he meant to identify the masc. *ātarš* with the neuter *x^varənō*.

The literal interpretation of the last two lines thus produces for the first time an Avestan reference‡ to the sacred fire which in Sasanian times was known under the name of *Ātur-farn-bag* or *Ātur-x^varrah*. 'The blazing Fire which is the strong Kavyan Fortune' is a terminology the author of our stanza would scarcely have used unless such a fire were an established object of worship at the time he was writing. That the chief sacred fire of later Zoroastrianism should have existed among Zoroastrians of the Achaemenian period was only to be expected, but no proof was thought to be available. [The proof is still missing, as the explanation

---

† Though Wn.–Debr., II, 2, 520, have a different explanation: a preposition corresponding to Gr. ἀνά, and a suffix *-īka-* abstracted from *abhíka, prátīka*.

‡ Apart from a possible allusion in the proper name *ātərəx^varənah-*, *Yt* 13.102.

offered above, p. 61, must be withdrawn: the Sasanian name of the fire contains °bāg 'share', not °bag 'god', cf. J. Harmatta, *Bull. du Musée hongrois des beaux-arts*, 1957, 9, and the Syriac and Arabic spellings with ā in G. Hoffmann, *Märtyrerakten*, 282 sqq. (who was probably right in assuming that the fire was named after a person).]

128[1]. Both here and in st. 129 two otherwise unknown words, θanvarə(i)tinąm (or θanvarətanąm, as even the scribe of F_1 first wrote), and srvī.stayąm respectively, are followed by what are generally held to be glosses. In the present passage Wo. translates: '...ein Tausend wohlgefertigter Bogen. ⟨Es kommt vor, (daß) die aus einer Tiersehne gefertigte Bogensehne von einem Gavasna (-Tier?) stammt⟩', in st. 129: '...Pfeile...mit hörnernem Widerhaken....⟨Es kommt vor, daß eiserne Sprossen (dran) sind⟩'. Lo. has, respectively, '...Bogen; (die Sehne der gutgemachten ist aus einer Darmsehne von...?)' and '...Pfeile, ...die hörnerne Widerhaken(?) haben. (Der Schaft der gutgemachten ist aus Eisen[?])'.

Both translators evidently regard yō and yā respectively as articles (cf. above, p. 154, n. †), yet assume that a following genitive or adjective is the predicate, respectively, of jya and sparəya. It seems incredible that asti yō...snāvya jya should mean 'the bow-string is sinewy', and not 'it (viz. the just-mentioned object) is the (yō) sinewy bow-string'. It is presumably with this consideration in mind that Htl. translated both glosses by 'was...bedeutet'; but he did not use our gloss for the purpose for which no doubt it was intended, viz. to explain the preceding θanvarə(i)tinąm (or θanvarətanąm). This word, the gloss tells us, means 'bow-string'; it is therefore a substantivized adjective θanvar-tan- 'stretching, i.e. bending, the bow'. Its masc. and neuter gen. pl. is represented by θanvar-tanąm, its fem. gen. pl. by θanvar-tinąm (which is what θanvarə(i)tinąm stands for). The latter is a haplological reduction of *θanvar-taninąm (fem. stem °tanī-). The fem. probably gained currency as an epithet of the fem. jyā-. The substantivized masc., it would appear, was used as a synonym of jyā. Either variant is justifiable, but the masc. θanvarətanąm has a better claim to stand in the text, as its gender may be the reason why instead of yā the masc. yō was used in the gloss.†

The above interpretation has the result that, contrary to Bth.'s expectation, we do not find Mithra's bow mentioned in the present enumeration (sts. 128–32) of his weapons. But the absence of the bow is only surprising if we disregard the fact that the weapons enumerated all lie on his chariot, including the mace (st. 132) which earlier (st. 96) the god was holding in

---

† The existence of the variant θanvarə(i)tinąm, which points to an original adjectival use of the compound, precludes what would otherwise have been an obvious alternative interpretation of θanvarətanąm: a tatpuruṣa compound of θanvar- and *tan(a)-, in which the latter corresponds to Oss. tæn 'string (of a musical instrument)'.

his hand. Clearly, then, at the moment the poet is here considering Mithra is free to hold the bow in his hand or carry it on his shoulder, as he sometimes does on Mithraic reliefs (cf. Cumont, *TMMM*, I, 183). The usual interpretation of *hazaṇrəm θanvarəitinąm* as 'a thousand bows' was in any case never convincing: why should Mithra clutter up his chariot with so many bows when he only needs one? He does not, for instance, take along more than one mace, obviously because, contrary to what happens with his clubs (st. 131), he never lets go the mace when striking it against his enemies (cf. Bth.'s note in *Wb. s.v. gaδā-*). It is arms that are *thrown*, and thereby lost, which he needs in large numbers: arrows, spears, knives, etc. In addition, we now learn, he requires a generous supply of bow-strings, as they would frequently snap with the constant use of the bow in the course of a long punitive expedition.

128². *gavasna-* was recognized by Bailey, cf. *BSOS*, VII, 69; VIII, 123, as belonging to NPers. *gavazn*, Khot. *ggūysna-*, etc.

128³. The last three lines are thought to have been added by an undiscerning scribe who wanted the stanza to end like sts. 129-31. However, in the case of discarded bits of snapped bow-string one might argue that they *could* 'fly and fall on the evil heads of the daēvas', and cause some havoc in view of their divine origin. Similarly in st. 132, where one may feel tempted to suppress the identical formula, I have chosen to retain it, on the assumption that the original *vazaiti and *pataiti were changed to plurals under the influence of the stereotyped wording of the preceding stanzas.

129¹. 'Die Spitze des Pfeils...wird auch sein Mund genannt, weil sie das Blut der Feinde trinkt', W. Geiger aptly remarked, *OK*, 447, referring to Av. *zaranyō.zafar-* and RV *yásyā áyo múkham* '(arrow) whose mouth is iron'. The same image is applied to the blade of the sword in Hebrew *hkh lpy ḥrb* 'to strike with the mouth of the sword', and *ḥrb pypywt* 'two-mouthed sword' Psalm cxlix.6; cf. Sogd. *d'r* 'blade': Syr. *pwm* 'mouth', above, note 96¹. Jackson's translation 'golden-notched' (*Classical Studies in honour of Henry Drisler*, 112), adopted by Lo. and Hzf., should be abandoned; so, too, Gdn.'s interpretation of °*zafar-* as 'neck or ring of the arrow's point' (based on NPers. *zurfīn*, *zufrīn* 'ring for holding the bolt of a door').

129². The likely meaning of *sparəya-*, viz. 'sprout' (cf., on the base, Henning, *BSOAS*, XII, 46 *sq.*), supports the common translation of °*stī-*, of which *sparəya-* is a gloss, by 'barb', lit. 'that which stands up'. We need not therefore think with Wi. and Spi. of the 'shaft' of the arrow, or with Gdn. of its 'point'. From our interpretation of the words *asti yā aṇhaēna sparəya* as a gloss to the preceding *srvī.stayąm* (cf. note 128¹) there follows that not only *sparəya-* and °*stī-*, but also *aṇhaēna* and *srvī*° should be synonymous, or nearly synonymous. The old explanation of *srvī.stayąm* as 'lead-balanced' or 'lead-based' (cf. Jackson, *loc. cit.*, Hzf.,

COMMENTARY   [129.2–136.1

Ap. I. 299, *Zor.* II, 435) breaks down over the *ī* of *srvī*-. The stem of the word for 'lead' is thematic both in *srum* < *srvəm and in the compound *srvō.zana-*, which Henning attractively translates 'with leaden jaws' (*Sogdica*, 50); such a stem offers no room for an ending *ī*. It is therefore clear that Bth. was right in identifying *srvī°* with the dual of *srū-* 'horn'; the compound means, as Duch., *Comp.* 149, aptly put it, 'ayant deux cornes pour arrêts'.

Once we feel assured of the meaning of *srvī°*, we can safely discard the emendation of *aṇhaēna* to *ayaṇhaēna* 'made of iron', which has held the field since Westergaard introduced it. For it becomes obvious that this emendation merely conceals the proof of the theory that the IE stem of the word for 'bone' is **os-*. This was urged by Meillet, *MSL*, 23, 259 *sq.*, on the strength of Lat. *os*, and taken for granted by F. Specht, *Der Ursprung der Indogermanischen Deklination*, 74: 'an die Wurzel *os-* ist ein Dentalsuffix in ai. *ásthi*, av. *ast*, heth. *ḫaštai* und griech. ὀστέον getreten, ein Gutturalsuffix in arm. *oskr*, kymr. *asgwrn*, korn. *ascorn*, ein *s*-Suffix wie in lat. *nāsus*, in lat. *os*'. If doubts continued to subsist it is because Lat. *os*, *ossis* shows an exceptional treatment, whether we derive it from **os-* or **ost-*.† Av. *aṇhaēna-* happily resolves all doubts. It can only mean 'made of **ah-*, or **aṇha-*', and this must be an alternative of *srū-*, 'horn', as the substance of which arrow-barbs are made. If, as is perhaps more likely, the stem of the noun was the unthematic **ah-*, it would be understandable if for lack of distinctiveness (nom. sg. *ō, **ə̄*, or **ə̊*!) it had been ousted by its doublet *ast-*. The adjective *aṇhaēna-*, no longer supported by the noun from which it had been derived, would then be gradually replaced by *astaēna-*, which was derived from *ast-*, presumably at a later period.

The dual ending which Bth. recognized in *srvī°* makes it clear that the words *yā aṇhaēna sparəya* are all three in the dual.

132¹. Cf. stanza 96.

134¹, 135¹. Cf. stanzas 97 *sq.*

136¹. The line *aēva čaxra zaranaēna* is generally thought to consist of three words, 'with one golden wheel'. As the next line contains no substantival instrumental which °*ča* might connect with *čaxra*, Lo. is forced to translate against all grammar: 'dem weiße Rosse den angeschirrten Wagen (instr. sg.) mit dem einen goldenen Rad ziehen, und ganz leuch-

† The evidence of Luvian *ḫašša* 'bone' cannot yet be evaluated with certainty, cf. H. Otten, *Zur grammatikalischen und lexikalischen Bestimmung des Luvischen* (*Veröffentlichungen des Instituts f. Orientforschung*, vol. 19), 64 (reference kindly supplied by H. W. Bailey). Wn.–Debr., II, 2, 722, still hold that *-thi* in OInd. *ásthi* 'bone' and *sákthi* 'thigh' (~ Av. *haxti-*) is not likely to be a suffix. Avestan not only proves the correctness of Meillet's analysis of the former as *as-thi*, but also supports the analysis of the latter as *sak-thi*, in view of the probable relation of Av. *hax-ti-* 'thigh' and *hax-a-* 'hollow of foot between heel and toes', see *BSOAS*, XIV, 488, n. 2.

281

tende Schleudersteine (acc.)'. Wo. resorted to the makeshift of implying a second, transitive, verb: 'dem die... Renner mittelst des einen goldnen Rades am geschirrten Wagen ziehen und die allglänzenden Schleudersteine (fahren)'. Other translators tacitly or explicitly emend or disregard the endings. To solve the difficulty we must recognize a bahuvrīhi in *aēva.čaxra*, thus setting free *zaranaēna* to act as second epithet of *vāša*, *vīspō.bāma* being the third. *yūxta*, which is generally held to agree with *vāša*, more likely governs it and agrees with *aurvanta*. Thus the irregularity of the otherwise transitive *θanj-* here taking its object in the instr. ('pull *at* the chariot'), disappears; the acc. *vāšəm is implied: 'runners yoked to the chariot pull (it)'. On the one-wheeled chariot see Introduction, pp. 35 *sq.* [See Addenda.]

136². The stem **bāma-* 'splendour, radiance' occurs in Av. only in *vīspō.bāma-* and *bāmaniva-* 'radiant'. The latter adj. was described by Bth. as a derivative of a stem **bāman-*, equivalent to **bāma-*. A different view was jotted down by Scheftelowitz in the margin of a second-hand book in my possession, according to which *bāmaniva-* means 'lichtähnlich', and corresponds to Skt. *bhāma-nibha-*. This suggestion, for which Scheftelowitz adduced no reason, deserves to be retained and developed, as it is supported by two independent equations. On the one hand, as Bth. explained *s.v.*, *bāmaniva-*, which is an epithet of clothes (*vastrā́sča kəšā́*† *bāmanivā́*), is rendered in Pahl. by *bāmīk tarāz* 'of splendid form'; on the other hand, the obvious etymological connection of Oss. (Iron) *niv* 'form, pattern, model, shape, picture' is with Skt. *nibha* 'appearance' (from *'shining', cf. Germ. *Schein*). The meaning of Pahl. (Arm., etc.) *tarāz* 'form, fashion, manner, way' assures us that Av. °*niva-* is identical with Oss. *niv*, hence is from **niba-* and corresponds to Skt. *nibha-*.

In Ossetic there is a complication in that Iron *niv* also means 'luck', which is the only meaning of Digor *nivæ* according to Miller-Freiman's Dictionary and Abayev, Осетинский язык и фольклор, I, 484. It therefore seems that Miller was mistaken when in Осетинские этюды, II, 32, he quoted both Digor *nivæ* and Iron *niv* in the meaning 'shape, appearance'. Moreover, the Digor form of the Iron adj. *nivdžin* 'shape-; lucky', is *nivgun*, which only means 'lucky'. For *nivæ/niv* 'luck' Miller, *op. cit.* II, 83, proposed a derivation from OPers. *naiba-* 'beautiful'. I should prefer to start from **hu-niba-* 'good shape (of circumstances), *or* good appearance = good omen'. The initial syllable **hu-* would be lost as is the *u-* of the verbal forms prefixed by *upa-*, or *us-/uz-*, cf. Miller, *Ossetisch*, 19.

nivæ is also found in the Digor compound *nivahæ* 'doom', whose second part is *ahæ* 'misfortune' (cf. Abayev, *op. cit.* 459, 439; neither *nivahæ* nor *ahæ* is to be found in Miller-Freiman's Dictionary). The

† On *kəša-* cf. Bailey, *Zor. Prob.* 8, n. 2.

literal meaning is 'misfortune of fortune'; for *ahæ* cf. Av. *aγa-*. Abayev's tentative comparison of Digor *nivesinæ* 'picture', *fes* 'luck', and Iron *niv* 'picture, luck', has only a bearing on the meaning, not on the etymology of these words. *nivesinæ* may represent **ni-paisanya-*; *fes* conceivably belongs to OInd. *páyate, pyáyate* 'to overflow, be exuberant', having developed from a verbal noun **fy-ā-sa-* based on an inchoative present stem comparable to OPers. *xšnāsa-* 'to know', MPers. *wygr's-* 'to wake up', Sogd. δm's- 'to blow', xw's- 'to tire'.

136³. For *asānas(ča)* Bth. and Lo. have 'sling-stones', Wi. and Spi. 'spokes', Gdn. 'axle', Da. 'hub', Hzf., 463, 'jewels', Htl. simply 'stones'. I side with Hzf., but take *asānas°* as acc. of relation depending on the instr. *vīspō.bāma*. Hzf. understood both words as being in the nom. pl.: 'and the jewels are containing-all-light'.

In support of the unusual meaning here assumed for *asan-* (normally 'stone') the following arguments can be adduced, none of which, admittedly, is decisive:

(1) In *RV*, v, 47.3 *áśman* seems to have the same meaning: *mádhye divó níhitaḥ pŕśnir aśmā ví cakrame rájasas pāty ántau* 'the variegated (precious) stone placed in the middle of the sky (= the sun?) paces about, protects the two ends of space'.

(2) Skt. *upala* is used for both 'stone' and 'jewel'.

(3) Khosrou Anōšarvān's 'car of Fortune', *despak paraç*, as described by Sebēos, was 'of gold' (cf. *zaranaēna* in the present stanza) and 'inlaid with precious stones and pearls', see Bailey, *Zor. Prob.* 46.

(4) In st. 143 Mithra's chariot is said to be *stəhrpaēsah-* 'star-decked'. What is meant can be gathered from the Pahl. Commentary to *Y* 57.21, where Sraoša's house is said to be *stəhrpaēsəm ništara.naēmāṯ* 'star-decked on the outside'; this means, the commentator explains, that it is 'adorned on the sides, that is, it is equipped on the sides with precious stones' (the passage is quoted by Bailey, *Zor. Prob.* 16).

136⁴. [Note H. Humbach's emendation of θanjasånte to θanjayånte, *Münchener Studien zur Sprachwissenschaft*, 9 (1956), 68 sq.]

137¹. On *mainya-* see above, p. 224.

137². *aṇhəuš dahmō*, Da. 'pieux du monde', Lo. 'ein weltweiser', Htl. 'für das (ewige) Leben, ein orthodoxer'; Spi. and Wo. connected *aṇhəuš* with *zaota ašava*: 'ein reiner Priester in der Welt', 'der ašagläubige Zaotar der Menschheit'.

137³. *miθrahe vāxš* seems to be an abbreviation of **raθwya vača miθrahe aoxtō.nāma yasnō* 'the prayer in which Mithra's name is mentioned with regular utterance', see sts. 30 *sq*.

137⁴. Quite differently other translators. Bth.–Wo., followed by Lo., took *anu.sastrāi* and *anu.mainyāi* as infinitives, and translated 'wenn er [der Mann] um seiner [Miθra's] Gunst willen seinen Befehl zur Ausführung bringt und seinen Befehl befolgt'. Benv., *Inf.* 68, tends to accept the infinitives, but considers the phrase 'de syntaxe très discutable'. Da. has 's'il en advient avec faveurs à lui instruit selon l'instruction, (à lui pensant selon l'instruction)'. Htl. translates 'wenn um seiner Gabe willen die Lehre zum Nachlehren, die Lehre zum Nachdenken wird', and comments: 'wenn erreicht ist, daß man der Lehre von Miθra entsprechend lehrt und denkt'.

What in my opinion has prevented the understanding of the last three lines is the failure to seek in them the previously mentioned *zaotar-* and his *miθrahe vāxš*. It is well known that 'the *hótṛ* and his assistants uttered verses of the Rig-Veda in a semi-musical recitative (*śastra*)' (L. D. Barnett, *Antiquities of India*, 155). In the present context an obscure *sastra-* mentioned soon after a *zaotar-* engaged in recitation, may be expected to have a meaning similar to that of OInd. *śastrá*. This at once suggests that *anu* is here a postposition governing the accusative. *manya-* is then a verbal noun derived from the Present stem *manya-* 'to think'; as it is contrasted with *sastra-* 'recitation (of prayers)' it probably means 'thinking (of prayers)'; the contrast is the same as between the *vačō.-marətanąm* '(prayers) recited aloud' and the *manō.marətanąm* '(prayers) recited in thought' of N 22. The subject of *bavaiti* will then be the *miθrahe vāxš*, the utterance of Mithra's name in prayer, which can take place both orally and mentally. The 'favour' (*yāna-*) which produces correct utterance on the part of the priest is evidently the salary he receives from the Truth-owning employer. It is not difficult to perceive the reason which prompted Zoroastrian priests to proclaim that Ahura Mazdāh had made the announcement quoted in st. 137. The stanza affords a glimpse of the organization of religious service in Aryana Vaējah, see Introduction, p. 17.

138[1]. *pərənəmča...darəyəmča...*: 'ein zu vollen Barəsman spreitend, und den Yasna zu sehr in die Länge ziehend' (Wo.); 'der...auch (noch so) viel...ausbreitet, auch (noch so) langes...darbringt' (Lo.); 'indem er ein volles...und ein langes...' (Htl.). The rendering I have adopted is Wi.'s.

The opportunity may be taken to revive Gdn.'s view that *pərətō* in the oft-quoted Gāthic stanza *Y* 51.12 belongs to *pərəna-* 'full'. The text, which presents several difficulties, is as follows:

*nōit tā.īm xšnāuš vaēpyō kəvīnō pərətō zəmō
zaraθuštrəm spitāməm hyat ahmī urūraost aštō
hyat hōi īm čaratasčā aodərəščā zōišənū vāzā*

'For this (reason) not pleasing-to him (was) the kavi's wanton at the gate of winter, to Zarathushtra Spitāma, (namely) that he prevented (his) stopping in that (place) when his two-horses both came to it and (were) shivering with cold' (Maria W. Smith). 'Le mignon du prince-sorcier, au Passage de l'hiver, a offensé Z.S. en lui refusant l'étape, à lui et à ses bêtes de trait grelottantes de froid en arrivant chez lui' (Duch., *Zor.* 277).

'Not listened to him, then, the Vēfiyo, the kāvian, in (month) *prta.-zimō* ["where the winter unfolds itself" = "December–January"], to him, Z.S., when he denied him hospitality, him and his attendants and the two-horse team that shivered with cold' (Hzf., *Zor.* I, 211).

Bth.'s interpretation of *aštō* as infinitive to *as-* 'to reach' was rightly rejected by Benv., *Inf.* 39, in whose opinion *aštō* is the acc. of **aštah-* 'lieu d'arrêt, halte', a noun he would also recognize in *vouru.ašta-* (on which see note 44[4] above). Benv. translated *urūraost aštō* 'il interdit la halte', but did not say what he thought of *ahmī*. That this loc. should stand for the dat., as Duch.'s rendering implies, is manifestly impossible. Bth.'s 'bei ihm' (meaning *chez soi*) (see *Wb.* 6, line 8 from bottom) is equally unbelievable. As to Hzf.'s 'him', it is based on an emendation of *ahmī* to *ā him*. Only Maria W. Smith's 'in that (place)' has a chance of being correct, since in Ossetic Digor *ami* < *ahmi* (see Miller, *Ossetisch*, 85) means 'here'. In Avestan, however, such use of *ahmi* would be exceptional (it is not recognized anywhere by Bth.) and one hesitates to admit it in the present passage, seeing that *aštō* may very well be a *-ti-* noun in the loc. with which *ahmī* agrees. From the point of view of word-formation a stem **ašti-* is more easily justified than **aštah-*, cf. OInd. *aṣṭi* 'reaching', against the rare appearance of the OInd. suffix *-tas-* (Wn.–Debr., II, 2, 615 *sq.*). The meaning will be 'point of reaching, place of arrival, goal', and *ahmī aštō* is either 'at this goal', or, more likely, 'at *his* goal', as we find *ahmi nmāne* 'in his house', etc., in YAv. (cf. Bth., *Wb.* 4 a).

In the case of *parətō zəmō* we must, I think, aim at a similar result, and strike a compromise between Bth.'s 'Gate of Winter' (as a place-name) and Hzf.'s 'December–January' (justified by the 'shivering' draught-horses).† It is enough to take *parətō* as the loc. of **parəti-* (cf. OInd. *pūrti* 'filling, completion'), instead of *parətu-* as Bth. had done, to obtain what is needed: 'in the fulness of winter'. This is what Gdn. meant in *KZ*, 30, 524, with his translation 'im härtesten Winter'. Bth. objected (*IF*, I, 191, n. 1) that *-ō* occurs as the loc. sg. ending of *-i-* stems only a few times in YAv., the usual ending being *-a*, which alone is attested in the Gāthās. This objection is invalid. As the alternation *-ā:au* in the loc. sg. ending of *-i-* stems is common in Vedic (cf. Wn.–Debr., III, 152 *sq.*), the YAv. cases of *-ō* (cf. *garō Vend.* 21.5, *haθra.jatō Vend.* 9.56, *azgatō* above, p. 228, n. ‡, *raptō* and *pairi.taxtō*, see note 45[2], *frayrātō*, see below, p. 293) are survivals, not innovations. We must therefore be prepared to find this ending in the older language, and not refuse on principle to recognize it even where it offers a better understanding than any alternative explanation.

In the last line Hzf. has the merit of having dispelled the nightmares

---

† The interpretation of *parətō.zəmō* as a compound meaning 'der weites Land besitzt' (Lo., *NGWG*, 1935, 159) fails to account for *parətō*.

which the two °čā-s had caused previous translators, cf. Bth., *Wb.* 575, n. 5. He took *īm čaratas°* as 'die ihn Begleitenden', (ἀμφί)πολοι. I would go further, and bring in also *hōi* (whose correctness Hzf. needlessly doubted) as a possessive dat. referring to both *čaratas°* and *vāzā*: 'sowohl (°čā) seine (*hōi*) ihn (*īm*) Begleitenden, als auch (°čā) (seine) beiden vor Kälte zitternden Zugtiere'. Accordingly, the whole stanza can be translated as follows:

'Not did the hymn-monger's (see note 34⁵) catamite please him (viz. Zarathuštra) in the fulness of winter, when he obstructed Zarathuštra Spitama at his (*scil.* Z.'s) goal, when both (°čā) his attendants (he obstructed), and his two draught-horses shivering with cold.'

139¹. As Bth. remarked in *Arische Forschungen*, III, 26, n. 1, the use of *anye* 'other' does not imply that the author counted Ahura Mazdāh among the Incremental Immortals. Cf. note 6⁴, last para., and Introduction, p. 12, n. †.

139². Arštāt is exclusively associated with Mithra in *Yt* 13.18 (quoted note 109⁶, p. 259), and accompanies him, together with the Religion, Rašnu, and Pārəndi, in *Visp.* 7.2 (see note 66¹), cf. also *Yt* 11.16, 21. The goddess can be identified by comparing the meaning of OPers. *arštā-*, which is thought to be a haplological shortening of **arštatā-*, as Av. *arštāt-* of **arštatāt-*. It occurs once in *Beh.* IV, 64 *upariy arštām upari-yāyam* 'I walked on (*or* in, = according to) *arštā-*', where the Akkadian word corresponding to *arštām* is *di-na-a-tú* 'laws', which elsewhere translates OPers. *dāta-* 'law', see Hzf., *Ap. I.* 285 *sq.* It is, however, unlikely that *arštā(t)-* is a mere synonym of *dāta-*; since *-tā-* and *-tāt-* are abstract suffixes, the evident meaning to conjecture is 'justice'. This conclusion meets a legitimate expectation arising from the nature of the Avestan vocabulary: it would be surprising if the word for 'justice' were absent in a collection of texts so rich in moral abstractions and their personifications as the Avesta.

As upholder of the social contract, Justice is a natural ally of Mithra's. In later Zoroastrian practice she is propitiated together with Rašnu, who is, perhaps, the 'Judge' (see above, p. 223). The Book of Ardā Virāf mentions her presence at the Činvat Bridge in the company of Mithra, Rašnu, Vayu, and Vərəθrayna; cf. Pavry, *Future Life*, 85 *sq.*, 103, n. 24; Da., *ZA*, II, 153 *sq.* The 26th day of the month was named after her, and an Avestan Yašt, of which unfortunately only a late substitute survives (*Yt* 18), was dedicated to her. Hzf., *loc. cit.*, failed to appreciate the vitality of the hypostasis of this virtue ('eine solche abstraction kann gar kein echtes *yasht* besessen haben'), because he did not believe what we take for granted, viz. that the virtue in question is neither more nor less than Justice: 'Die bedeutung ist εὐνομία. *Iustitia* kommt noch näher als *rectitudo*, und es ist auf dem wege zur "wahrheit"'. It seems that Hzf., though he had correctly understood the OPers. passage (*op. cit.*, p. 290, he translates 'ich wandelte nach der gerechtigkeit'), was still influenced by the earlier interpretations of Av. *arštāt-* as 'Aufrichtigkeit' (Spi., Justi, Bth., Lo.) and 'Wahrheit' (Wi.). Kent translated OPers. *arštā-* by 'rectitude'.

140[1]. *yazāi...spitama*. Since Zarathuštra is addressed we must assume, as in st. 143, that Ahura Mazdāh is the speaker, and resumes here the remarks he had made in sts. 137 *sq.* That Ahura Mazdāh worships Mithra is said of him in the third person in st. 123.

140[2]. *hvāmaržḍika-*, lit. 'well, very, merciful'; differently Gdn. ('gern sich erbarmend') and Spi. ('von selbst verzeihend'). Cf. *maržḍikāi* in st. 5.

140[3]. *amiθwa-*, according to Bth. 'unermeßlich, unvergleichlich', to *mā(y)-* 'to measure'. Htl. has 'dem Unermeßlichen (oder: nicht Gepaarten, d. i. Unvergleichlichen?)', obviously thinking of *miθwa-*, etc.; Lo. translates 'ohne gleichen', Spi. 'ohne Genossen', Da. 'incomparable', Duch. (*Comp.* 122) 'incommensurable'. Gdn. connects the word with Ved. *mī-* 'to alter', and translates 'unwandelbar'; Wi. has 'ohne Lüge', comparing OInd. *mithyā*.

140[4]. *uparō.nmāna-*, lit. 'whose house is above', namely in Paradise, see note 50[1]. Two comparable compounds occur unrecognized in *Vend.* 13.45 *sq.*, in a context dealing with dogs which has occupied us before (notes 45[2], 105[3]). In *Vend.* 13.46 we read:

(*a*) *parō pasča nmānahe yaθa vāstryō fšuyąs*
(*b*) *pasča parō nmānahe yaθa vāstryō fšuyąs*

Wo. translates 'vor (und) hinter dem Haus (ist er [*scil.* the dog]) wie der viehzüchtende Bauer, hinter (und) vor dem Haus (ist er) wie der v. B.', while Da. has 'il est le premier, il est le dernier de la maison, comme un laboureur; il est le dernier, il est le premier de la maison, comme un laboureur'. The respective Pahlavi explanations are:

(*a*) '*kā gōspand* '*hač pahast* '*bē-āyēt aš pēš* '*andar* '*ēstēt*
(*b*) '*kā gōspand* '*andar* '*ō pahast* '*šavēt aš* '*apāč ō* '*pas* '*estēt*

'when the cattle goes out of the stable he is in front of it; when the cattle goes into the stable he is back behind it'.

The Avestan text will be seen to convey exactly what the Pahlavi commentator says if *pasča.nmānahe* and *parō.nmānahe* are taken as compounds governed respectively by the preceding *parō* and *pasča*. *pasča.nmāna-* is the cattle 'which has the house behind', hence walks away from it; *parō.nmāna-* is the cattle 'which has the house in front', hence goes towards it.

In *Vend.* 13.45 only the (*a*) formula occurs: *parō pasča nmānahe yaθa raθaēštā̊* 'he is in front of what has the house behind, like a warrior'. Here the Pahl. Comm. has '*kā gōspand* '*hač pahast* '*bē-āyēt aš* '*pas* '*andar* '*ēstēt*; Da. was evidently right in proposing that we should read *pēš* instead of '*pas*.

141.1–142.2]   COMMENTARY

141¹. *aš.xraθwastəmō.* Cf. Mithra's *āsnō xratuš* in st. 107, and Introduction, p. 52, on the qualities in which Mithra is supreme. On *xᵛarəna hačimnō* see notes 67², 84¹.

142¹. Up to here I follow Bth.–Wo., as everybody has done for the whole stanza since 1905. Previous opinion differed: 'der als erster Verkünder stark mehrt des heiligen Geistes Geschöpfe...' (Wi.); 'der als der erste Verkünder das Starke fördert unter den Geschöpfen des Çpenta-mainyus...' (Spi.); 'welcher zuerst den morgen verkündend den geschöpfen des heiligen geistes bringt' (Gdn.); 'qui accroît puissamment en connaissances multiples la création du Bon Esprit...' (Da., commenting: 'son apparition illumine l'âme'). It must be admitted that, since *vaēidiš* only occurs here and *sūrəm* is ambiguous, Da.'s rendering is as plausible as Bth.'s. If the latter is to be preferred, the reason is that it provides Mithra with more convincing circumstances in which to 'light up' his body.

Following up Gdn.'s excellent suggestion Bth. made out a good case for taking *sūrəm* to mean 'in the morning'. However, he omitted an argument tentatively adduced by Gdn., which deserves to be retained, viz. that *ušåṇhəm* (or *ušąm*) *sūrąm*, occurring in *Vend.* 18.15 and *F* 27*b*, means 'the morning (adj.) dawn'. *ušå* 'dawn', which is the name of the period from midnight till sunrise, is divided into *ušå sūra*, which lasts from midnight till daybreak, and *raočaṇhąm fragatiš* ('the approach of daylight'), which lasts from daybreak till sunrise. It would seem that Gdn.'s interpretation suits the definition of *ušå sūra* better than the earlier rendering 'the strong dawn', to which Bth. reverted; one might even argue against the latter that up to daybreak dawn is at its weakest. If we link the meaning of the adverbial *sūrəm* in our passage with the definition of *ušå sūra*, we shall have an additional reason for arguing that the light Mithra brings is that which pervades the earth at daybreak, as against the later burst of light which the rising sun produces, cf. Introduction, p. 31.

vaēidi- is the noun with an eye on which Henning, *BBB*, p. 93, reconstructed an OIr. stem **vidah-* 'shape' which might serve as etymon to Sogd. *yδw*. Unfortunately neither the meaning nor the origin of the Sogd. word are sufficiently assured to serve as confirmation of the meaning Bth. had assigned to *vaēidi-*.

142². *måṇhō hvā.raoxšnō* was analysed as gen. of *māh- hvā.raoxšan-* by Bth., who translated the last two lines 'sobald er (seinen) Leib zum Leuchten bringt wie (den Leib) des eigenlichtigen Mondes', assuming that the object of the comparison is *raočayeiti*. The thought that Mithra should light up not only his own body but also the moon's, and this despite her being provided with 'own light', is incongruous. It is, moreover, wrong, as a close look at *hvā.raoxšnō* shows. The only other attested form of this adjective is *xᵛā.raoxšnəm*; no stem **raoxšan-* occurs anywhere, while *raoxšna-* is extremely common. The very strong reasons one expects to see produced before *hvā.raoxšnō* is taken as anything but the nom. sg. masc. of a them. stem *hvā.raoxšna-* have never been offered. Wi. was simply observing normal Avestan grammar and lexicology when he translated 'wenn er den Leib erleuchtet wie der Mond selbst leuchtet' (Spi., more clearly, has 'wie der von selbst leuchtende Mond leuchtet'); so did Justi, *Handbuch*, when he quoted *hvā.raoxšnō* and °*nəm* under the entry *hvā.raoxšna-* (pp. 89 and 334), and *måṇhō* as nom. sg. masc., with dat. *måṇhāi*, gen. *måṇhahe*, and voc. *måṇha*, under the entry *måṇha-*, thematic doublet of *māh-*, and equivalent of Ved. *mā́sa*, them. doublet of *mā́s* (cf. Wn.–Debr., III, 322, and Kent's interpretation of OPers. *māhyā* as gen. of *māha-*, *Old Persian*, s.v.). Bth.'s obliteration of the stem *måṇha-* and far-fetched analysis of *hvā.raoxšnō*, by which he overrode and consigned to oblivion previously reached conclusions of plain common sense, deserve to be remembered as deterrents to over-sophistication in Avestan philology.

It will have been noticed that with Bth.'s predecessors also the comparison revolved around *raočayeiti*. As *raočayeiti* governs *tanūm*, and is therefore necessarily a causative, this interpretation implies the poet's belief that the moon lights up her own body. Such a theory he may well have held, but it is not necessary to assume that he expressed it in the present stanza. I should prefer to shift the stress to *hvā.raoxšnō*, and take *it* as the pivot of the comparison: Mithra, *hvā.raoxšnō* like the moon, lights up his body. On the background of the epithet *hvā.raoxšnō*, and its reflex in Western Mithraism see Introduction, pp. 31 *sq.*, 61.

With regard to the order of words it may be noted that *hvā.raoxšnō*, as understood by us, occupies the same position in respect of *yaθa måṇhō* as *bāmya* does in st. 143 in respect of *yaθa dāmąn sraēštāiš hū*: 'self-lighted like the moon' ~ 'shining like the sun's most beautiful creature(s)'.

143[1]. *yeṇhå* seems to stand for *yeṇhe* by attraction to the gender of *tanūm*, Mithra and his body being identical, just as the gender of *kəhrpō* is responsible for *yeṇhå* (referring to the masc. Haoma) in st. 90. On Da.'s interpretation see note 143[5]. All other translators are agreed that the *ainikō* in question is Mithra's.

143[2]. The voc. *spitama* would seem to indicate that at least the line in which it occurs is spoken by Ahura Mazdāh. We were driven to a similar conclusion in regard to st. 140 (see note 140[1]), where, moreover, Mazdāh, speaking of himself, says *yazāi* 'I worship'. To penetrate the haze which surrounds the present stanza, easily the most intricate of the whole hymn, we shall do well to cling to what analogy with st. 140 can be detected, and assume that it is again Mazdāh who says: *spitama...yazāi*. The next step will be to decide where Mazdāh's speech begins. It would be quite feasible to consider the whole stanza as spoken by him. I am not doing so, because the first two lines form a better sequel to st. 142 than introduction to what follows. The reference to Mithra's shining face goes well with his lit-up body which 'brings into evidence' the many shapes. [See Addenda.]

143³. *yeṅhe vāšəm hangrəwnāiti...yazāi* could, of course, be translated 'whose chariot she guides...; it (chariot) I worship'. But although the analogy with *Yt* 6.5 (quoted note 79², p. 227) could be invoked, where Mithra's mace is an object of worship, it seems less than likely that, at what is virtually the conclusion of a solemn hymn, Ahura Mazdāh himself should be worshipping not Mithra, but his chariot.

143⁴. *paoiriš* cannot, of course, belong to *paurva-*, or *paoirya-*, 'first, in front', as oddly enough everybody except Bth., Htl. and Hzf. assumed. Hzf.'s transformation of *paoiriš* into *pārandīča (p. 461) need not detain us. Htl. suggested that *paoiri-* is a proper name, and compared *upa.- paoirīm*, the name of the stars 'in front of the Pleiades' (see Henning, *JRAS*, 1942, 247). This is a possibility which cannot be ruled out, but as there is nothing to confirm it I prefer Da.'s identification of the subject of *hangrəwnāiti* with Aši (see next note), which goes well with Bth.'s understanding of *paoiriš* as nom. sing. (as if of an *-i-* stem) of *paoirī-*, fem. of *paru-* 'much, frequent', in the present context best translated by an adverb 'frequently'.

143⁵. The discovery that Aši is meant by 'she (who) guides' Mithra's chariot (cf. above, notes 68¹ and 68²), was made by Da., and promptly forgotten by his successors. The reasons for rejecting this obvious identification (cf. st. 68: *yeṅhe vāšəm hangrəwnāiti ašiš vaṅuhi...*) seem to have been two. One was presumably the more than subtle construction of Da.'s period: 'de qui soulève le char, ô Spitama, la première de ceux qui ne trompent pas, celle de qui le visage étincelle comme l'étoile Tishtrya'. It does seem unlikely that the poet would obscure his meaning to the extent of interposing between *yeṅhe* and its remotely situated antecedent *miθrəm* a relative clause that anticipates a different 'antecedent' mentioned only *after yeṅhe*. But dismissal of this construction need not entail rejection of the identification of the subject of *hangrəwnāiti* with Aši. A more weighty reason for disbelieving Da. must have been the assumption, first made by Gdn., that *bāmya* meant 'dawn', which is discussed in the next note.

The description 'who shines like the most beautiful creature(s) = like daylight' (see note 143⁸) would scarcely suit a star *paoirī (see note 143⁴). It does, however, suit Aši, both on the score of (1) radiance (*bāmya*), and (2) beauty (*sraēštāiš*): (1) in *Yt* 17.6 we read *aši bānumaiti...vyāvaiti bānubyō* 'radiant Aši, effulgent with rays' (cf. Lo., *Yäšt's*, 117, n. 1, on *vyāvaiti*); (2) in the same passage Aši is addressed as *srīre* 'beautiful', and in *Yt* 17.15 as *huδāta-* 'well-made', and *hučiθra-* 'of beautiful appearance' (cf. note 64⁶).

143⁶. Both Gdn., who took *hū.bāmya* as a compound, and Bth., who separated the two words, considered that *bāmya* is used here and in *Vend.*

19.28 as a substantive meaning 'dawn'. The *Vend.* passage reads: θrityą̊ xšapō vīusaiti uši raočaiti bāmya gairinąm ašaxᵛāθranąm āsənaoiti miθrəm huzaēnəm hvarəxšaētəm uzyōraiti. Bth.–Wo. translate: 'Aufleuchtet in der dritten Nacht, aufflammt die Morgenröte; die das Behagen des Aša gewährenden Gebirge ersteigt der gut bewaffnete Miθra; die Sonne geht auf'. 'Aufflammt' was obtained by falling back on the strange reading *usi.raočaiti*. To install *bāmya* as 'dawn' in this passage, and suppress the obvious word for 'dawn', *uši*, by changing it into a non-existent preposition, does strain credulity more, it would seem, than if in a passage where the acc. *miθrəm* fulfils the function of a nom., the apparent loc. *uši* were held to do likewise. In support of the latter alternative the following argument can be advanced. Pahl. *ušbām* 'daybreak, morning' presupposes what one would put down in the nom. as OIr. **uš bāmya* 'the shining dawn'. However, the Man. MPers. spelling of the word is *'wšyb'm*, and this invites the reconstruction of an OIr. **uši bāmya* as attested in our passage. One may then feel inclined to postulate for OIr. a stem **ušī-* 'dawn', coexisting with *uš-* and *ušah-*, of which *uši* in *Vend.* 19.28 would be the correct nom. sing. Alternatively the stem to postulate may be **uši-*, with analogical *-ī-* stem nom. ending. The latter assumption has the advantage of permitting the guess that this stem was secondarily abstracted from compounds like *uši.dam-*, *uši.darəna-*, cf. Gr. ἠϊ-κανός, in which *uši°* is loc.† Conceivably, too, a playful contamination with the dual *uši* 'intelligence' may have helped to bring about this stem, cf. the English figure of speech *it dawned upon him*, meaning 'he began to understand'.

We may then translate *Vend.* 19.28 as follows, taking *vīusaiti* as present participle, nom. sg. fem. (cf. *usaitīm ušå̄ŋhəm* in *Yt* 14.20), and *(hu)zaēna-* as belonging to *zaēnahvant-*‡ (cf. note 61¹): 'The flashing (*vīusaiti*), shining (*bāmya*) dawn (*uši*) of the third night flares up; Mithra, the keeper of good watch, approaches the mountains where Truth breathes freely; the sun rises.'

With the exclusion of *Vend.* 19.28 our stanza remains the only passage where *bāmya* has been taken as a substantive, but only at the cost of serious disadvantages. Gdn., to achieve his translation 'dessen Wagen... die schönste unter den Geschöpfen, die Strahlende, lenkt', was forced to sacrifice *yaθa* and emend 'das sinnlose *sraēštāiš*' to **sraēšta*. Wo., on the other hand, translates: 'dessen...Wagen die lichte (Göttin)—(schön) wie die schönsten Geschöpfe—...lenkt'; that the essential element of the comparison has to be supplied in brackets should suffice to condemn

---

† Cf. also the variant *ušistaire* of *ušastaire* in our Yašt, st. 104.

‡ This is no more than a matter of preference, as Bth.'s rendering of *huzaēnəm* by 'gut bewaffnet' is fully justified by st. 141 of the Mithra Yašt. As Mithra has only one epithet in the context of *Vend.* 19.28, where the keeping of time would seem to matter more than the smiting of *miθrō.drug-*s, one might expect this epithet to refer to his watch rather than armour.

this interpretation. No such difficulty is encountered if we agree that *bāmya-* is here, as everywhere else, an adjective; as far as *bāmya* is concerned we must therefore adopt Lo.'s solution: 'die wie die schönsten Geschöpfe leuchtet'.

143⁷. The early translators took *hū.bāmya-* as a compound:† Wi. ('Glanz'), Spi. ('Sonnenglanz'), Justi ('schöner Glanz'), Gdn. ('die Strahlende'), probably also Da. ('resplendissante de lumière'). Later it was realized that *hū...xšaētāi* is a case-form of the common *hvarəxšaēta-* 'sun, *lit.* majestic sun' in which both members of the juxtaposition are inflected, as in *nairyehe saṅhahe*, gen. of *nairyō.saṅha-*. The Av. dative having largely taken over the function of the possessive genitive, it became possible for the dat. *xšaētāi* to stand in apposition to the gen. *hū*; a parallel is found in the formula *θraētaonō jánta ažōiš dahākāi* 'Thraētaona, the slayer of Aži Dahāka', *Vend.* 1.17. Accordingly, Bth. has 'dessen—des strahlenden Sonnen(gotts)—Wagen...', Lo. 'dessen Wagen...die nie trügende Erste (=Morgenröte) des Königs Sonne lenkt', and Hzf. (p. 461) '[in beauty] like (that, *scil.* face) of the most beautiful of creatures, radiant, of Hvar χšēto [the Sun]'. Hzf. obviously took *sraēštāiš dāmąn* as apposition of *hū...xšaētāi* (cf. also his *Zor.*, II, p. 522), and thought that the comparison was between Mithra's and the sun's faces (*ainikō*, line 1). If this were so one would expect to find a gen. sing. instead of *dāmąn sraēštāiš*; also, *bāmya* can hardly be the attribute of an implied masc. *ainikō*. Lo.'s 'Erste', as we saw (note 143⁴), is impossible; in any case *hū...xšaētāi*, as a possessive genitive, can hardly be separated from its governing noun in the manner assumed by Lo. The latter objection applies even more forcefully to Bth.'s rendering, since the separation of *hū* from *xšaētāi* by a noun (as *bāmya* is in his opinion) which is *not* the one governing the two halves of the juxtaposition renders the construction of the period intolerable. There is no doubt that if *hū...xšaētāi* is a possessive gen.—and no other interpretation seems possible—there is only one noun on which we can assume it depends without having to force the syntax of the period, and this is *dāmąn*. If therefore it can be shown that 'the most beautiful creatures of the sun' is not a meaningless expression, the main difficulty of construction will be solved.

143⁸. We now come to the crux of the stanza, the 'most beautiful creatures'. It need hardly be stressed that by YAv. grammatical standards *dāmąn sraēštāiš* is the correct nom.-acc. plur. of the neuter **dāma sraēštəm*, on a par with *vīspāiš avi karšvąn yāiš hapta* (sts. 64, 89), etc., cf. E. Schwyzer, *IF*, XLVII (1929), 248 *sqq.* The expression occurs again in *H*, II, 9, where the good man's soul in the other world meets his Daēnā (=his individual Religion) 'in the *shape* of the beauty of such a one as

† Thus also Htl. ('die Schönstrahlende').

(are) the most beautiful creatures' (*kəhrpa avavatō srayå yaθa dāmąn sraēštāiš*, cf. Bth., *Wb. s.v. sray-*). From this wording we gather what is in any case obvious, namely that the 'most beautiful creatures' have the 'most beautiful shape'. It will therefore repay to consider Y 36.6, where the 'most beautiful shape' is defined: *sraēštąm aṯ tōi kəhrpąm kəhrpąm āvaēdayamahī mazdā ahurā imā raocå barəzištəm barəzimanąm avaṯ yāṯ hvarə avācī*. 'The most beautiful shape of shapes we attribute to you, O Mazdāh Ahura, these lights here, yonder highest of the high, which is called sun'. By *imā* and *avaṯ* the contrast between things 'here (on earth)' and 'yonder (in the sky)' is expressed, a contrast which is familiar from Darius' tribute to Ahura Mazdāh: *hya{h} imām būmim adā{t}, hya{h} avam asmānam adā{t}* 'who created this earth, who created yonder sky', NR, a, 1-3, cf. O. G. von Wesendonk, *Die religionsgeschichtliche Bedeutung des Yasna haptaŋhāti*, 35, W. B. Henning, *Asiatica (Festschrift Weller)*, 290, and above, note 118[2].

By *imā raocå* 'these lights (here on earth)' daylight is meant. This was clearly understood by Bth., *Wb.* 1490, sect. (3), and a glance at the passages there quoted leaves no doubt on the point. In *Yt* 15.55 we read *frataraēibyō raocå, vītaraēibyō ušåŋhəm* 'earlier than (full) daylight (*lit.* "the lights"), later than the dawn'. In *Vend.* 18.16 it is said of Būšyąstā: *hā vīspəm ahūm astvantəm hakaṯ raocaŋhąm frayrātō nix{v}abdayeiti: x{v}afsa darəyō mašyāka nōiṯ tē sačaite* 'At the awakening of daylight (*lit.* "of the lights") she persuades all the material world at once, to sleep on: sleep long, man, your time is not (yet) up' (cf. note 109[1]). Here *raocaŋhąm frayrātō* could simply be translated 'at the awakening of the day'. It would not be wrong to say that in Avestan the equivalent of OPers. *raučah-*, NPers. *rōz* (etc.), 'day', is the plural of *raočah-* 'light'.

In the mind of the author of Y 36.6 *imā raocå* is identical with *barəzištəm...avaṯ...hvarə*, daylight being, as it were, the sun on earth. Here, as on several other occasions (cf. notes 126[3], 127[5], 140[4]), Bth. was wrong in supplying an 'and' ('das Licht hier (und) jenes Höchste... dort', *Wb.* 163), thereby obliterating an intentional identification.

It is then clear that the 'most beautiful *shape*' is daylight, which in Avestan is expressed by the *nomen plurale tantum raočå*. Any apposition of such a noun may be expected to take plural endings. Accordingly, if we are told in *H*, II, 9 that the Daēnā is *in shape* like the most beautiful *creatures*, in the plural, there can be no doubt that the Daēnā is compared to daylight. Thus the absurd looking plural endings of *dāmąn sraēštāiš* turn out to be mere formal signs of grammatical concordance. From the wording of *H*, II, 9 and our stanza we may infer that, because there is only one 'most beautiful creature', namely daylight, Avestan poets were free, at least in comparisons affecting beauty, to replace the word *raočå* by the metonymic expression *dāmąn sraēštāiš*.

On the word order *yaθa dāmąn...bāmya* see note 142[2], end.

143⁹. *stəhrpaēsaṇhəm mainyu.tāštəm* is a stereotyped combination which is found elsewhere in apposition to *asman-* 'sky', Haoma's girdle *(aiwyāṇhana-)*, and the Haoma-stalks (st. 90). That Mithra's chariot is not improperly said to be *mainyu.tāštəm* is clear from the epithet *mainyu.- hąm.tāšta*, which it has in st. 67. The latter summarizes, as it were, the words *hąm.taštəm yō daδvā̊ spəntō mainyuš* 'fashioned by the creator Spənta Mainyu', which we read in the present stanza; these explain in what respect the chariot can be said to be *mainyu.tāšta-*. It would be wrong, in my opinion, to argue that in introducing with good reason the adjective *mainyu.tāštəm* the poet was induced by the above stereotyped formula to prefix it by *stəhrpaēsaṇhəm*, disregarding its unsuitability in the present context. I prefer to think that the chariot was really imagined as being studded with 'stars', these being precious stones as explained in note 136³.

143¹⁰. *hąm.taštəm yō daδvā̊*, instead of **hąm.taštəm ahmāi yō (asti) daδvā̊* is harsh, but not really harsher than *nmānəm...yasə.θwā... yazaite* in st. 30 (cf. above, p. 161, n. †). A close parallel is provided by *hunivixtəm...miθrō yō vouru.gaoyaoitiš* 'well-brandished by grass-land magnate Mithra' in *Yt* 6.5 (see above, p. 227, n. †). Before realizing that this is a possible solution I was prepared to follow Wi. and Gdn. in emending *yazāi* to *yazatāi*, and propose a revised translation of the syntactically faultless six lines which would thus emerge:

> *yeṅhe vāšəm hangrəwnāiti*
> *aδaviš paoirīš spitama*
> *yaθa dāmąn sraēštāiš hū bāmya*
> *xšaētāi yazatāi hąm.taštəm*
> *yō daδvā̊ spəntō mainyuš*
> *stəhrpaēsaṇhəm mainyu.tāštəm*

'of whom frequently she, the undeceiving—who shines like the sun's most beautiful creature—guides the star-decked, supernaturally fashioned chariot, O Spitamid, built by the majestic god who (is) the creative Incremental Spirit'. What finally deterred me from resorting to this course is the presence of the voc. *spitama*, which invites the conclusion set forth in note 143², and the improbability that *xšaēta-* should be here, and nowhere else, an epithet of Spənta Mainyu, and not of the sun, mentioned two words earlier, with which elsewhere it forms the juxtaposition *hvarə-xšaēta-*. [See Addenda.]

143¹¹. The previous versions of this difficult stanza are here quoted for the reader's convenience. In all but Da.'s (and, in a minor detail, Wo.'s) the first two lines are rendered as by us.

'...; dessen Wagen mitergreift, der nicht Irrende, Erste, o heiliger, wie die schönsten Geschöpfe mit Glanz dem leuchtenden Yazata bereitete ihn der Schöpfer der heilig-geistige den sterngeschmückten, geist-

COMMENTARY [143.11–144.1

gebildeten (Wagen) der Zehntausendseher, der starke, allwissende, unbeirrte' (Wi.).

'..., dessen Wagen mit ergreift der unbetrügliche, erste, o Heiliger, nämlich unter den schönsten der G., den mit Sonnenglanz für den glänzenden Yazata geschaffenen, den sternenglänzenden vom Schöpfer Ahura-mazda auf himmlische Weise geschaffenen (Wagen). Er wacht mit zehntausend (Augen), etc.' (Spi.).

'...; dessen wagen nie irrend vorne stehend, o Çpitama, die schönste unter den g., die Strahlende, lenkt, den für den mächtigen gott gezimmerten sternbesetzten, gottgeschaffenen. Er, etc.' (Gdn., suppressing *yō daδvå spəntō mainyuš*).

'...(see note 143⁵)...Tishtrya; aussi belle créature qui soit, resplendissante de lumière. Je veux adorer le [char] fabriqué par le Créateur Ahura Mazda, brodé d'étoiles, fait dans le ciel; [le char] du Dieu aux dix mille espions, etc.' (Da.).

'(so daß) dessen Antlitz strahlt wie (das) des Tištrya-Sterns; dessen [des Mithra]—des strahlenden Sonnen(gotts)—Wagen die lichte (Göttin) —(schön) wie die schönsten G.—sich untrüglich immer wieder (einstellend) lenkt, o Spitama. Ich will den (Wagen) verehren, (den) er der schaffende heilige Geist gezimmert (hat), den sternengeschmückten, von Geistern gezimmerten; (den Wagen des Miθra), der zehntausend, etc.' (Wo.).

'..., dessen Wagen, o Spitāma, die nie trügende Erste [nämlich: Morgenröte] des Königs Sonne lenkt, die wie die schönsten G. leuchtet. Ich will ihn anbeten, etc. [as Wo.]' (Lo.).

'...; dessen Wagen die untrügliche Paoiri lenkt, Spitama, wie die leuchtendsten G., die Schönstrahlende, für den Strahlenden. Ich will dem Gefertigten [= dem Wagen Mithras?] opfern, der Spender [= Ahura Mazdāh], der erleuchtete Geist; dem sternengeschmückten, geistgefertigten, der 10000 Späher hat, etc.' (Htl.).

'whose face shines forth like (that) of the Tištrya star; [in beauty] like (that) of the most beautiful of creatures (reading [*sraya*] *yaθa dāmān*), radiant, of Hvar χšētō [the Sun]', and, as if belonging to a different context, 'whose chariot mount with him Rtiš and Pārandī (*aδaviš paoiriš*)' (Hzf., pp. 453, 461).

144¹.

	aiwi.d°	*antarə.d°*	*ā.d°*	*upairi.d°*	*aδairi.d°*	*pairi.d°*	*aipi.d°*
Wi.	bei (dem Land)	innerhalb	am	über	unter	um	auf
Spi.	über	in	an	über	unter	vor	hinter
Gdn.	in der Nähe	mitten in	auf	über	unter	um	in der Nähe
Da.	autour	à l'intérieur	dans	au-dessus	au-dessous	devant	derrière
Wo.	rings um	inmitten	innerhalb	über	unter	vor	hinter
Lo.	rings um	inmitten	innerhalb	über or nördlich	unter or südlich	vor or östlich	hinter or westlich
Htl.	um	in	?	über	unter	um	über hin(?)
Wn.†	—	inmitten	—	—	—	um	—
I.G.	facing	between two	inside	above	below	around	behind

† *Altindische Grammatik*, II, 1, 312 sq.

The interpretation of some of these prepositional compounds, as introduced by Spi. on the strength of the NPers. translation of the identical stanza in *Ny* 2.11, cannot be right, even though it was adopted by Da., Bth., and Duch., *Comp.* 194 *sqq.* In particular, *pairi.dahyu-* cannot mean 'who is in front of the country', an interpretation which no doubt gained credit because the compound was thought to be contrasted with *aipi.dahyu-*, as *upairi°* and *aδairi.dahyu-* are in contrast with each other. The sequence of Mithra's positions in relation to the country may be less erratic than would appear from the usual translations of this stanza. Instead of jumping (with Wo., etc.) from 'around' to 'inmidst' and 'inside' (what is the difference?), then 'up and down' and 'forward and backward', it may be suggested that the following sequence is both more logical and more in keeping with the meaning of the prepositions involved: facing the speaker's country from the outside, that is, from a neighbouring country, Mithra approaches, crosses the frontier and enters; this is expressed by the prepositions 'facing', 'between (the two countries)', and 'inside'; once in the country Mithra inspects it, looking for violators of the contract, from above, below, and all round; finally, proceeding along the direction in which he had entered, he leaves the country at the opposite end; he is then 'behind the country' from the point of view of the man who had faced him at his arrival.

145[1]. The formula *vīspanąm dahyunąm daiṅhupaitīm*, which also occurs in *Ny* 1.7, the condensed hymn to Mithra quoted above, p. 263, has been translated 'Landesherr aller Länder' (Wi., Spi., Wo., Lo., Htl.), 'könig aller völker' (Gdn.), 'maître de tous les pays' (Da.). What exactly these scholars thought was the meaning of the title is not known, but Hzf., who on p. 434 translates it by 'the dahyupatiš of all provinces', and on p. 444 by 'ruler-of-the-country of all countries', goes on to explain on p. 445 that the formula implies 'the union of many satrapies in one empire, not as a hope, but as a projection of worldly facts into myth'. The 'worldly facts' Hzf. had in mind is the Median empire: 'vispānām dahyūnām dahyupatiš is the sovereign Median title, and Mithra, thereby, is saluted...as sovereign ruler of Ērānšahr' (p. 447). We shall presently see that there are more reliable hints than Mithra's title under discussion, on the political set-up which the poet had in mind. The title itself, 'Landesherr aller Länder', is at least as likely to mean 'master in each of all countries' as 'master of a state consisting of all countries'. The former interpretation is so consistent with Mithra's ubiquity (on which cf. note 95[1], end), and with his being identified simultaneously with all the religious chiefs of the Zoroastrian hierarchy (st. 115), that one would hesitate to exchange it for the latter even if it were safe to assume that the poet who coined it (who very likely was a remote ancestor of the author of our hymn) was acquainted with the notion of a Median Chief of chiefs.

COMMENTARY [145.1

As it happens, our Yašt contains indications which do not favour such an assumption.
We have seen that Da. had already recognized in the term *daiṅhusasti-* of st. 87 (also occurring in *Y* 62.5) a word meaning 'empire'. It was Htl. who associated with *daiṅhusasti-* the *dahyunąm fratəmaδātō* of st. 18, which he rendered by 'Oberherrschaften der Länder'. Although he was, I think, wrong in considering (p. 138) *fratəmatāt-* a *synonym* of *daiṅhusasti-*, his observation would have provided a firmer basis for Hzf.'s 'imperial' speculations than Mithra's title in our stanza. Hzf., however, was unable to take advantage of Htl.'s combination, because he mistook *fratəmaδātō* for an imaginary OPers. word meaning 'first-born' (p. 469), although the spelling *fratəmatātō* in *Yt* 13.95 (see above, p. 27) should have sufficed to discourage such a vagary.†
Let us take up Htl.'s hint and compare the two series in sts. 18 and 87:

18	87
nmānanąm nmānō.paitiš	nmānəm
vīsąm vīspaitiš	vīsəm
zantunąm zantupaitiš	zantūm
dahyunąm daiṅhupaitiš	dahyūm
dahyunąm fratəmaδātō	daiṅhusastīm

Although all the items in the accusative of the left-hand column are inconveniently in the plural, one notices a discrepancy between the first four and the last. All must indicate persons, but for the first four °*pati-* is used, a word which in the singular refers to one person, whereas the singular of the abstract *fratəmatāt-* can only with some hesitation be taken as denoting a single person, by comparing, e.g., Italian *podestà* 'chief magistrate' < Lat. *potestatem*. More readily one would infer from the abstract suffix that even in the singular a body of chiefs was meant, who held authority collectively.
Not only the persons at the head of the last unit are described by an abstract noun, but also, in the right-hand column, the unit itself, *daiṅhusasti-*. As *sasti* means 'command', one is reminded of the twofold meaning of Lat. *imperium*. In Iranian a similar semantic development took place with *xšaθra-*: as the abstract of a base meaning 'to rule' it first meant 'rule, power', then 'kingship', 'kingdom', 'empire' (in MIr. *Ērānšahr*), and, on a smaller scale, 'estate' (cf. above, p. 208), and 'town' (NPers. *šahr*). As the context of st. 87 does suggest that *daiṅhusasti-* is used in the sense of 'empire', it will be appropriate to interpret the compound, which theoretically could be taken in the sense of 'command of the country (sing.)', as meaning *imperium regionum* (plur.)'.
It is then likely that *dahyunąm* in line 5 of the left-hand column is to be interpreted differently from *dahyunąm* in line 4. The singular of *dahyunąm*

† On the coexistence of *-tāt-* and *-δāt-* after a vowel, cf. note 44³.

297

daiṅhupaitiš is *daiṅhāuš daiṅhupaitīm*, cf. st. 17. But the singular of *dahyunąm fratəmaδātō* would appear to have been **dahyunąm fratəma-δātəm*, *dahyunąm* 'countries' being the equivalent of **daiṅhusastōiš* 'empire'. We may say that *dahyunąm* in line 5 has the same meaning as *daiṅhusasti-*, and the same function (viz. of indicating the area over which sway is held) as *daiṅhu°* in the compound *daiṅhupaiti-*.

A confirmation of the existence of a fivefold division of temporal authority is provided by the evidence of a fivefold division of religious authority, which we examined in note 115[1]:

To the *ratu-* called *nmānya-* corresponds the *nmānō.paiti-*,
to the *ratu-* called *vīsya-* corresponds the *vīspaiti-*,
to the *ratu-* called *zantuma-* corresponds the *zantupaiti-*,
to the *ratu-* called *dāhyuma-* corresponds the *daiṅhupaiti-*;

it is then clear that the Ratu General called *zaraθuštrō.təma-*, the first holder of whose office was Zarathuštra himself, corresponds in the ecclesiastical hierarchy to what in the political organization was called **dahyunąm fratəmatāt-*.

A priori it would be legitimate to assume that the dignitaries who constituted the body called *fratəmatāt-* individually bore the title *fratama-*, just as the *karapō.tāt-* was the society of *karapan-*s, the *kəvītāt-* the *kavi-*s as a body, and the *daēvō.tāt-* the daēva-class; cf. also the abstract *ašavasta-* in the collective meaning the word sometimes seems to have, above, note 5[2]. It may be presumed that each country of the 'empire' sent a representative called *fratama-* 'premier' to a central council called *fratəmatāt-* 'Council of premiers', which controlled even the power of the individual *daiṅhupaiti-*s. It is in this sense that we may understand the OPers. title **fratama-*, which Eilers has recognized in its Elamite disguise, *pir-ra-tam-ma*, *Zeitschrift für Assyriologie*, N.F., 17 (51), 225 *sqq.*: the **fratama* Bagadāta who is mentioned in the Elamite tablets may have been a member of Xerxes' **fratamatāt-*, a Council of State on which sat the representatives of the countries that had been integrated in the Achaemenian empire.†

The glimpse we obtain of the political organization which the writer of sts. 18 and 87 had in mind may accordingly be said to reveal a federation of countries, each having its own head (*daiṅhupati-*), under the supreme control of a central council of 'premiers'. Some, or all, of these 'premiers' may have been *daiṅhupati-*s of the countries they represented. Had there been anything like a 'King of kings' at the head of this federation, one would have expected his downfall to be contemplated in st. 18. The state

† There is no need to seek this title also in the expression *fratamā*[b] *anušiyā*[b] of the Behistun inscription, as Eilers does on pp. 233 *sq.*, since *fratama-* is obviously also the ordinal 'first', and the translation 'foremost supporters' (on *anušiya-* see Henning's explanation quoted in *GMS*, p. 250) is entirely satisfactory.

in question was then neither the Achaemenian nor the Median empire, as Htl. and Hzf. believed, but rather the Greater Chorasmia of whose importance and relevance to the background of the Avesta Henning has made us aware (see his *Zor.* 42 *sq.*). We have no means of telling to what period of pre-Achaemenian Chorasmian history the scanty references we have examined apply. The division of religious authority which places Zarathuštra, or the high priests who succeeded him, above the Ratus of each country is, of course, no indication that the fivefold division dates only from the time when Zarathuštra's doctrine became the state religion of Aryana Vaējah. For the organization of the Zoroastrian Church was very likely modelled on the priesthood it had displaced. If Vištāspa was a Chief of chiefs, sts. 18 and 87 must refer to a political organization which preceded his accession by an unknown number of years. It is, however, possible that Vištāspa was merely the *daiṅhupati* of one of the countries that belonged to the Chorasmian federation, and sat at the latter's *fratəmatāt-* as the *fratəma* representative of his country.

145^2. Here the Mithra Yašt ends. There follow the usual prayers and formulae (constituting the 146th and last paragraph of the hymn in Gdn.'s edition), which are recited at the end of every Yašt. These are identical for each Hymn, except that after the words

yasnəmča vahməmča aojasča zavarəča āfrīnāmi

'I praise the worship, prayer, might, and strength' there follows in the genitive the name of the divinity to whom the Hymn is dedicated. In the present case we read after *āfrīnāmi*:

miθrahe vouru.gaoyaoitōiš rāmanō x^{v}āstrahe.

In a similar formula embedded in the paragraph which precedes the beginning of our hymn (paragraph 0 of Gdn.'s edition), we find the same four words, except that *rāmanō* is joined to what precedes by the enclitic °*ča*:

miθrahe vouru.gaoyaoitōiš rāmanasča x^{v}āstrahe.

We may therefore without hesitation translate both in 0 and in 146: 'of grass-land magnate Mithra *and* Rāman who provides good pastures'.

The genius Rāman x^{v}āstra is so frequently invoked together with Mithra in the liturgical parts of the Avesta that it is surprising not to find him even mentioned in the body of the long hymn to Mithra. Contrary to Nyberg, *Rel.* 241, I would infer from this situation that the association of the two gods is late and secondary, having perhaps arisen from the semantic affinity of their respective constant epithets *vouru.gaoyaoiti-* and *x^{v}āstra-*. In Middle Iranian times Rām was closely associated with Vayu, to the extent of being identified with him in the Bundahišn (cf. Zaehner, *Zurvan*, 338 and *passim*); hence the title of the Avestan Yašt to Vayu appears as *Rām Yašt* in most MSS., though not in the excellent MS. F$_1$; cf. Bth., *Wb. s.v. rāman-*. [See Addenda.]

CRITICAL APPARATUS

KARL F. GELDNER'S CRITICAL APPARATUS TO YAŠT X†

1. (1) F_1 Pt_1; *yim miθrəm* L_{18} H_3 J_{10} Ml_2; *yim m° yim* P_{13}. — (2) F_1 Pt_1 E_1 P_{13}; *fradaδąm* L_{18} J_{10}; in K_{15} *a* appears to have been corrected to *ā*. — (3) *dīm* F_1 Pt_1 E_1 L_{18} P_{13} J_{10}; *dym* K_{15}. — (4) F_1 Pt_1 E_1 P_{13} K_{15}; *yasnyača, vahmyača* L_{18}; cf. Yt. 8^{50}.

2. (1) F_1 E_1; *mərantaite* Pt_1 L_{18} P_{13}; *mərančaiti* $K_{40,12}$; *marančaiti* H_4; *məračaiti* K_{15}; *marəntiti* J_{10}. — (2) so in H_4 as a correction of *vīspanąm*; all other MSS. *vīspanąm*. — (3) F_1 Pt_1 E_1 P_{13} K_{15}; *daṅhaōm* L_{18}; *daṅhum* J_{10} K_{40}. — (4) L_{18} has only *druš* instead of *m° dr°*. — (5) F_1 Pt_1 K_{15}; *avaṯ* L_{18} P_{13}. — (6) F_1 Pt_1 L_{18} P_{13} H_3; *jačaṯ* K_{15} J_{10} Ml_2; *janaṯ* K_{12}; deest K_{40}. — (7) F_1 K_{15} H_3 Ml_2; deest Pt_1 L_{18} P_{13} J_{10}. — (8) F_1 Pt_1 E_1 K_{15}; *daenāṯ* without $x^v ā$ L_{18} P_{13}. — (9) F_1 Pt_1 E_1 P_{13}; *avayå* L_{18} H_4.

3. (1) all MSS. *yō*. — (2) F_1 Pt_1 E_1 L_{18}; *družintim* P_{13} (the next word wanting); *drujinti* K_{40}; *daružənti* K_{15}. — (3) *družənti* F_1 Pt_1 P_{13}; thus and *drujənti* L_{18}; °*žənta* and °*žənti* E_1; °*zənta* and °*žənti* K_{15}. — (4) all MSS.

4. (1) Parr. 4–6 = Ny. 2^{13-15}. — (2) F_1 Pt_1 E_1; *haoxšayanəm* L_{18}; *haoxšyanəm* P_{13}. — (3) F_1 Pt_1 E_1 K_{15} P_{13}; *daṅhu°* $H_{3,4}$; *daṅhu°* L_{18}.

5. (1) F_1 Pt_1 E_1 K_{15} L_{18} P_{13}; *avaṅhe* J_{10}. — (2) F_1 Pt_1 E_1 K_{15} L_{18} P_{13}; *ravaṅhe* J_{10}. — (3) F_1 P_{13} J_{10}; *rafnaiṅhe* L_{18}; *rafnaiṅhe* Pt_1; *rafnanuhe* E_1. — (4) Ml_2; *baešazyāi* F_1 Pt_1 E_1 $K_{15,40}$ L_{18} P_{13} J_{10} H_4. — (5) F_1 Pt_1 E_1 Ml_2 (Mf_3); °*γnyāi* K_{15} L_{18} P_{13} J_{10} K_{40} H_4. — (6) L_{18} P_{13}; *aiwi.θrō* F_1 Pt_1 E_1 K_{15}. — (7) F_1 Pt_1 E_1 K_{15} P_{13}; *yasnyō* L_{18} K_{40} H_4 J_{10}. — (8) F_1 Pt_1; *druxtō* L_{18} P_{13} H_4 Ml_2; *draoxtō* J_{10}. — (9) emended: *vīspəm māi aṇuhe* F_1 Pt_1; *vīspəm māiaṇuhe* E_1 K_{15}; *vīspəm mā aṇuhe* P_{13} (the second *m* above the line); *vīspəm måṇhe* L_{18}; *vīspa måṇhe* H_4; *vīspa māṇhe* K_{40}; *vīspamåṇhe* J_{10}; (*vīspəmāi aṅhe* Mf_3 in Mih. Ny.). — (10) (Mf_3;) *astvaiti* F_1 Pt_1 E_1 K_{15} L_{18} P_{13}.

6. (1) F_1 Pt_1; *dāmō hu* E_1 K_{15}; *dāmō* L_{18} P_{13} J_{10} Ml_2; *dāmōi* H_4. — (2) F_1 Pt_1 E_1; *səvištīm* K_{15}; *husəvištəm* L_{18} P_{13} Ml_2. — (3) F_1 Pt_1 E_1 K_{15}; *vanatača* L_{18} P_{13}. — (4) desunt Pt_1. — (5) K_{15}; *haoma Yō* F_1 Pt_1 E_1 P_{13}; *haomaya* L_{18}. — (6) cf. Ny. 1^{16}. — (7) $F_{2,1}$ Mf_3 J_9 H_2 $L_{18,9,11,25}$; *gavō* L_{12} Pt_1 O_3; P_{13} has *ō* written above *a*. — (8) all MSS., exc. F_1 L_9 *hizva*. — (9) F_2 Pt_1 L_{25} Jm_4; *daṅhaṇha* Mf_3; *daiṅhaṇha* K_{18a} F_1 L_{18} O_3; *daiṅhaṇhe* L_{12}. — (10) F_2 Mf_3 L_{18} J_{15}; *vāyžəbyō* J_9 H_2 Pt_1 P_{13} Jm_4. — (11) cf. Y. 4^{26}; F_2 L_{12} divide after *paiti*, J_9 H_2 L_9 after *ahurō*, L_{12} connects *vaṇhō* sec. m.

† In the transcription of Avestan words no notice has been taken of the difference between the simple and the complex signs for *š* and *n* or between internal and initial *v*. The sign for initial *y* is rendered by *Y*.

APPARATUS

with what precedes. — (12) J_9 H_2 $L_{9,12,18,25}$ F_1 Pt_1 P_{13} Jm_4 Mb_2; vaiṅhō F_2.

7. (1) cf. Ny. 1^6. The Par. recurs at the beginning of each Karde. The MSS. abbreviate. — (2) F_1 (the first a is corrected to i); jayāurvå̇nhəm L_{18} P_{13} $H_{3,4}$; jiyaurvå̇nhəm Pt_1 E_1 K_{15}.

8. (1) F_1 Pt_1 E_1; yazənti K_{15}; °zate P_{13}; °zaiti L_{18}. — (2) F_1 Pt_1 K_{15} L_{18} P_{13}. — (3) L_{18} P_{13} J_{10} K_{12} H_3; arəzahi F_1 Pt_1 E_1 K_{15}; arəzaδa H_4; arəzaiδi K_{40}. — (4) F_1 Pt_1 E_1 K_{15} L_{18}; ava P_{13} H_4. — (5) F_1 Pt_1 E_1 K_{15} P_{13} L_{18} (in the two last š a correction of m; L_{18} has, in its erroneous repetition of these words, xrūišyeitīšavi); cf. Yt. 10^{48}. — (6) F_1 Pt_1 E_1 K_{40}; daiṅhu K_{15}; daiṅhō P_{13}; daṅhō L_{18}. — (7) F_1 Pt_1 E_1 L_{18} P_{13}; °tāni K_{15}; °tāna H_4.

9. (1) cf. Yt. 13^{47}. — (2) F_1 Pt_1; vādim K_{15}; vādəm L_{18}; vāδəm P_{13}. — (3) Pt_1 E_1 $K_{15,40}$ L_{18} P_{13} $H_{3,4}$; F_1 has frāyazāinti, but n struck out. — (4) fraxšni F_1 Pt_1 E_1 $K_{15,40}$ L_{18} P_{13} H_3; fraxšne J_{10}; °xšna H_4. — (5) L_{18} P_{13} H_4; zarid° Pt_1; zrizd° F_1 E_1 K_{15}; zarəždātōiṯ J_{10}; zaražd° H_3. — (6) F_1 Pt_1 E_1 $K_{15,40}$ P_{13} J_{10}; aṅuhīaṯ L_{18} (u above the line). — (7) Pt_1 H_4 F_1 (sec. m.); ātaθra F_1 (pr. m.); ā araθra K_{15} E_1; āaθra K_{40}; āθra H_3; ātaraš Ml_2; ātarə J_{10}; ātarafraoisyeiti L_{18} P_{13}. — (8) F_1; fraōraisyaeiti H_4; fraosyeti K_{15} E_1; fraōisyeiti Pt_1; frōiris° K_{40}; fraošyənte K_{12}. — (9) F_1 Pt_1; vāča L_{18} P_{13}.

10. (1) Yt. 10^7. — (2) jiyaurv° F_1 E_1 K_{15}; jayāurv° L_{18} P_{13}; jiyāurv° Pt_1.

11. (1) F_1 Pt_1; Yazante E_1 K_{15} L_{18} P_{13}. — (2) H_4 K_{40}; aspašəm F_1 Pt_1 E_1 K_{15} L_{18} P_{13} H_3 Ml_2; aspå K_{12}. — (3) J_{10} H_4; spaxštim F_1 Pt_1 E_1 Ml_2; spaxštəm K_{15} L_{18}; paxštīm P_{13}; paoruš paxštīm H_3; cf. Y. 57^{26}. — (4) P_{13}; δbišayatąm F_1 Pt_1 E_1; dabišayatąm K_{15} (the first a is corrected sec. m. to i); ṯbaēšyantąm H_3 J_{10}; ṯbaišayantąm L_{18}; ṯbaešvantąm H_4; ṯbaešavantəm K_{40}. — (5) dušmainyūnąm J_{10}; dušmainyavanąm F_1 Pt_1 E_1 K_{15} L_{18} P_{13} H_4. — (6) Ml_2; haθrā navāitīm P_{13} H_4 K_{40}; haθrānavaitīm L_{18}; haθra navaitīm F_1 Pt_1 E_1 K_{15}; haθravānavāitīm J_{10}. — (7) K_{15}; hamrə° F_1 Pt_1 E_1 L_{18}; hamrə° P_{13}. — (8) P_{13}; ṯbašayatąm K_{15}; ṯbaešayantąm H_4 K_{40}; ṯbaešyantąm H_3; ṯbišavatąm F_1 Pt_1; ṯbaešavatąm L_{18}; ṯbašavatąm E_1.

12. (1) jiyaurv° F_1 Pt_1 E_1; jayāurv° L_{18}; jiyā urv° P_{13}.

13. (1) F_1 Pt_1 E_1 L_{18} H_4; pisō P_{13}; vīsō J_{10}. — (2) F_1 Pt_1 E_1 L_{18} K_{15} H_4; āaṯ P_{13}. — (3) F_1 Pt_1 E_1 L_{18}; ādidāiti K_{15}; ādaδāiti P_{13} H_4. — (4) F_1 Pt_1 E_1 P_{13}; šayamanəm L_{18}.

14. (1) F_1 (a correction of Yahmō) Pt_1 K_{40} H_3; ahmya L_{18} P_{13} (in this Y added above the line) J_{10}. — (2) P_{13} inserts here paorva. — (3) F_1 Pt_1 E_1 K_{15} L_{18} P_{13} H_3. — (4) F_1 Pt_1 L_{18} P_{13} $H_{3,4}$; ūrå $K_{15,40}$ E_1; yā rå J_{10}. — (5) F_1 Pt_1 E_1 K_{15}; rāzayenti P_{13}; rāzāyenti L_{18}. — (6) F_1 Pt_1 L_{18} K_{15} Ml_2; E_1 has θatairō corrected sec. m. to θātairyō; θātirō H_3; θā tirō K_{12}; θātairyō K_{40}; θrātārō H_4; θrātairyō J_{10}; xāθrō P_{13}: qu. θātārō? — (7) frāδayənte P_{13}; frāδayene H_3; frāδayəne F_1 Pt_1 E_1; °dayəne K_{15};

APPARATUS [14-20

°dayəna K_{40}; frādayən L_{18}; frādayanti H_4; frāvayanti J_{10}. — (8) F_1 Pt_1 K_{15} H_4; yaṭ P_{13} L_{18} J_{10} H_3 Ml_2. — (9) F_1 Pt_1 E_1; °ti K_{15} L_{18} P_{13} J_{10}. — (10) F_1 Pt_1 E_1 L_{18} P_{13} H_3; °wəš K_{15}. — (11) F_1 Pt_1 E_1; °ṇhā L_{18}; °ṇhå̄ P_{13}. — (12) F_1 E_1 K_{15}; °ti Pt_1 L_{18} P_{13}; °šanta H_3. — (13) H_3; āiškatəm F_1 Pt_1 E_1 P_{13}; āikatəm L_{18}; āu katəm K_{15}. — (14) F_1 Pt_1 E_1 K_{15} P_{13} H_4; pōurvtəmča L_{18}; pōurvatəmča H_3. — (15) F_1 Pt_1 E_1 L_{18} P_{13} $H_{3,4}$ K_{40}; āroyum K_{15}; J_{10} Ml_2 append ča. — (16) F_1 Pt_1 E_1 K_{15} H_4 Ml_2; suxδəmča J_{10}; saoxδəm K_{40}; sauxδəm P_{13}; suδəmča L_{18}. — (17) F_1 Pt_1 E_1 L_{18} K_{15} $H_{3,4}$; hāiriząmča P_{13}.

15. (1) F_1 Pt_1 E_1; arəzahe L_{18} P_{13} $H_{3,4}$. —(2) F_1 Pt_1 E_1 L_{18} P_{13}; in K_{15} a correction of °he; savahe $H_{3,4}$. — (3) deest P_{13}. — (4) P_{13} appends ča and omits the following word. — (5) surō F_1 Pt_1 E_1 K_{15} L_{18} P_{13}. — (6) F_1 Pt_1 E_1 L_{18}; ādaδāiti P_{13} H_3.

16. (1) F_1; karšvō hu Pt_1 L_{18} P_{13}. — (2) F_1 Pt_1; vazaiti L_{18}; P_{13} both; E_1 °zaēte and °zaite; K_{15} °zaēte and °zaiti. — (3) P_{13}; δå̄ Pt_1 L_{18} K_{15}; F_1 E_1.δå̄ and °δå̄. — (4) F_1; gūnōti Pt_1 L_{18}; °nōiti P_{13}. — (5) dīm F_1 Pt_1 E_1 K_{15} P_{13} H_3 Ml_2. — (6) F_1 Pt_1 E_1 H_4; dīdaha L_{18} (after ī an m is inserted); dahi H_3; δaha P_{13}.

17. (1) jiyavåṇhəm Pt_1; jiyāur° L_{18}; jigāur° P_{13}. — (2) F_1 Pt_1 E_1 L_{18} P_{13} K_{15} H_3; daruxδō H_4 Ml_2; druxtō J_{10} K_{40}. — (3) P_{13}; daṇhōuš F_1 E_1; daṇhōuš Pt_1 L_{18} K_{15}. — (4) daṇhupatēe F_1 Pt_1 L_{18} P_{13}; °paitēē K_{15} E_1.

18. (1) F_1 Pt_1 E_1; vādəm L_{18} P_{13}. — (2) F_1 Pt_1 E_1 L_{18}; daraoẑaiti $K_{15,40}$. — (3) F_1 Pt_1 L_{18} P_{13}; vīsō paitiš E_1 K_{15}. — (4) F_1 Pt_1 E_1 P_{13}; daṇhōuš, daṇhup° L_{18}; in K_{15} these three words are wanting. — (5) sčandayeiti F_1 Pt_1 E_1 L_{18} P_{13} K_{15} H_4; upasčən day° K_{40}; upasčin day° H_3; upastand° J_{10}. — (6) zantum, dahyum F_1 Pt_1 E_1 K_{15}; K_{40} zantūm and dahyum. — (7) F_1 Pt_1 E_1 L_{18} P_{13} $K_{15,40}$ H_4; patōiš J_{10} Ml_2; K_{12} H_3 the first time paitīm; the words uta vīsąm v° are wanting in K_{15}. — (8) F_1 Pt_1 P_{13} L_{18}; E_1 has dahyunąm corrected to daiṇhunąm; K_{15} daiṇhōnąm pr. m. to daiṇhūnąm. — (9) P_{13}; daihupaitiš F_1 Pt_1 E_1; daiṇhupaitiš K_{15}; daṇhu p° $H_{3,4}$; daṇhu p° L_{18}. — (10) F_1 Pt_1 E_1 P_{13} H_3; °dātō L_{18}; fratma dātō K_{15}; frātəm dātō J_{10} Ml_2; fratəm tātō H_4 K_{40}.

19. (1) The next seven words are wanting in K_{15}. — (2) F_1 Pt_1 E_1; pāiti L_{18} P_{13} J_{10} $H_{3,4}$ K_{40} Ml_2; pāite deest K_{15}.

20. (1) F_1 Pt_1 L_{18} J_{10}; spačiṭ P_{13}; aspi čiṭ H_4; aspaičaṭ K_{40}; aspačaṭ K_{15} E_1 H_3. — (2) all MSS. yō. — (3) F_1 Pt_1 E_1 K_{15} P_{13} H_3 Ml_2; drujəm L_{18}; J_{10} has drvjantō and leaves out the next nine words. — (4) Pt_1 L_{18} P_{13} $K_{15,40}$ $H_{3,4}$; F_1 has vazyąstra corrected to °stra; vazyastra K_{40}; Westergaard emends vazyąstara. — (5) L_{18}; bavaiti F_1 Pt_1 E_1 P_{13} $K_{15,40}$ $H_{3,4}$ Ml_2. — (6) an emendation of Westergaard; apayeintō F_1, but ō is a correction pr. m. (of i?); apayentō E_1 K_{15}; apayantō Pt_1 L_{18} P_{13} Ml_2 H_3 K_{40}; apayanta H_4; apayanō K_{12}; the next four words are wanting in F_1 Pt_1 L_{18} P_{13} K_{12} Ml_2, and are appended in E_1 in marg. In F_1 the line ends with apayein and the next begins with tō nōiṭ framanyente, Pt_1 ends with

apayantō and begins with *nōiṭfram°*. It seems that in F_1 or the archetype a line has fallen out, which began with *ti* and ended with *vazən*. — (7) *frastanviti* K_{15}; *fraxšta navanti* K_{40}; *fraxšti navanti* H_4; *frastanyå̊ nta* H_3. — (8) H_3; *vazantō* K_{40} H_4; *vazinti* K_{15}. — (9) F_1 Pt_1 E_1 L_{18}; °*mainyaeti* P_{13}; °*mainyanti* H_4 J_{10}. — (10) F_1 Pt_1 E_1 L_{18} P_{13} K_{15} $H_{3,4}$; *apasə* J_{10}; *apaiš* K_{40}. — (11) F_1 Pt_1 E_1 P_{13}; *vazaiti* L_{18} H_3 J_{10}; *vazinti* K_{15}. — (12) all MSS.

21. (1) P_{13}; *Yaδćiṭ* F_1 Pt_1; L_{18} both. — (2) F_1 Pt_1 E_1 P_{13} K_{15} Ml_2; *hvastəməm* L_{18}; *hvaspəm* H_4. — (3) K_{40} H_4; *aiṅhyeiti* F_1 Pt_1 (in this *ṅh* conjoined) E_1; *aiṅhayaeta* H_3; *aiwyeiti* L_{18} Ml_2 J_{10}; *āwyeiti* P_{13}; *yaṅhayaiti* K_{12}; *avi məθriš* K_{15}. — (4) *tanum* F_1 Pt_1 E_1 L_{18} P_{13} K_{15}. — (5) *aδćiṭ* F_1 Pt_1 L_{18} P_{13}; *aδčïṭ* E_1; *aδaćiṭ* K_{15}. — (6) F_1 Pt_1 E_1 K_{15} P_{13}; *rāšayanti* L_{18}; *rāšyanti* J_{10} K_{40}; *rāšyante* H_4; *rāšayaeta* H_3. — (7) F_1 Pt_1 E_1 P_{13}; *ava* L_{18}. — (8) *aiṅhayeiti* F_1 Pt_1 L_{18} (in this *ṅh* conjoined) H_3; *aiṅhaye* E_1; *aṇhayeti* H_4; *aṇhyaeiti* K_{40} J_{10}; *aṅh* P_{13} (at *h* the word and line break off); K_{15} has instead of this word: *aiṅhayearetīm bariti Yạm aiṅhaye*. — (9) F_1 Pt_1 L_{18} P_{13}; *frəne* E_1.

22. (1) *jaγaōur°* F_1 (*jaγ°* corrected to *jiγ°*); *jiγaōur°* Pt_1; *jaγā ur°* L_{18} H_4; *jiγā ur°* P_{13}. — (2) F_1 Pt_1 E_1 L_{18}; *aiθyaj°* J_{10} H_3; *aiθyazaṅhaṭ* P_{13}.

23. (1) F_1 Pt_1 E_1 P_{13} K_{12}; *ạzaṇhaibyō* J_{10}; *ạzaṇhubyō* K_{15} L_{18} H_4; *ạzaṇhbyō* K_{40}; *ạzaṇhṭ byō* H_3. — (2) F_1 Pt_1; °*θyå̊ sə* E_1; *haepaθyå̊sća* L_{18}. — (3) F_1 Pt_1 E_1 L_{18}; *θwayạm* P_{13}. — (4) F_1 Pt_1 P_{13}; *barahe* L_{18} H_3 K_{40}. — (5) cf. Yt. 10^{63}. — (6) here all MSS. *tanūm*, exc. H_4 *tavm*. — (7) F_1 (sec. m. *b* struck out) E_1; *arahi* Pt_1 P_{13}; *arahe* L_{18}.

24. (1) F_1 Pt_1 L_{18} P_{13}; *hixšnut°* K_{15}. — (2) *išōiš* H_4; *ušaōš* F_1 Pt_1 E_1 L_{18} $K_{15,12}$; *ušaōiš* P_{13}; *saoš* H_3; *aošōiš* K_{40} J_{10}. — (3) F_1 Pt_1 E_1 K_{15}; *parapaθạm watō* P_{13}; *para paxạm vatō* L_{18}. — (4) F_1 Pt_1 E_1 L_{18} Ml_2; *avi . snōiti* P_{13}; *āsnōiti* J_{10}. — (5) *šanmaōyō* F_1 Pt_1 Ml_2; *šana maōyō* P_{13} H_4; *šanamayō* L_{18}; *šnamanōyō* E_1 H_3 K_{40}; *ašnaōitšnamanōyō* K_{15}; *ava asarōiš namnōyō* K_{12}; *sn moyō* J_{10}. — (6) all MSS., exc. J_{10} *fraxšne*. — (7) F_1 Pt_1 E_1 P_{13}; *aviṅhe* L_{18}. — (8) *aδōyamnō* F_1 Pt_1 E_1 L_{18} P_{13} K_{15}.

25. (1) *jiγaurv°* Pt_1; *jaγā ur°* L_{18} P_{13}. — (2) F_1 Pt_1 E_1 K_{40}; *vyāxnəm* K_{15}; *vayāxnəm* L_{18}; *vīāxnīm* P_{13}. — (3) F_1 Pt_1 L_{18}; *ašahe narəm* P_{13}. — (4) *aojaṇuhạm* F_1 Pt_1 E_1 P_{13} K_{15}; in L_{18} *u* sec. m. above the line.

26. (1) F_1 Pt_1 E_1 K_{15} H_3; *katarəm* L_{18} J_{10}; *akutarəm* P_{13}. — (2) F_1 Pt_1 E_1 L_{18} K_{15}; *ačištārəm* P_{13} J_{10}. — (3) Pt_1; *daiṅhəaom* E_1 K_{15}; *daiṅhəm* F_1 (above *ə* is *ao* as correction); *daṇhəm* L_{18}. The sentence is wanting in P_{13}; L_{18} has the clause following first, then this, and then again that following. — (4) F_1 E_1 K_{15}; *daṇhaōm* Pt_1; *daiṅhum* P_{13}; *daṇhum* and *daṅhōm* L_{18}.

27. (1) Pt_1 P_{13} K_{15}; *daṇhəuš* L_{18}; *daiṅhəuš* F_1 E_1 — (2) F_1 Pt_1 L_{18} P_{13}; *rạxšyạθyå̊* H_3; *rạxšayạiθyå̊* E_1; *rạxštayạiθyå̊* K_{15}. — (3) F_1 Pt_1 E_1 K_{15}; *apavaiti* K_{40}; *apayaiti* H_3; *hišapiv°* P_{13}; *hīša piv°* L_{18}; *hīšipavaiti* H_4; *hīšə pavaiti* J_{10}. — (4) F_1 Pt_1 E_1 L_{18} P_{13} K_{15} Ml_2; *γanānå̊* K_{40}; °*γnānå̊* H_4; *gainānå̊* J_{10}; *gənanå̊* K_{12}.

APPARATUS [28-37

28. (1) F_1 Pt_1 E_1 L_{18}; *staonå* P_{13} K_{40}; *asti nå* K_{15}. — (2) L_{18}; *viδ°* F_1 Pt_1 E_1 P_{13}; *vīidār°* K_{15} (*ī* struck out). — (3) F_1 Pt_1 E_1 L_{18}; *stawarå* K_{15}. — (4) P_{13} H_4 J_{10}; *qiθayå* F_1 Pt_1 L_{18} $K_{40,15}$ E_1 (in the last two *a* above the line). — (5) F_1 Pt_1 E_1 L_{18} P_{13} K_{15} H_3 Ml_2; *yahvō* K_{40} H_4; *yāhvō* J_{10}: qu. *yahmi*? — (6) P_{13}; *xšnutō* F_1 Pt_1 E_1 L_{18} K_{15}. — (7) J_{10} H_4; *navaiti* F_1 Pt_1 L_{18} $K_{15,40}$ Ml_2; *navaeti* P_{13}; the three next words are wanting in P_{13}. — (8) *sčindayeiti* F_1 Pt_1 E_1 K_{15}; *sčandayate* K_{40}; *sti dayeti* L_{18}.

29. (1) *tum* F_1 Pt_1 P_{13}. — (2) F_1 Pt_1 E_1; *daṇhubyō* L_{18}. — (3) all MSS., exc. K_{15} *āxštōišča*. — (4) deest K_{15} P_{13}. — (5) F_1 E_1 L_{18} K_{15} J_{10}; *°hi* Pt_1 P_{13}.

30. (1) deest F_1 Pt_1 E_1 L_{18} P_{13} J_{10} Ml_2; stands $K_{15,40}$ H_3; *tvm* H_4. — (2) all MSS., qu. *aspayå*? — (3) F_1 Pt_1 P_{13}; *mašitå* L_{18} H_4; *misitå* K_{15} E_1; P_{13} has before this the words *mazištå nmānå*. — (4) L_{18}; *sraōganǝm* F_1 Pt_1 E_1 P_{13} K_{15}. — (5) K_{40} H_4; F_1 has *aspǝm* corrected pr. m. to *aspaēm*; *aspaē* E_1 K_{15}; *aspǝm* Pt_1 L_{18} P_{13} H_3 J_{10} Ml_2; *aspi* K_{12}.

32. (1) *āhiša* F_1 E_1 K_{15}; *āhīš* Ml_2; *āhiš* Pt_1 H_3 J_{10}; *āheš* L_{18} K_{40}; *āhuiš* P_{13}. — (2) F_1 Pt_1; *zaoθra* L_{18} P_{13}. — (3) These four words are wanting in K_{15} H_4. — (4) Ml_2; *hiš* F_1 Pt_1 E_1 L_{18} P_{13} K_{15}. — (5) F_1 Pt_1 E_1 L_{18} P_{13} K_{15} J_{10} Ml_2; *čimāni* H_4; *hiščamānae* H_3.

33. (1) F_1; *°mahe* Pt_1 L_{18} P_{13} K_{15} H_3. — (2) F_1 Pt_1 E_1 L_{18} P_{13}; K_{15} has *urvata* corrected to *urvaiti*. — (3) *ča* deest K_{15}; stands F_1 Pt_1 E_1 L_{18} P_{13} Ml_2. — (4) L_{18}; *vaiδīmča* J_{10}; *vaēiδīnmča* F_1 Pt_1 E_1 Ml_2; *vaē.δinmča* K_{15}; *vaēiδīnǝmča* P_{13}. — (5) F_1 Pt_1; *ča* deest K_{15}; *paiti* deest L_{18} P_{13}.

34. (1) F_1 Pt_1 E_1; L_{18} has *frā man°* and *frāman°* (so J_{10}); P_{13} *fram°* and *frām°*. — (2) F_1 Pt_1 E_1 L_{18}; K_{15} has *urvāzumna* and *°zimna*. — (3) F_1 Pt_1 E_1 L_{18} and K_{15} the second time; *haomanaṅhamna* K_{15}; *humanaṅhimna* and *haomanaṅhimǝna* P_{13} L_{18}. — (4) F_1 Pt_1 L_{18} P_{13}; *°mi* E_1 K_{15}. — (5) *haomnaṅhimna* F_1 Pt_1 E_1; *haomanaṅhimana* L_{18}; *humanaṅhimna* K_{15}. — (6) K_{15}; *karafanǝmča* F_1 Pt_1 E_1 L_{18}.

35. (1) *jiyaōur°* F_1 Pt_1 E_1; *jiyåur°* P_{13}; *jiyā ur°* L_{18}. — (2) Parr. 35-41 in L_{18} are written double by mistake. — (3) F_1 Pt_1; *spādim* K_{15} H_3; *vendaṯ* L_{18}; *vaendaṯ* P_{13}; (*vaenadaṯ* L_{18}). — (4) F_1 Pt_1 E_1 P_{13}; *hazaṇhrā* L_{18}.

36. (1) F_1 Pt_1 E_1 L_{18}; *frasāvayetei* P_{13}. — (2) desunt L_{18} (L_{18}); in E_1 sec. m. in marg. — (3) F_1 Pt_1 E_1 P_{13}; *arǝzǝm* L_{18}. — (4) *frārasmanō* F_1 L_{18} $K_{15,40}$; *frārāsmanō* Pt_1 P_{13}. — (5) *sčandayeiti* F_1 Pt_1 E_1 (L_{18}) P_{13} $K_{15,40}$ $H_{3,4}$; *sča dayaeiti* L_{18}; *sčindaiti* J_{10}. — (6) F_1 E_1; *vīspa* K_{15}; *vīspi* Pt_1 L_{18} P_{13}. — (7) *frāmaiδyǝnǝm* F_1 E_1 P_{13} H_3; *framaidyǝnǝm* Pt_1 L_{18}. — (8) F_1 Pt_1 E_1 P_{13} Ml_2; *xraṇhayeti* L_{18} $K_{15,40}$ H_3; *θaraṇhyeti* H_4; *haraṇhyeti* J_{10}. — (9) emended: *xrvīšyentahe* F_1 E_1; *xrūīšyentahe* K_{15}; *xrvī šyentahe* P_{13}; *xrvī išyentahe* Pt_1 L_{18}.

37. (1) P_{13} (L_{18}) J_{10}; *barāti* F_1 Pt_1 L_{18} K_{15} Ml_2; *baraiti* K_{40}. — (2) F_1 Pt_1 E_1 P_{13}; *θwayąmča* L_{18}. — (3) F_1 Pt_1 P_{13} L_{18} H_3 Ml_2; *spāyeiti* K_{15} E_1. — (4) The rest of the Par. wanting in L_{18}. — (5) F_1 Pt_1 E_1 P_{13}; *vazaiti* K_{15}; (*vaziti* L_{18}).

38. (1) *šaitayō* F_1 Pt_1 E_1 P_{13} K_{15} Ml_2; *išityō* K_{40}; *saitayō* L_{18} (in the repetition ..*itayō*, the first letter not clear, perhaps *s*); *saityō* J_{10}. — (2) F_1 E_1 K_{15}; *frazinti* Pt_1 P_{13}; *frāzinti* L_{18} (L_{18}); *frazənti* H_3; *frazanti* K_{40} J_{10}. — (3) *šyete* P_{13}; (*šyeite* L_{18};) *šyeiti* K_{15}; *šyeiti* F_1 Pt_1 E_1 J_{10} Ml_2; *šyeti* L_{18}; *šiaeti* H_4; *šayeti* K_{40}; *sayaeti* H_3. — (4) *haiθim* F_1 E_1 K_{15}; *haiθəm* Pt_1 L_{18} P_{13} J_{10}. — (5) F_1 Pt_1 E_1 P_{13}; *varəθīm* L_{18}. — (6) F_1 Pt_1 E_1 H_4; *azate* P_{13}; *azaiti* $K_{15,40}$ L_{18} H_3 J_{10}; (*azāiti* L_{18}). — (7) F_1 Pt_1 E_1 P_{13} Ml_2; *frazašta* L_{18}; *frajasaiti* J_{10}; (*frazišta* L_{18}). — (8) F_1 (a correction of *raēθya*) Pt_1 E_1 P_{13} L_{18}; (*raθwya* L_{18};) *raiθwīm* J_{10}. — (9) F_1 Pt_1 E_1 P_{13}; °*ti* K_{15} L_{18}. — (10) *anuzafnō* K_{40}.

39. (1) K_{15} H_3; *išvasčiṯ* F_1 Pt_1 L_{18} (L_{18}) P_{13}. — (2) F_1 Pt_1 E_1 L_{18} P_{13}; *hutaxtāṯ* H_3. — (3) *ašəmnō* F_1 Pt_1 E_1 L_{18}; *ašəmanō* $K_{15,40}$ (L_{18}); *ašə manō* J_{10}; *ašmanō* P_{13}; *asmanō* H_3; *išamanō vanō vīdō* H_4. — (4) all MSS. here and in the next Kardes. — (5) K_{40}; *aipiti* F_1 Pt_1 E_1 H_3; *aipaiti* L_{18} P_{13}. — (6) F_1; *zaraštayasčaṯ* P_{13}; the sentence wanting in K_{15}. — (7) *ašəmnō* F_1 Pt_1 E_1; *ašəmanō* L_{18} (L_{18}) H_3 K_{40}; *ašmanō* P_{13}; *asimanō* H_4; *ašəm nō* J_{10}. — (8) here J_{10} *bavanti*. — (9) Pt_1 F_1 E_1 P_{13}; *zaraštvačiṯ* L_{18}; *zastavačiṯ* K_{15}. — (10) L_{18}; *ašəmnō* F_1 Pt_1 E_1; *ašəmanō* $K_{15,40}$ P_{13} H_3.

40. (1) The first sentence wanting in Pt_1 L_{18} (L_{18}) P_{13} J_{10} Ml_2. — (2) F_1; *Yō* E_1 K_{15} H_3. — (3) F_1; *srahu* E_1. — (4) F_1; *ašəmō* E_1 K_{15}. — (5) *vazaračiṯ* F_1 Pt_1 E_1 L_{18} P_{13} K_{15}. — (6) *Yō* F_1 Pt_1 E_1 P_{13} K_{15}; deest L_{18} (L_{18}) J_{10}; cf. n. 2. — (7) *srahu* F_1 E_1 L_{18} P_{13} K_{15} J_{10}; *srvahu* Pt_1; cf. n. 3. — (8) *ašəmnō* F_1 Pt_1 E_1 P_{13}; *ašəmanō* L_{18} K_{15} H_3; *aša manō* J_{10}.

41. (1) F_1 Pt_1 E_1 K_{15}; L_{18} has °*yeti* and °*yete*; *ava θrånhaiti* P_{13}; the three following words wanting in P_{13} $K_{15,12}$. — (2) F_1 Pt_1 E_1 L_{18} P_{13} K_{15} J_{10}; *vāite* $H_{3,4}$. — (3) *Yazatatē* E_1 K_{15}. — (4) Ml_2; *raēčayeiti* F_1 E_1 L_{18} K_{15} H_3; *raečayeite* Pt_1 P_{13}.

42. (1) *uityōjanå* F_1 E_1 K_{15}. — (2) These four words are wanting L_{18} J_{10} Ml_2. — (3) °*tǰe* F_1 Pt_1 E_1 K_{15} P_{13}. — (4) F_1 E_1; *miθrāiδa* Pt_1 L_{18}; *miθrāiδi* H_3; *miθrāiṭa* P_{13}; *miθrāṭ* $K_{15,40}$ H_4; *miθrāide* J_{10}. — (5) F_1 Pt_1 E_1 P_{13}; *nayeinti* K_{15}; *nayenti* L_{18}; *nayanti* J_{10}. — (6) F_1 Pt_1 E_1 L_{18} P_{13} H_3; *bāzva* $K_{15,40}$. — (7) F_1 Pt_1 E_1 K_{15} H_3; *sčinδyanti* J_{10}; *sčindayeti* L_{18}; *sča nadayeinti* P_{13}; *sčandayaiti* H_4 K_{40}.

43. (1) F_1 Pt_1 E_1; *fraspayeinti* K_{15}; *frā spaēiti* P_{13}. — (2) L_{18} P_{13} K_{15}; *baēvaraγnāi* F_1 Pt_1 E_1.

44. (1) *jiγaur*° F_1 E_1 K_{15}; *jiγāur*° Pt_1; *jaγāur*° P_{13} L_{18}. — (2) an emendation of Westergaard; *mazāṯ* H_4 K_{12}; *mazāδa* F_1 E_1 (in this *dāt* appended sec. m.); *mazdaδa* L_{18}; *mazdāδa* Pt_1; *mazdāda* K_{15} (the first *d* struck out); *mazdāṭa* P_{13}; *mazdaδāṯ* K_{40}; *mazda δāta* H_3 Ml_2; *mazdaδātanąm* J_{10}. — (3) *anąm zō* H_4; all the rest *nązō*; deest J_{10}. — (4) F_1 Pt_1 L_{18} P_{13}; *vouruaštəm* E_1 K_{15}; *aštīm* Ml_2; *astəm* H_3 J_{10}; *hištəm* K_{40}; *hastəm* H_4; *vō urustąm* K_{12}.

45. (1) F_1 Pt_1 E_1 L_{18} P_{13} K_{15}; *asta* K_{40} Ml_2 — (2) desunt L_{18}. — (3) F_1 Pt_1 E_1 K_{15}; *spašō* P_{13}; *spō* L_{18}. — (4) L_{18} P_{13}; *drujīm* F_1 Pt_1 E_1 K_{15}.

— (5) hišpō səmana L_{18}; hispō səmna F_1 Pt_1 E_1 P_{13} K_{15}. — (6) avī Pt_1 E_1 L_{18} P_{13} $K_{15,40}$ $H_{3,4}$; avū F_1; aēvī J_{10}. — (7) F_1 E_1; družənti K_{15}; drujinti Pt_1 L_{18}; °jinte P_{13}. — (8) F_1 Pt_1 E_1 L_{18} K_{15} J_{10}; pāθō K_{40} $H_{3,4}$; miθō P_{13}. — (9) F_1 Pt_1 E_1 L_{18} P_{13} J_{10}; asənti K_{15}.

46. (1) F_1 Pt_1 E_1 L_{18} H_4; spš K_{15}; spašə J_{10}; ašpaš P_{13}. — (2) F_1 Pt_1 E_1 L_{18}; vīdaiti K_{15} H_4; vīdaeti J_{10}. — (3) K_{15}; aṅhe F_1 Pt_1 E_1 L_{18} P_{13} H_3; frå̇ṅhe H_4; manaṅhe J_{10}. — (4) F_1 Pt_1 E_1 L_{18} P_{13} $K_{15,40}$ Ml_2; fraxšni H_4; frašne J_{10}. — (5) F_1 Pt_1 E_1 K_{15}; avaṅhe L_{18} P_{13} J_{10} H_3.

47. (1) F_1 Pt_1 E_1; jaγāur° L_{18}. — (2) Ml_2 J_{10}; frasrutəm F_1 Pt_1 E_1 K_{15} L_{18} H_3; frasraotīm P_{13}. — (3) F_1 Pt_1 E_1 P_{13}; zaranimanəm H_3; zaraninəm L_{18}; zaranyamanəm K_{40}; zaraniumanəm H_4; zaranumanəm K_{15}; zairinəm J_{10}. — (4) F_1 Pt_1 E_1 K_{15} L_{18}; xrvīšyeitiš P_{13}; cf. Yt. 10⁸. — (5) F_1 Pt_1 E_1 K_{15} L_{18}; daiṅhō P_{13}.

48. (1) F_1 E_1; °ti Pt_1 L_{18} P_{13} K_{15}; °vazāiti J_{10}. — (2) P_{13}; the rest as Par. 47 n. 4 and 5. — (3) cf. Yt. 14⁶³. — (4) F_1 Pt_1 E_1 L_{18} P_{13} $K_{15,40}$ H_4; pairə H_3; pairi K_{12} Ml_2; pərədəm J_{10}. — (5) F_1 Pt_1 L_{18} H_4; gaošayaeiti E_1 P_{13} (in this gaoša is wanting). $K_{15,40}$ H_3 J_{10} Ml_2. — (6) F_1 Pt_1 E_1 K_{15} H_3 Ml_2; tā deest P_{13}; tå̇ J_{10}; tādaiṅh° L_{18}. — (7) təhamərəθə̄ K_{40} H_4; hamarəθe F_1 Pt_1 E_1 P_{13} K_{15} H_3; marəθe L_{18}. — (8) F_1 Pt_1 E_1 P_{13} K_{15} Ml_2; °barəntō J_{10}; duža barətō K_{40}; dužbərənte H_3.

49. (1) jiγaur° F_1 Pt_1 E_1 K_{15}; jaγāur° L_{18} P_{13}.

50. (1) F_1 Pt_1 E_1; fraourvaēsyąm L_{18}; fraōrvaēšyąm P_{13}; fraōryaēsyąm K_{15}. — (2) F_1 Pt_1 E_1; dunąn P_{13}; dunąm L_{18}; K_{15} has dunmąm, but the final m corrected sec. m. to n. — (3) all MSS., except J_{10} uzgasənti. — (4) F_1 E_1; harəθyō Pt_1 P_{13}; deest L_{18} J_{10}. — (5) deest F_1 Pt_1 E_1 L_{18} P_{13} K_{15} H_3 J_{10} Ml_2; stands H_4 K_{40}.

51. (1) F_1 Pt_1 L_{18} P_{13}; kərənå̇n E_1 K_{15}. — (2) F_1 E_1 H_4; aməšāspənta Pt_1 L_{18} P_{13} H_3. — (3) havarə L_{18}; havara F_1 Pt_1 E_1 P_{13}. — (4) F_1 Pt_1 E_1 L_{18}; frašni K_{15}; fraxšayəni P_{13}; fraxšne J_{10}. — (5) F_1 Pt_1 E_1; zarazd° K_{15} L_{18} P_{13} J_{10}. — (6) F_1 Pt_1 E_1 L_{18} P_{13} K_{15}; aṅhuyaṱ H_3. — (7) F_1 Pt_1 E_1 L_{18}; ādaδāiti P_{13}.

52. (1) F_1 K_{15} Ml_2 H_4; deest Pt_1 P_{13} L_{18} J_{10} H_3. — (2) Pt_1 L_{18} P_{13}; °te F_1 E_1. — (3) F_1 Pt_1 E_1 L_{18} $K_{15,40}$ H_4 J_{10}; mā̇uš P_{13}; māyaoš K_{12} H_3. — (4) F_1 Pt_1 E_1 K_{40}; vā dəm J_{10}; vādim $H_{3,4}$; vādəm L_{18} Ml_2; vā P_{13}; in P_{13} K_{15} the rest of the Par., from dim on, is wanting.

53. (1) F_1 Pt_1 E_1 L_{18} P_{13} H_3; °ti K_{15}.

54. (1) H_4; nipāiti K_{40}; nipāpa F_1 Pt_1 L_{18} P_{13} K_{15} H_3 Ml_2. — (2) all MSS., exc. H_4 hvāpō. — (3) F_1 Pt_1 E_1 H_3; yazənti K_{15} L_{18} P_{13}. — (4) J_{10} H_4 Ml_2; deest F_1 Pt_1 E_1 K_{15} L_{18} P_{13}. — (5) F_1 Pt_1 E_1 L_{18} P_{13} K_{15}; yazinti J_{10}; yazanta H_4; the clause is wanting H_3.

55. (1) cf. Yt. 8¹¹, 10⁷⁴; Yeiδi zī F_1 Pt_1 E_1 L_{18}; yaṱ zī K_{40} H_4 J_{10}. — (2) Yezayanta F_1 Pt_1 E_1 K_{15}; yazīnti H_4; yezənti L_{18}; yazənti P_{13}; yazinti J_{10}; P_{13} has the clause repeated by mistake, with Yeizayanta the second time. — (3) L_{18} P_{13}; Yazainti F_1 E_1 K_{15} (P_{13}); Yazanti Pt_1; °te H_3. —

(4) F_1 Pt_1 E_1 L_{18} K_{15} $H_{3,4}$; *ašavōyō* P_{13} H_4 K_{40}. — (5) F_1 Pt_1; *zrūāiu* K_{15}; *zrūāyušušyąm* L_{18}; *zrūānahe* ə P_{13} (but *nahe* struck out again); *zrvāṯ* H_3; *zrū āyīda* Ml_2; *zarvnāyušū* H_4; *zaurvāyušū* K_{40}; *zravane āyaṯ* J_{10}. — (6) F_1 Pt_1 E_1 K_{15} H_3 Ml_2; *šaošyąm* P_{13}; *šušyąm* K_{40}; *šuyąm* H_4. — (7) F_1 E_1 $H_{3,4}$ J_{10}; *upaθwarštahe* Pt_1 L_{18} P_{13} $K_{15,40}$.

60. (1) F_1 E_1 $K_{15,40}$ H_3; *frāsasti* Pt_1; *frasasta* J_{10}; *frāsasta* L_{18} P_{13}. — (2) F_1 Pt_1 K_{15} P_{13} H_3 Ml_2; L_{18} has *gaōiδa* instead of *g°* *at°* *iδa*. — (3) F_1 Pt_1 E_1 H_3 Ml_2; *ataurvōyō* K_{40}; *ataurvōiδa* P_{13}; *ata urvyō* H_4; *atourvayaṯ* J_{10} (*iδa* deest); this and the two next words are wanting in K_{15}. — (4) *Yaō nāi intąm* F_1; *Yō nāi intąm* Pt_1 (sec. m. before *ō* an *a* inserted); *Yō nāyntąm* E_1 K_{15} H_3; *yō nāintąm* J_{10}; *yaō nāi intąm* L_{18}; *yaō nāi inatąm* P_{13}; *yō tā intąm* Ml_2; *yō nāyatąm* K_{12}; *yaonāi uatąm* K_{40}; *yaonāi gatąm* H_4; the passage is corrupt. — (5) F_1 Pt_1 E_1 $K_{15,12}$ H_3 J_{10} Ml_2; *spāsānō* L_{18}; *sāsānō* P_{13}; *spasnō* K_{40}.

61. (1) F_1 Pt_1 E_1 L_{18} P_{13} H_3; *zəntəm* K_{15}. — (2) *zaēnavhuntəm* F_1 Pt_1 L_{18} P_{13}; *zaēnavhutəm* K_{15} E_1. — (3) F_1 E_1 K_{15} H_3; *aspasəm* Pt_1 L_{18} P_{13}; *aspəm* Ml_2. — (4) F_1 Pt_1; *vyāxnəm* L_{18} P_{13} H_3 J_{10}. — (5) *fraṯ* F_1 $Pt_1E_1L_{18}$ K_{15}; *fraδa* P_{13}; *frada* J_{10} K_{40}; *fradāpəm* H_4. — (6) all MSS. — (7) F_1 Pt_1 Ml_2; *vyāxnəm* E_1 L_{18} P_{13} H_3 K_{40} J_{10}; *vyānəm* K_{15}. — (8) F_1 Pt_1 L_{18} K_{15}; *yaoxštavantəm* P_{13}. — (9) F_1 Pt_1 L_{18}; *dāmi dātəm* P_{13}.

62. (1) *daδańti* F_1 Pt_1 L_{18}; *dadaniti* P_{13}. — (2) F_1 Pt_1 E_1; *mīždəm* L_{18} P_{13}.

63. (1) F_1 Pt_1 E_1 P_{13} K_{15} $H_{3,4}$ Ml_2; *tanūm* L_{18} K_{40}.

64. (1) F_1 Pt_1 E_1 K_{15} J_{10} Ml_2; *ahmi* P_{13}; *Yaṯ ahmi* L_{18}; *Yahi* H_3; *yahmāi* K_{40} H_4. — (2) F_1 Pt_1 E_1 P_{13} H_3 J_{10}; *vyāni* K_{15}; *vīdāni* L_{18}; *vīāyaeini* H_4; *vīsāne* K_{12}. — (3) F_1 Pt_1 E_1 P_{13}; *frakayāi* L_{18}. — (4) F_1 Pt_1 E_1 L_{18} P_{13} $K_{15,40}$ Ml_2; *Yahmi* H_3 J_{10}; *yahmāi* H_4. — (5) F_1 Pt_1 E_1 P_{13} K_{15}; *ava* L_{18}.

65. (1) F_1 Pt_1 E_1 L_{18} K_{15}; *vīāxnanąm* P_{13}. — (2) F_1 Pt_1; *vyāxnō* E_1 K_{15} P_{13}. — (3) F_1 Pt_1 E_1 P_{13} K_{15} H_3; *frašti* L_{18}. — (4) F_1 Pt_1 E_1; *azuiti* H_3; *āzviti* L_{18}; *āzvaiti* P_{13}; *āzyti* K_{15}.

66. (1) F_1 Pt_1 E_1; *°ti* L_{18} P_{13} K_{15}. — (2) F_1 Pt_1 L_{18} P_{13}; *pārəndača* E_1 K_{15} H_3 J_{10}. — (3) F_1 Pt_1 E_1 L_{18} H_3; *vərətiš* P_{13}; in K_{15} the words from *uyrača* to *var°* are wanting. — (4) all MSS., exc. $H_{3,4}$ *ašaunąm*.

67. (1) F_1 Pt_1 L_{18} P_{13} Ml_2; *tāšti* J_{10}; *tāša* K_{15} E_1 (in this *t* appended sec. m.) H_3. — (2) *bərəza* K_{15} E_1 H_3; *čaxra* F_1 Pt_1 E_1 L_{18} P_{13}; *čaiθra* K_{15}; *čiθra* H_4 Ml_2. — (3) These four words are wanting in F_1 Pt_1 L_{18} P_{13} H_3 J_{10} Ml_2; in E_1 they stand sec. m. in marg.; cf. Yt. 10[20] n. 6. — (4) *arəzahe* K_{15} E_1; *ərəzihi* H_4. — (5) H_4 $K_{40,12}$ J_{10} Ml_2; *aṯ* F_1 Pt_1 E_1 L_{18} P_{13} K_{15} H_3. — (6) *čaθrahe* Ml_2; *čixra* F_1 Pt_1 E_1 L_{18} P_{13} K_{15}; *čixrə* J_{10}; *čaxra* H_3; *raiθyača xra* H_4. — (7) desunt K_{15}.

68. (1) *hangrəwnāiti* F_1 Pt_1 E_1 P_{13} K_{15} (in this *a* appended). — (2) L_{18} P_{13}; *māzdayesniš* F_1 Pt_1 E_1 K_{15}. — (3) F_1 Pt_1 E_1 L_{18} P_{13} J_{10}; *hīti* K_{15} H_3. — (4) F_1 Pt_1 E_1 L_{18} P_{13}; *rādaiti* K_{15}. — (5) *auruš* F_1 Pt_1 E_1 L_{18} P_{13} K_{15} $H_{3,4}$ Ml_2; *uruš* J_{10} K_{40}. — (6) all MSS., exc. L_{18} *raoxšni*. — (7) F_1 Pt_1 E_1 L_{18}

APPARATUS [68–77

P_{13} H_3; *āsaya* Ml_2; *ašaya* K_{40} H_4 J_{10}. — (8) F_1 Pt_1; *mainiv°* E_1 K_{15}; *manaiv°* L_{18} P_{13}. — (9) *dīm* F_1 Pt_1 E_1 L_{18} P_{13} H_3 Ml_2. — (10) *hu īrixtəm* F_1 E_1 P_{13} K_{15} H_3; *huīrixtəm* Pt_1; *hu īrəxtəm* L_{18}. — (11) F_1 Pt_1 L_{18} P_{13} H_3 Ml_2; *yahmāṯ* K_{15}.

69. (1) F_1 E_1 K_{15} H_3; *mōitū* Pt_1 P_{13}; *mōitū θra* L_{18} Ml_2. — (2) P_{13} H_4; *haməraθāi* F_1 Pt_1 L_{18}; *haməraθrāi* E_1 $K_{15,12}$. — (3) F_1 Pt_1 E_1 K_{15} P_{13}; *jasāiti* L_{18} J_{10}. — (4) F_1 Pt_1 E_1 L_{18} P_{13} K_{15} H_3 Ml_2.

70. (1) F_1 Pt_1 E_1; °*ti* K_{15} L_{18} P_{13} H_3 J_{10}. — (2) F_1 Pt_1 E_1 L_{18} P_{13} K_{15} H_3 J_{10} Ml_2; *hūkəhrpa* K_{40}. — (3) J_{10}; *paētərənō* F_1 Pt_1 E_1 L_{18} P_{13} K_{15}; *paiti rənō* K_{40}; *paiti ranō* H_4; *paiti ərašnō* H_3. — (4) desunt K_{15}. — (5) F_1 Pt_1 E_1; *jayehe* P_{13}; *jayaihe* L_{18}; *jyahe* H_3; deest J_{10}; this and the following compound wanting in K_{15}.

71. (1) F_1 Pt_1 E_1; *haməraθāṯa* K_{15} P_{13}; *haməraθāṯ* L_{18} H_3. — (2) F_1 Pt_1 E_1 $K_{15,40}$ H_3; *upaš* P_{13} J_{10} Ml_2; *aupaš* L_{18}. — (3) F_1; *āmanaṅha* Pt_1 L_{18} P_{13} K_{40}; *ā manaṅhi* E_1 K_{15}. — (4) all MSS. — (5) F_1 Pt_1 E_1 L_{18} P_{13} K_{15} H_3; *varəəta* H_4; *varəti* J_{10}. — (6) *haməraθə̄* F_1 Pt_1 E_1 L_{18} P_{13} H_3; *haməraθə̄* K_{15}. — (7) F_1 Pt_1 E_1 P_{13}; *jaγnyā̊* K_{15}; *nijaiγnvā̊* L_{18}.

72. (1) L_{18}; *mastrəynasča* F_1 P_{13}; *mastrə̄ynasča* Pt_1 E_1; *mastraγanasča* H_3; *mastarə̄γanasča* K_{15}. — (2) F_1 Pt_1 E_1 P_{13}; °*nasča* L_{18}. — (3) F_1 P_{13}; *zəmāṯa* Pt_1 E_1 L_{18} K_{15}; *zəmāṯ* H_3. — (4) F_1 Pt_1 E_1 L_{18} K_{15}; *raθwayaeiti* P_{13}. — (5) K_{40}; *druzəm* H_4; *drujəm* F_1 Pt_1 E_1 L_{18} P_{13} K_{15} H_3 Ml_2.

73. (1) deest L_{18}. — (2) F_1 Pt_1 E_1 P_{13}; *vāčəm* L_{18}, and K_{15} a correction of *vāčim*. — (3) L_{18} P_{13}; *mainyu* F_1 Pt_1 E_1 K_{15}.

74. (1) cf. Yt. 10⁵⁵, 8¹¹; *Yēδi* F_1 Pt_1 E_1 K_{15} P_{13} L_{18}; *yaṯ zī* $H_{3,4}$ K_{40} J_{10}. — (2) *Yazanti* F_1 Pt_1 E_1 P_{13} K_{15} H_3 J_{10}; *Yazatai* L_{18}. — (3) F_1 Pt_1 E_1 P_{13}; *Yazinti* L_{18}; *Yazinta* K_{15}. — (4) *frānurvyō* H_4 K_{40}; *frānōyō* F_1 Pt_1 E_1 L_{18} P_{13} K_{15} H_3; *naravayō* J_{10}. — (5) Ml_2; *ašāvaōyō* F_1; *šāvaōyō* Pt_1; *šāvōyō* E_1 K_{15}; *šāvayō* P_{13}; *švaōyaō* L_{18} (sec. m. *ā* appended). — (6) F_1 Pt_1 E_1 L_{18} P_{13}; *zrū āyada* Ml_2. — (7) *šūšyąm* F_1 Pt_1 E_1 K_{15} H_3; *sušyąm* P_{13}; *sušyąm* L_{18}. — (8) Pt_1 Ml_2; *upa aθwar°* F_1 E_1 K_{15}; *upaθwar°* L_{18} P_{13} H_3.

75. (1) K_{15} inserts here *tē* which is wanting in the rest. — (2) P_{13}; *višō* F_1 Pt_1 E_1 L_{18} K_{15} H_3. — (3) F_1 Pt_1 E_1 K_{15}; *daiṅhō* P_{13}; *daṅhu* L_{18}. — (4) F_1 Pt_1 E_1 P_{13} K_{15} H_3; *bāzuš* L_{18}. — (5) F_1 Pt_1 E_1 L_{18} K_{15} Ml_2; *nivanāṯ* P_{13}. — (6) *ṯbišynbyō* E_1 H_3; *ṯbišainbyō* F_1 Pt_1; δ*bišyntōbyō* K_{15} (*tō* struck out); *ṯbaešanibyō* K_{40} H_4; this and the two next words wanting in L_{18} P_{13} J_{10} Ml_2.

76. (1) *ṯbišayatąm* F_1 Pt_1 E_1 L_{18}; *ṯbišyantąm* P_{13} H_3; δ*bišiatąm* K_{15}. — (2) F_1 Pt_1 E_1 H_3; °*ṅuhantąm* L_{18} P_{13} Ml_2 H_4; *ṯbiēšaṅuhatąm* K_{15}. — (3) F_1 Pt_1 E_1; °*he* L_{18}, P_{13} K_{15} H_4; *sčandayehe* H_3. — (4) Pt_1 E_1 L_{18}; *sčandaya* P_{13}; F_1 has *sčandaya*, but *sča* corrected sec. m. to *sči*. — (5) $K_{15,40}$; *hariθyō* K_{12}; *hiriθyō* H_3; *hurāθvō* F_1 Pt_1 E_1 L_{18} P_{13} H_4 J_{10} K_{40} Ml_2. — (6) *ahī* F_1 Pt_1 E_1 L_{18} P_{13} K_{15} Ml_2.

77. (1) F_1 Pt_1 E_1 L_{18} P_{13}; *avaṅhe* and *avaṅhe* K_{15}. — (2) F_1 Pt_1; E_1 has *aiwišayama*, with *an* appended later. [*an*: Gdn.'s misprint for *na*?]

APPARATUS

78. (1) so all MSS. — (2) all MSS. — (3) all MSS. $y\bar{a}$; L_{18} $y\bar{a}hub\partial r\partial it\bar{\imath}m$. — (4) $hub\partial r\partial itim$ F_1 Pt_1 E_1; hu $b\partial r\partial t\bar{\imath}m$ P_{13}; hu $b\partial r\partial tim$ K_{15}. — (5) all MSS. — (6) F_1 Pt_1 E_1; $frasč\partial n°$ K_{40}; $frasčindayehe$ L_{18} P_{13} K_{15}; $frasčandayaehi$ H_3. — (7) F_1 Pt_1 E_1 L_{18} P_{13} K_{15}. — (8) F_1 Pt_1 E_1 K_{15} insert here avi, $H_{3,4}$ K_{40} ava; deest L_{18} P_{13} J_{10} Ml_2. — (9) F_1 Pt_1 L_{18}; $Yasny\bar{o}$ P_{13}.

79. (1) all MSS. — (2) F_1 Pt_1 E_1 P_{13}; $d\bar{a}\delta e$ K_{15}; $dai\delta i$ L_{18}. — (3) F_1 Pt_1 E_1 L_{18} P_{13} H_4; $raš\partial uš$ K_{15} H_3; $ršn\bar{o}š$ J_{10}. — (4) F_1 Pt_1 E_1 K_{15} H_3; $mainavaintym$ L_{18}; $manait\bar{\imath}m$ P_{13}.

80. (1) H_4 K_{40}; tum F_1 Pt_1 E_1 L_{18} P_{13} K_{15} H_3. — (2) F_1 Pt_1 E_1 K_{15} L_{18} H_3; $druj\d{q}m$ P_{13}. — (3) F_1 Pt_1 L_{18} P_{13} K_{15}. — (4) so F_1; ahe H_4; deest $K_{15,40}$ H_3; in Pt_1 H_3 J_{10} Ml_2 the words $var°$ to $har\partial ta$ are wanting; L_{18} P_{13} have instead of $var\partial z\bar{a}nahe\ldots adruž\d{q}m$ merely $varuž\d{q}m$. — (5) F_1 Pt_1 E_1 P_{13} H_3; $s\bar{o}iri$ L_{18}; $š\bar{o}iri$ K_{40}; $s\bar{o}rai$ K_{15}; $s\bar{u}re$ J_{10}. — (6) F_1 Pt_1 E_1 L_{18} P_{13} K_{15} H_3; $v\bar{\imath}\theta iš$ H_4; $v\bar{\imath}\theta\bar{u}š$ Ml_2; $v\bar{\imath}\theta\bar{a}iš$ J_{10}.

81. (1) cf. Par. 80; $daie\delta i$ F_1 Pt_1 E_1 K_{15}; $dai\delta i$ L_{18}; deest P_{13}. — (2) $ba\bar{e}var\partial$ F_1 Pt_1 E_1 L_{18} P_{13} $K_{15,40}$ H_4 J_{10} Ml_2; $baevar\partial m$ H_3. — (3) $manavaint\partial m$ F_1 Pt_1 E_1 L_{18} P_{13} K_{15}.

82. (1) F_1 Pt_1 E_1 K_{15}; $v\bar{\imath}\delta\bar{o}i\theta re$ L_{18} P_{13}. — (2) F_1 Pt_1 E_1 K_{15}; $spayaeti$ P_{13}; $paiti$ L_{18}. — (3) L_{18} P_{13}; $druj\bar{\imath}mča$ F_1 Pt_1 E_1 H_3; $daroj\bar{\imath}mča$ K_{15}. — (4) J_{10}; $aiwyača$ F_1 Pt_1 L_{18} P_{13} K_{15} H_3 Ml_2. — (5) F_1 (sec. m.) Pt_1 E_1; $spasn\bar{o}$ F_1 (pr. m.) L_{18}; $aspasn\bar{o}$ P_{13}.

83. (1) F_1 Pt_1 E_1 L_{18} P_{13} K_{15}. — (2) F_1 Pt_1 E_1 P_{13} K_{15} and L_{18}. H_3 below; $ava\dot{n}he$ L_{18} $H_{3,4}$; $ava\dot{n}he$ J_{10}.

84. (1) Pt_1 E_1 L_{18} P_{13} K_{15} H_3; in F_1 stands e pr. m. above the line; below it an erased letter (perhaps a?); $pai\theta e$ H_4; $pa\theta a$ J_{10}; $pai\theta i$ K_{40}. — (2) F_1 Pt_1 L_{18}; $hačamna$ E_1 K_{15}. — (3) L_{18}; $dri\gamma\bar{u}šči\d{t}$ F_1 Pt_1 E_1 H_3; $dru\gamma\bar{u}šči\d{t}$ K_{15}; $dr\partial yušč\d{t}$ P_{13}. — (4) F_1 Pt_1 E_1 P_{13}; $hva\bar{\imath}š$ L_{18} H_4.

85. (1) deest L_{18}. — (2) F_1 E_1 H_3; $vaxš$ Pt_1 L_{18} P_{13}. — (3) $vihapta$ F_1 Pt_1 E_1 K_{15} H_3. — (4) F_1 Pt_1 E_1 L_{18} P_{13} H_4; $gaošači\d{t}$ K_{15}; $gaošča\d{t}$ K_{40}; $gaoči\d{t}$ H_3; $gušči\d{t}$ J_{10}.

86. (1) all MSS., exc. J_{10} Ml_2 $var\partial ti$. — (2) F_1 Pt_1 E_1 P_{13}; $azimana$ L_{18} H_3; $az\partial mna$ K_{15}. — (3) $paitišm°$ L_{18}; $paiti$ $šm°$ F_1 Pt_1 E_1; $paiti$ $mar\partial mna$ K_{15}. — (4) J_{10}; $ka\d{t}$ $n\bar{o}$ H_4; $kapa$ $n\bar{o}$ F_1 Pt_1 E_1 L_{18} P_{13}; $kafa$ $n\bar{o}$ K_{40}; $kap\bar{o}$ $n\bar{o}$ K_{15} P_{13}. — (5) H_3; $fra\bar{o}urva\bar{e}sy\bar{a}ite$ F_1 Pt_1 E_1; $fra\bar{o}urva\bar{e}sy\bar{a}iti$ K_{15} L_{18}; $°va\bar{e}isyaiti$ P_{13}. — (6) $az\partial m$ $n\d{q}m$ F_1 Pt_1 E_1 L_{18} P_{13} $K_{15,40}$ $H_{3,4}$ Ml_2; $azimanqm$ J_{10}.

87. (1) F_1 Pt_1 E_1 K_{15}; $a\d{t}$ L_{18} P_{13}. — (2) P_{13}; $xšnut\bar{o}$ F_1 Pt_1 E_1 L_{18} K_{15} H_3.

88. (1) The next six words desunt K_{15}. — (2) $b\partial r\partial zi$ Pt_1 E_1; $b\partial r\partial za$ L_{18} P_{13}; F_1 has the words $paiti$ $b\partial r\partial zi$ appended sec. m.; in H_4 the text runs: $bar\partial zasta$ $v\bar{o}h\bar{u}$ $man\bar{o}$ $paiti$ $bir\partial zaihi$. — (3) H_3; $v\bar{o}če$ F_1 Pt_1 E_1 K_{15} L_{18}; $vače$ P_{13}. — (4) $h\bar{u}$ $kair\bar{\imath}m$ L_{18}; $h\bar{u}kairim$ F_1 Pt_1 E_1 K_{15} H_3; $hukairim$ P_{13}. — (5) $bar\partial smn$ Pt_1 E_1 P_{13}; $bar\partial s\partial mna$ K_{15} E_1 H_3; $bar\partial s\partial m$ L_{18}. — (6) L_{18}; $v\bar{a}\gamma žiby\bar{o}$ J_{10}; $va\gamma ž\partial by\bar{o}$ F_1 Pt_1 E_1 P_{13} K_{15}.

APPARATUS [89–98

89. (1) $Pt_1 E_1 L_{18} P_{13}$; F_1 has *Yasnīm* corrected to $°nəm$. — (2) $F_1 Pt_1 P_{13}$; *bərəza* $E_1 K_{15} H_3$; *bərəzō* L_{18}. — (3) desunt $K_{15,12}$.
90. (1) F_1; *haōmąn* $K_{15} E_1 H_3$; *haōmąm* $Pt_1 L_{18} P_{13}$. — (2) $F_1 L_{18} E_1$; *mānyut°* Pt_1; *mainyū tāsta* P_{13}. — (3) $Pt_1 P_{13}$; *bərəzayaṯ* F_1 (above *z* is written *j*); *bərəzjayaṯ* E_1; *bərəza jayaṯ* K_{15}; in L_{18} the clause is wanting. — (4) F_1; *bərəjayan* $Pt_1 L_{18} P_{13}$; *bərəzayən* $K_{15} E_1$. — (5) *durāṯ* $F_1 K_{15}$; *drāṯ* $Pt_1 L_{18}$; *darāṯ* P_{13}.
91. (1) *čašmaini* $F_1 Pt_1 L_{18} P_{13} K_{15}$. — (2) $F_1 Pt_1 E_1$; *Yasnyō* $L_{18} P_{13}$. — (3) $F_1 Pt_1 E_1 L_{18} J_{10}$; *frastarətāṯ* $P_{13} K_{15} H_4$. — (4) $F_1 E_1 K_{15}$; *barəsmənə* P_{13}; *barəsman* $Pt_1 L_{18}$.
92. (1) cf. Y 57^{24}. — (2) $F_1 Pt_1 E_1 K_{15} P_{13} H_3$; *fraorənti* $L_{18} J_{10}$. — (3) $Pt_1 L_{18} P_{13} J_{10}$; *ārmaite* $F_1 E_1 K_{15} H_3$; *ārmaitiš* H_4. — (4) $F_1 Pt_1 L_{18}$; *bərəza* $P_{13} K_{15} E_1$. — (5) H_3 inserts here *ahurō* which is wanting in $F_1 Pt_1 L_{18} P_{13} H_4 J_{10}$. — (6) $F_1 E_1 K_{15,40} H_3$; *hvapå* $Pt_1 L_{18} P_{13} J_{10} Ml_2$. — (7) *Yōi* $F_1 Pt_1 L_{18} P_{13}$; *Yō* $E_1 K_{15} H_3$. — (8) $K_{40} Ml_2$ append *ča* which is wanting in $F_1 Pt_1 E_1 L_{18} P_{13} K_{15} H_{3,4}$. — (9) all MSS.; qu. *vahištəm*?
93. (1) cf. Y $57^{25 sq.}$. — (2) $Pt_1 E_1 L_{18} P_{13} H_3$; F_1 has *drvaδaṯēibyō*, but *ṯ* is struck out and δ appears to have been corrected to *t*; *drvaδibyō* H_4. — (3) $F_1 Pt_1$; *haēnəbyō* $E_1 L_{18} P_{13} K_{15}$; *hinaēbyō* H_3. — (4) *usixrūrəm* $F_1 Pt_1 E_1 L_{18} K_{15}$. — (5) $F_1 Pt_1 E_1 K_{15}$; *gərəwąn* $L_{18} P_{13} H_3$. — (6) $F_1 Pt_1 E_1$; *darō məbyō* L_{18}; *dru maēibyō* P_{13}. — (7) $F_1 Pt_1 E_1 K_{15} L_{18}$.
94. (1) $F_1 Pt_1 E_1$; *dāyå* $L_{18} P_{13}$. — (2) $F_1 E_1 K_{15} H_3$; *pouru paxštīm* $Pt_1 L_{18} P_{13} K_{40}$; *spaxštəm* J_{10}; *xšpaxstīm* H_4. — (3) $J_{10} Ml_2$; *ṯbišayatąm* F_1; *ṯbišayatąm* Pt_1; *ṯbaišyantąm* $L_{18} H_4$; *dabaišyantąm* P_{13}; *ṯbaešyantąm* K_{40}; *ṯbaešayatąm* E_1; *ṯbaešayntąm* H_3 (and below). — (4) $F_1 Pt_1 E_1 L_{18} K_{15}$; *duš mainyūnąm* $P_{13} J_{10}$; *°yavanąm* Ml_2. — (5) Ml_2; *haθrānavāitīm* $H_{3,4} P_{13} J_{10}$; *haθra* $F_1 Pt_1 E_1 L_{18} K_{15}$; *navāitīm* L_{18}; *navāitim* $E_1 K_{15}$; *navaitīm* Pt_1; *navaitim* F_1. — (6) $L_{18} P_{13} H_4$; *hamarəθanąm* $F_1 Pt_1 K_{15}$. — (7) *ṯbišyantąm* P_{13}; *ṯbišayatąm* $F_1 Pt_1$; *ṯbašayatąm* $E_1 K_{15}$; *ṯbaišayantąm* L_{18}; *ṯbaešyantąm* J_{10}.
95. (1) L_{18}; *zim* $F_1 Pt_1 E_1 P_{13} K_{15}$. — (2) *avakərəna* K_{40}; *kərəni* H_4; *darana* $F_1 Pt_1 E_1 L_{18} P_{13} K_{15} H_3 Ml_2$; *darane* J_{10}. — (3) $F_1 Pt_1 E_1 L_{18} P_{13} K_{15}$; *aṇhå* H_3.
96. (1) *adrazimanō* K_{40}; *družimnō* $F_1 Pt_1 E_1$; *drujimanō* $L_{18} P_{13} H_3$; *druzimanō* K_{15}; *draoj°* J_{10}. — (2) $Pt_1 L_{18}$; *vīrōnayåṇčim* $F_1 E_1 K_{15}$; cf. Yt. 10^{132}. — (3) all MSS. — (4) $F_1 Pt_1 E_1 L_{18} K_{15}$; *frahixštəm* P_{13}. — (5) $F_1 Pt_1 E_1 L_{18} K_{15} H_3$; P_{13} *zayanąm* and *zaenąm*.
97. (1) $F_1 Pt_1 E_1 L_{18} P_{13} H_3$; *yahmāṯ* K_{15}. — (2) $F_1 Pt_1 L_{18}$; *vīspa* $E_1 K_{15}$.
98. (1) all MSS. — (2) $Pt_1 E_1 P_{13} K_{15}$; *miθri* L_{18}; *miθrō* $H_{3,4} K_{40}$; *miθrahe* $J_{10} Ml_2$; F_1 has *miθra Yō*, but *a Yō* sec. m.; the original reading is not recognizable. — (3) deest $L_{18} P_{13} J_{10} Ml_2$. — (4) $F_1 Pt_1 E_1 L_{18} P_{13}$; *gaoyōtōiš* $J_{10} Ml_2$. — (5) desunt K_{15}. — (6) $F_1 Pt_1 E_1 K_{15}$; *fraxštaye* P_{13}; *fraxštayaēhaiti* L_{18} instead of *fr° p°*. — (7) *āyazəmā* $F_1 E_1$; *ayazəmā* $Pt_1 P_{13}$; *ayazəm* L_{18}.

APPARATUS

99. (1) $Pt_1 E_1 K_{15}$; *fratərəspənta* F_1; °*tərəta* L_{18}; °*tarəta* P_{13}. — (2) $F_1 Pt_1 E_1 K_{15}$; °*ti* $L_{18} P_{13}$; *frvāzāti* H_3.

100. (1) $F_1 Pt_1 E_1 K_{15}$; °*ti* $L_{18} P_{13}$. — (2) $F_1 Pt_1 E_1 L_{18} K_{15} H_3 Ml_2$; *vairyastārəm* P_{13}; *vairyāstārəm* H_4. — (3) $F_1 E_1$; °*ti* $Pt_1 L_{18} P_{13} K_{15}$. — (4) all MSS. — (5) all MSS., exc. H_3 *ašāunąm*.

101. (1) $Pt_1 P_{13}$; *avidīš* $F_1 E_1 K_{15}$; *avadīš* L_{18}. — (2) all MSS. *ašavō*. — (3) *ərəzfyō* $F_1 Pt_1 E_1 P_{13}$; *parana* $F_1 Pt_1 E_1 L_{18} P_{13} K_{15}$. — (4) $F_1 Pt_1 E_1 L_{18}$; *miθranayā̊* $P_{13} K_{40}$; *ava* $P_{13} K_{40}$. — (5) $F_1 Pt_1 E_1 K_{15}$; °*yeti* $L_{18} P_{13}$. — (6) $F_1 E_1 K_{15}$; *avaya* Pt_1; *avayaspa* $L_{18} P_{13}$. — (7) all MSS., exc. Ml_2 *virača*.

102. (1) $F_1 Pt_1 E_1 K_{15} L_{18}$; *arəštīm* $P_{13} J_{10}$. — (2) Ml_2; *išum* $F_1 Pt_1 E_1 K_{15} L_{18} P_{13}$.

103. (1) cf. Y. 57^{15} seq.. — (2) *anavaɴuha bədəmnō* F_1 (in this *ə* struck out) $Pt_1 K_{15}$.

104. (1) *fragrəwənti* $F_1 E_1 K_{15}$; *fratrəwənti* $Pt_1 L_{18} P_{13}$. — (2) cf. Y. 57^{29}; *ušistaire* $F_1 E_1$; *ušistairi* K_{15}; *ušistūaire* $Pt_1 P_{13}$; *ušištvaire* L_{18}. — (3) *hindvō* $F_1 Pt_1 E_1 H_4 K_{40}$ and K_{15} pr. m. as a correction of *həndvō*; *həndvō* $L_{18} J_{10}$; *hənδvō* P_{13}. — (4) *āgə̄urvayeite* $F_1 Pt_1 E_1$; *āgə̄urvaireiti* K_{15}; *āgə̄urvaite* L_{18}; °*rvaiti* P_{13}. — (5) $F_1 Pt_1 E_1 K_{15}$; *naɣne* P_{13}; in L_{18} this and the two preceding words are wanting. — (6) $F_1 Pt_1 L_{18} P_{13}$; *sanike* $E_1 K_{15}$. — (7) $F_1 Pt_1 L_{18}$; *vīmaēδīm* $E_1 K_{15}$; *vī maiδəm* P_{13}; $E_1 K_{15}$ insert before this word *vīmiṭ*, H_3 *vīmaṭ*, K_{40} *vīmaiṭ*, K_{12} *vīmaiδe*.

105. (1) $F_1 Pt_1 E_1 L_{18} P_{13} H_{3,4} J_{10} K_{40}$; *təmča* Ml_2; *tumčiṭ* K_{15}. — (2) $F_1 Pt_1 E_1 K_{15}$; *hanagr*° P_{13}; *hangarəm fršəmanō* L_{18}; *hangərəfšəmnō* H_3. — (3) F_1 (in this *e* seems to be a correction of *i*) $Pt_1 E_1 P_{13}$; *bāzuwi* L_{18}; *bāzividuš* K_{15}. — (4) $F_1 Pt_1 E_1 P_{13} K_{15} H_3 Ml_2$; *nastō* $L_{18} J_{10}$; *nistō* H_4. — (5) all MSS., exc. L_{18} *aštitō*. — (6) $F_1 Pt_1 E_1 L_{18}$; *aɴuhayaθa* P_{13}; *aɴuhya* K_{15}; *aɴhuya* H_3. — (7) $F_1 Pt_1 E_1 L_{18} K_{15}$; *aθa* $H_{3,4} K_{40}$. — (8) $F_1 Pt_1 E_1 L_{18} P_{13} K_{15} H_3$; *druxtə̄* H_4; *druxta* $K_{40} Ml_2$; *draoxti* J_{10}. — (9) $F_1 Pt_1 E_1 K_{15} L_{18} P_{13}$; *apašma* H_3; *apišmi* K_{40}.

106. (1) $F_1 Pt_1 E_1 K_{15}$; *mainya* $L_{18} H_3 J_{10}$; *ainya* P_{13}. — (2) $F_1 Pt_1 E_1 K_{15}$; *satē* $L_{18} P_{13}$; *stə̄* $H_4 K_{40}$; *stə* H_3; *sti* J_{10}. — (3) $F_1 Pt_1 K_{15}$; *mainyete* $E_1 L_{18} P_{13}$. — (4) $F_1 Pt_1 E_1 L_{18} P_{13} H_3$; *stə̄* $K_{15,40}$. — (5) $F_1 Pt_1 E_1 L_{18}$; *dužvxtəm* $P_{13} K_{15}$. — (6) $F_1 Pt_1 L_{18} K_{15}$; *satē* P_{13}.

107. (1) $F_1 Pt_1 L_{18}$; *satē* P_{13}; *stə̄* $E_1 K_{15}$. — (2) $F_1 E_1$; °*ti* $K_{15} P_{13}$; $Pt_1 L_{18}$ have both. — (3) $F_1 Pt_1$; *mainyum* $K_{15} E_1$. — (4) $F_1 Pt_1 E_1 L_{18} K_{15}$; *satē* P_{13}. — (5) $F_1 Pt_1 L_{18}$; *Yaoxtiš* P_{13}. — (6) $F_1 Pt_1 E_1 H_3$; *drujintəm* K_{15}; *drujintim* L_{18}. — (7) $F_1 Pt_1 E_1 P_{13}$; °*ti* K_{15}; *fraxšaθrahe* L_{18} (the next three words are wanting). — (8) $F_1 Pt_1 E_1 H_3$; °*ti* $K_{15} P_{13}$. — (9) all MSS., qu. *diδāiti*? — (10) $F_1 E_1 H_{3,4}$; *dōiθrābyō* $Pt_1 L_{18} P_{13} K_{15,40} J_{10} Ml_2$.

108. (1) *raēsča* $F_1 Pt_1 E_1 P_{13} K_{15} H_4$; *raeisča* L_{18}; *raēščiṭ* Ml_2. — (2) $F_1 Pt_1 L_{18}$; *baxšāne* $E_1 P_{13} K_{15}$. — (3) $F_1 Pt_1 K_{15}$; °*šāne* $E_1 L_{18} P_{13}$. — (4) $F_1 Pt_1 L_{18} P_{13}$; *usapara* H_3; *usa para* $E_1 K_{15}$; *uš para* K_{40}; *us upara* $Ml_2 H_4$. — (5) $F_1 Pt_1 E_1 L_{18} K_{15} H_4$; *barəzayən* P_{13}.

APPARATUS [109–117

109. (1) F_1 Pt_1 E_1 L_{18} K_{15} H_3; *hanisaxtəm* P_{13}. — (2) *ništāti* F_1 Pt_1 E_1 L_{18} K_{15} $H_{3,4}$; P_{13} has °*ta* corrected to °*te*; cf. Par. 111. — (3) F_1 Pt_1 L_{18} P_{13} H_3; *kiryeti* K_{15}; *kiryeite* E_1. — (4) *ištayēti* L_{18} $K_{15,40}$ J_{10} Ml_2; *ištayete* F_1 Pt_1 E_1 P_{13} $H_{3,4}$; deest K_{12}. — (5) F_1 Pt_1 E_1 P_{13}; *axšništahe* K_{15} $H_{3,4}$; *axīštahe* L_{18}; *axšnūtahe* J_{10}. — (6) F_1 Pt_1 E_1 L_{18} P_{13} K_{15} H_3; *miθramanō* K_{40}; *miθrəmanō* H_4; *miθrō* J_{10} K_{12}; in Ml_2 this and the four next words are wanting. — (7) *hūxšnūitīm* L_{18}; *hūxšnūitim* F_1 Pt_1 E_1; *huxšnūitəm* P_{13}.

110. (1) $K_{15,40}$ P_{13} $H_{3,4}$ J_{10} Ml_2; *dučiθrim* F_1 Pt_1 E_1; *dučiθrīm* L_{18}; *dužiθrəm* K_{12}. — (2) F_1 Pt_1 E_1 P_{13} K_{15} H_3; *jatā* L_{18}; *jati* J_{10} K_{12}.

111. (1) $x^vainisθaxtəm$ F_1 Pt_1 E_1 P_{13} K_{15} H_3; *hainisiθaxtəm* L_{18}. — (2) all MSS. — (3) F_1 Pt_1 E_1 L_{18} P_{13} K_{15} H_3. — (4) *miθrəmnō* J_{10}; the rest as Par. 109 n. 6.

112. (1) F_1 E_1; °*hātəm* Pt_1; °*haδəm* L_{18}. — (2) *vīsūpatīm* K_{15}; *vīsō paitīm* P_{13} Ml_2; *vīsūptīm* F_1 Pt_1 E_1 $H_{3,4}$; *vīspō patīm* L_{18} J_{10}. — (3) F_1 Pt_1 E_1 L_{18} P_{13} K_{15}; *fryanå* J_{10}; *frāyanå* H_4. — (4) F_1 Pt_1 E_1; *yesətąm* L_{18} P_{13}. — (5) *daḣiyum* F_1 Pt_1 E_1 K_{15} L_{18} P_{13}. — (6) F_1 Pt_1 E_1 K_{15} H_3; *ačaraeti* P_{13}; *ā* deest L_{18}. — (7) F_1 Pt_1 E_1 L_{18} P_{13} $K_{15,40}$ $H_{3,4}$ Ml_2; *gaoyōiti* J_{10}. — (8) F_1 Pt_1 E_1 L_{18} P_{13} $K_{15,40}$ H_3 J_{10} Ml_2; *āhva* H_4. — (9) F_1 H_3; *fračraite* Pt_1 L_{18} P_{13}; °*čaraiti* H_4; *frača rāiti* J_{10}.

113. (1) F_1 J_{10}; *tada* P_{13}; *taṭa* Pt_1; *taṭ* L_{18} $H_{3,4}$. — (2) F_1 Pt_1 E_1 P_{13} K_{15}; *nōiṭ* L_{18}. — (3) F_1 Pt_1 E_1 L_{18} K_{15}; *avaṅhe* P_{13}. — (4) K_{15} H_3; *bərəzanti* F_1 Pt_1 L_{18}; *bərəzənti* P_{13}. — (5) F_1 Pt_1 E_1 H_4; *vāčəm* L_{18} P_{13} H_3. — (6) *ča* deest K_{15}. — (7) F_1 Pt_1 E_1 L_{18} K_{15} H_3; *navaiθyąn* P_{13} J_{10}; *navaθyąn* H_4 K_{40}. — (8) F_1 Pt_1 E_1 L_{18} P_{13} K_{15} H_3 Ml_2; *astayō* K_{12}; *astyō* K_{40}; *aštryō* J_{10}. — (9) F_1 Pt_1 E_1 L_{18} J_{10} Ml_2; *tata* P_{13}; *taṭ* $K_{15,12}$. — (10) F_1 Pt_1 E_1 L_{18} P_{13} H_3; °*ti* K_{15} J_{10} H_4. — (11) F_1 Pt_1 E_1 L_{18} P_{13} K_{15} H_3.

114. (1) a repetition from Yt 10[94]. — (2) F_1 Pt_1 P_{13}; *spaxtīm* L_{18}. — (3) *ṭbišyantąm* P_{13} J_{10}; *ṭbišayatąm* F_1 Pt_1 E_1 L_{18}; *δbišayatąm* K_{15}. — (4) F_1 Pt_1 E_1 L_{18} K_{15}; *duš mainyūnąm* P_{13}. — (5) *haθra navāitīm* F_1 Pt_1 L_{18}; *haθrā nūāitīm* P_{13}; K_{15} has *haθrā* corrected to °*ra*. — (6) K_{15}; *hamərəθranąm* F_1 Pt_1 E_1 P_{13}; *hamərətanąm* L_{18}.

115. (1) F_1 Pt_1 E_1 L_{18} K_{15}; *ratūō* H_3; *ratavō* P_{13} H_4.

116. (1) F_1 Pt_1 E_1 L_{18}; *supti* δ° K_{15}; *su paitiδ*° P_{13}. — (2) F_1 Pt_1 E_1 K_{15} P_{13} H_3; *varəzahāna* L_{18}; *vərəzāna* H_4. — (3) E_1 K_{15} H_3; *huyāyana* F_1 Pt_1; *hvyāyana* L_{18}; *huyå āyana* P_{13}; *huyāyin* H_4; *huyāiyine* K_{40}; *hūyå gane* J_{10}. — (4) F_1 Pt_1 E_1 P_{13} K_{15}; deest L_{18} J_{10}.

117. (1) F_1 Pt_1 E_1 P_{13} K_{15} H_3 Ml_2; *paitarə* L_{18}; *patare* J_{10}. — (2) F_1 Pt_1 E_1 L_{18} K_{15} $H_{3,4}$ J_{10}; *avi* P_{13} K_{40}. — (3) *hačate* F_1 Pt_1 E_1 L_{18} P_{13} K_{15} H_3; *hačaiti* H_4 J_{10}; *hačāiti* K_{40}. The four next words are wanting in L_{18} J_{10}. — (4) F_1 Pt_1 E_1 $K_{15,40}$ $H_{3,4}$; *ahmi* P_{13}: qu. *hamahe*? — (5) $H_{3,4}$; *ayąnaθa* F_1 Pt_1 E_1 K_{15}; *ayanaθa* P_{13}. — (6) Pt_1 E_1 K_{15}; *aṇhāiθi* P_{13}; F_1 has *aṇhaθāiti*, but *aθ* struck out. — (7) F_1 Pt_1 E_1 L_{18} P_{13} J_{10}; Ml_2 °*γne* or °*γna*.

118. (1) P_{13}; *āitī* Pt_1 E_1 L_{18} K_{15}; *āitū* F_1. — (2) F_1 Pt_1 E_1 P_{13}; °*ti* L_{18} K_{15}. — (3) desunt P_{13}. — (4) F_1 Pt_1 E_1 L_{18} P_{13} Ml_2; *zaošqm* $K_{15,12}$ $H_{3,4}$.
119. (1) F_1 Pt_1 L_{18} P_{13}; *Yazāiš* E_1 K_{15}. — (2) F_1 Pt_1 E_1 L_{18}; *framrviš* P_{13}. — (3) F_1 Pt_1 E_1 P_{13} K_{15}; *yazayata* L_{18}. — (4) *māzdayasna* L_{18}. — (5) F_1 Pt_1 E_1 L_{18} K_{15} $H_{3,4}$; *patara taēibya* P_{13}. — (6) all MSS. — (7) F_1 Pt_1 L_{18} H_4 K_{40} J_{10}; *parīnīnō* P_{13}; *pərənīnō* E_1 K_{15} H_3. — (8) Pt_1 L_{18}; °*ti* P_{13}; *fravzǻnte* E_1; *fravzvǻnte* K_{15}; *frazǻnte* F_1 (sec. m. *v* appended); *fravāzǻnte* H_3.
120. (1) F_1 Pt_1 E_1 K_{15}; *vīspa* H_3; *vīspe* J_{10}; *vīspəm* L_{18}; *vīspi* P_{13}. — (2) P_{13} H_4 J_{10}; *mazdayasnąm* F_1 Pt_1 E_1 L_{18} K_{15} H_3 Ml_2. — (3) F_1 Pt_1 H_4; *āvīstō* P_{13}; *āvstō* L_{18}; *āvištō* E_1 K_{15} Ml_2. — (4) F_1 Pt_1 E_1 P_{13} H_3; °*ti* L_{18}; *vaidyǻnti* K_{15}. — (5) F_1 Pt_1 E_1 P_{13}; °*ti* L_{18} K_{15}. — (6) F_1 Pt_1 E_1 K_{15}; *franharāṯ* L_{18}; *franurāṯ* P_{13}. — (7) L_{18} P_{13}; *xšnutō* F_1 Pt_1 E_1 H_4. — (8) K_{40}; the rest *hyaṯ*; cf. Par. 121.
121. (1) F_1 Pt_1 P_{13}; *zaoθranąmana* L_{18}. — (2) F_1 E_1 $K_{15,40}$; *hyaṯ* Pt_1 L_{18} P_{13} H_3 J_{10} Ml_2.
122. (1) F_1 E_1 K_{15} H_3; *frasnayanta* Pt_1 L_{18} P_{13}. — (2) F_1 Pt_1 E_1 L_{18} K_{15} H_3; *ākayanta* P_{13}; *ākayanti* J_{10}; *ādayante* H_4; *ādayanti* K_{12}. — (3) L_{18}; *tanum* F_1 Pt_1 E_1 P_{13} K_{15}. — (4) F_1 Pt_1 E_1 P_{13}; *frasnayanta* L_{18} K_{15}. — (5) *visaiti* F_1 Pt_1 E_1 L_{18} P_{13} K_{15} H_3. — (6) F_1 Pt_1 E_1 L_{18} (in this *ā* appended sec. m.) K_{15} H_3; *ākayanta* P_{13} Ml_2. — (7) F_1 Pt_1; *yānōiṯ* L_{18} P_{13} K_{40}.
124. (1) F_1 Pt_1 E_1 P_{13} Ml_2; *uz bāzåš* H_3; *uzāuš* K_{15}; *uzbāzuš* L_{18}. — (2) F_1 Pt_1 E_1 H_4; °*ti* L_{18} P_{13} K_{15}; *fra avazāti* H_3. — (3) L_{18} P_{13}; *srirəm* F_1 Pt_1 E_1 K_{15} H_4. — (4) F_1 Pt_1 E_1 K_{15} Ml_2; *paisəm* L_{18}; *paesanhəm* P_{13}.
125. (1) F_1 Pt_1 E_1 P_{13} K_{15} H_4; °*ti* L_{18}. — (2) *hama* $K_{12,40}$; *hami* F_1 Pt_1 E_1 P_{13}; *spaēti tahami* L_{18} H_3; *spaēti tahmi* K_{15}; *spitatahmi* H_4. — (3) *mainyuš* F_1 Pt_1 E_1 L_{18} K_{15}; *mainyə̄uš* P_{13} J_{10}; *mainyaoš* $H_{3,4}$. — (4) $K_{15,40,12}$ E_1 H_3 append *āaṯ* which is wanting in F_1 Pt_1 L_{18} P_{13} H_4 Ml_2. — (5) F_1 Pt_1 E_1 L_{18} P_{13} K_{15}; *pāiti šmuxta* H_4. — (6) F_1 Pt_1 E_1 L_{18} P_{13} K_{15} $H_{3,4}$. — (7) F_1 Pt_1 E_1 K_{15}; *Yaoxta* L_{18} P_{13} H_4. — (8) Pt_1 L_{18} H_4 Ml_2; F_1 has *imąmča* corrected pr. m. to *ivąmča*; *əvąmča* J_{10}; *aevąmča* P_{13} (the first *a* struck out); *isąmča* E_1 K_{15} H_3; *išąmča* K_{40}. — (9) F_1 Pt_1 E_1 $K_{15,40}$ H_3 Ml_2; *upairəspātā* P_{13}; °*spāta* H_4; *uairispātā* L_{18}.
126. (1) F_1 E_1; °*ti* Pt_1 L_{18} P_{13} K_{15}. — (2) F_1 Pt_1 E_1 K_{15} H_3; *hāvaya* L_{18} K_{40}; P_{13} has *hāvaya* corrected to *hāvōya*. — (3) H_4; *čištəm* F_1 Pt_1 E_1 L_{18} P_{13} K_{15} J_{10} H_3 Ml_2. — (4) Ml_2; *vanhvaiti* J_{10}; *vanuhaita* F_1 Pt_1 E_1 L_{18} K_{15}; *vanuhiti* K_{40}; *vanuhita* P_{13} H_3; deest H_4; *vanhasaite* instead of *v° sp°* K_{12}. — (5) F_1 L_{18} P_{13} K_{15}; *mazd°* Pt_1.
127. (1) F_1 Pt_1 E_1 L_{18} P_{13} H_3; *hukərpa* K_{15}; cf. Yt. 10[70], 14[15]. — (2) so J_{10} as correction of *paiti ərətō*; *paiti rənō* $H_{3,4}$ K_{40} Ml_2; *paētərənō* F_1 Pt_1 E_1 L_{18} K_{15}; *paētaranō* P_{13}. — (3) *asvrahe* F_1 Pt_1 E_1 P_{13} K_{15}; L_{18} abbreviates. — (4) F_1 Pt_1 E_1 P_{13} K_{15}; *yuxtahe* Ml_2. — (5) F_1 Pt_1 E_1 $K_{15,40}$ H_3;

pāirəv° P_{13}; pairivāzahe L_{18}. — (6) F_1 Pt_1 E_1 K_{15}; nixšta P_{13} H_4 K_{40}; nixta L_{18}; naxšata H_3. — (7) deest K_{12}, but is found in the rest.

128. (1) hištaēte L_{18} P_{13}; °taēite F_1 Pt_1 E_1 K_{15}. — (2) Pt_1, and F_1 as a correction pr. m. of θanvarətanąm; θanvarətinąm E_1 L_{18} K_{15} H_4; θanvarə tanąm P_{13} H_3. — (3) all MSS.; asti yō desunt P_{13}. — (4) F_1 Pt_1 E_1 L_{18} P_{13} K_{15} Ml_2; gavainehe H_4; gavašti J_{10}; daēvayasna K_{12}. — (5) F_1 E_1; snāvya Pt_1 P_{13}; šnāuya H_4; snāya K_{15} H_3; snāvhe snāvya L_{18}; snāva snāvayō J_{10}. — (6) F_1 Pt_1 E_1 L_{18} P_{13} K_{15} J_{10}; jaya H_3; ǰyā K_{12}. — (7) L_{18}; mainyava så F_1 Pt_1 E_1 P_{13} K_{15}.

129. (1) F_1 Pt_1 E_1; °ti K_{15} L_{18} P_{13}. — (2) ašunąm F_1 Pt_1 E_1 K_{15}; ašaonąm L_{18} P_{13} $H_{3,4}$ J_{10} K_{40}. — (3) paranąm all MSS. — (4) F_1 Pt_1 E_1 K_{15} L_{18}; zafąm P_{13}. — (5) deest P_{13}; aršti J_{10}. — (6) F_1 Pt_1 E_1 L_{18} K_{15} $H_{3,4}$; yå aŋhən K_{12}; yā aŋhən J_{10} Ml_2: qu. yā ayaŋhaēna?

130. (1) F_1 Pt_1 E_1 H_3; barōiθrō P_{13} K_{15} H_4 J_{10}; barōθrō L_{18}; tižinąm H_3 K_{40} J_{10}. — (2) K_{15} repeats this sentence, also K_{12} with zarastanąm for arštinąm. — (3) čakušanąm F_1 Pt_1 (here k appears to be a correction of γ); čakušnąm E_1 $K_{15,40}$ H_3; ča gušanąm L_{18}; ča γušanąm P_{13}; ča ušnąm H_4; deest Ml_2. — (4) all MSS.

131. (1) F_1 Pt_1 E_1; avō d° P_{13}; dātranąm L_{18}.

132. (1) vazarəm F_1 Pt_1 E_1 L_{18} K_{15} H_4; hazaŋrəm P_{13}. — (2) F_1 Ml_2; °fšānəm Pt_1 L_{18} P_{13} K_{15}; cf. Yt. 10^{96}. — (3) F_1 Pt_1 L_{18} P_{13} Ml_2; fravaēδəm E_1 K_{15} H_3. — (4) nayåncim F_1 Pt_1 E_1 K_{15}; anyåntim P_{13}; deest L_{18}. — (5) $H_{3,4}$ Ml_2; zarōišiayaŋhəm F_1 (in this °əm corrected pr. m. to °ō) P_{13}; °yaŋhəōm Pt_1; išiayaŋhəm L_{18} (in this ō too appended to m sec. m.). — (6) P_{13}; zarənyehe F_1 Pt_1 E_1 L_{18} K_{15}. — (7) all MSS. — (8) vazanti P_{13}; vazainte F_1 Pt_1 E_1; the clause is wanting in L_{18} K_{15}.

133. (1) ne γninti J_{10}; niγaninti F_1 Pt_1 E_1 L_{18} P_{13} K_{15} H_3; niγainti H_4; niγainte K_{40}; in P_{13} the rest of this Par., and Par. 134, are wanting. — (2) F_1 Pt_1 E_1; °ti L_{18} K_{15}. — (3) F_1; arəzahe Pt_1 E_1 L_{18} K_{15} H_3. — (4) F_1 Pt_1 E_1 K_{15}; savahe H_3; savaŋhe L_{18}. — (5) barəšta F_1 Pt_1 E_1 L_{18} K_{15}. — (6) F_1 Pt_1; jarəšta E_1 L_{18} K_{15}.

134. (1) aēsmō F_1 Pt_1 E_1 K_{15} H_3 Ml_2; cf. Yt. 10^{97}. — (2) H_4; pišō° F_1 Pt_1 L_{18}; pašō E_1 K_{15}.

135. (1) repeated from Yt. 10^{98}. — (2) all MSS. — (3) desunt K_{15}. — (4) F_1 Pt_1 L_{18} P_{13} K_{15}; aš H_4 Ml_2; deest H_3. — (5) F_1 Pt_1; fraxštāte E_1 K_{15} H_3; fraxštaiti L_{18}. — (6) F_1 Pt_1 L_{18} P_{13} J_{10}.

136. (1) F_1 Pt_1 E_1 L_{18} P_{13} K_{15} $H_{3,4}$; °tō K_{40} Ml_2. — (2) Ml_2; Yuxta F_1 Pt_1 E_1 P_{13} H_4. — (3) H_3; θan jasånti K_{40}; θanjaysånte F_1 Pt_1 E_1 L_{18} Ml_2; θanjaysånti K_{15}; θanjayasånte H_4; θanjyånti J_{10}. — (4) F_1 Pt_1 E_1 L_{18} K_{15} H_4; ašava P_{13}; ava K_{12}. — (5) F_1 Pt_1 P_{13}; čixra L_{18}; čaθra E_1; čaθara K_{15}; čiθra $H_{3,4}$. — (6) F_1 Pt_1 E_1 K_{15} H_4; ašā nasča K_{40}; aspānasča P_{13}; spānasča L_{18} Ml_2; spānišču J_{10}. — (7) yezišē J_{10}; Yezī F_1 Pt_1 E_1 L_{18} P_{13} K_{15} H_3. — (8) F_1 E_1 K_{15} H_3; avi še desunt P_{13}; še deest Pt_1 L_{18} J_{10}; ava J_{10}.

137. (1) H_3; ušte F_1 Pt_1 E_1 L_{18} P_{13} H_4; ušti K_{15} J_{10}. — (2) H_3; uitī F_1 Pt_1 E_1 L_{18} P_{13} K_{15} H_4. — (3) K_{15} (pr. m.); zaotō F_1 Pt_1 E_1 L_{18} P_{13} $H_{3,4}$ $K_{40,15}$ (sec. m.); in Ml_2 these words are wanting. — (4) F_1 Pt_1 E_1 L_{18} P_{13} K_{15} H_4; frastarətāṯ H_3 K_{40}. — (5) F_1 Pt_1 E_1 P_{13} K_{15}; barəšmən L_{18}; barəsman H_4. — (6) F_1 Pt_1 E_1 P_{13}; °ti L_{18} K_{15} H_3. — (7) F_1 Pt_1 E_1 L_{18} K_{15} H_4 J_{10} Ml_2; rāstəm P_{13} H_3. — (8) E_1 K_{15} H_3; manyāi Pt_1 L_{18} P_{13}; F_1 has mānyāi corrected to manyāi. — (9) yazišē J_{10}; Yezī F_1 Pt_1 E_1 L_{18} P_{13} K_{15} H_4. — (10) Pt_1 E_1 L_{18} Ml_2; Y̨ąnāδa F_1 (a̯ corrected sec. m. to ā); yānāṯa K_{15} H_3; yānāṯ P_{13} H_4. — (11) F_1 Pt_1 E_1 L_{18}; sastarāi K_{15}; anusastarāi P_{13} H_4 J_{10}. — (12) manyāi F_1 Pt_1 E_1 L_{18} K_{15}; anumanyāi P_{13} H_4 K_{40}.

138. (1) F_1 Pt_1 E_1; uitī L_{18} P_{13}. — (2) L_{18}; zaotō F_1 E_1 $H_{3,4}$ Ml_2; zaōto Pt_1 P_{13}; K_{15} as Par. 137 n. 3.

139. (1) F_1 Pt_1 E_1 K_{15}; aməšāspənta L_{18}. — (2) rašnumča F_1 Pt_1 E_1 L_{18} P_{13} K_{15}.

140. (1) F_1 Pt_1 E_1 L_{18} $K_{15,40}$ H_3 Ml_2; vaṇhə̄uš P_{13} K_{12} J_{10}. — (2) all MSS., exc. Ml_2 uγrīm. — (3) F_1 Pt_1 E_1 L_{18} P_{13}; aməθwəm $K_{15,40}$ $H_{3,4}$.

141. (1) F_1; təmaṇhāṯa Pt_1; təm aṇhāδa E_1; təm aṇhāṯa H_3; təmaṇhāδaṯ L_{18}. — (2) F_1 E_1 K_{15} H_4; jiγaurvm Pt_1 P_{13} L_{18}; jaγāurūm H_3. — (3) all MSS. — (4) F_1 Pt_1 E_1 K_{15}; tanjištəm Ml_2; tančištəm L_{18} P_{13}. — (5) aš F_1 Pt_1 E_1 K_{15} L_{18} $H_{3,4}$; ašahe P_{13}. — (6) L_{18} P_{13} J_{10} Ml_2; x^varən F_1 Pt_1 E_1 H_4; haranaṇhčamnō H_3; deest K_{15}. — (7) hazaṇrā F_1 Pt_1 L_{18} P_{13} H_4; °rāi E_1 $K_{15,12}$ H_3.

142. (1) F_1 Pt_1 E_1 L_{18} P_{13} K_{15} H_4 J_{10}; paōuruiš H_3; paouruš K_{12} Ml_2. — (2) F_1 Pt_1 E_1 L_{18} P_{13} K_{15} H_4; vīduš J_{10} Ml_2. — (3) all MSS., exc. J_{10} sarəm. — (4) F_1 Pt_1 E_1 P_{13} $K_{15,40}$ $H_{3,4}$; dāmanąm L_{18} J_{10} Ml_2. — (5) Pt_1 E_1 L_{18} P_{13} K_{15} H_3; må̇ṇhəm H_4; F_1 has må̇ṇhəm corrected pr. m. to må̇ṇhō. — (6) F_1 Pt_1 E_1 P_{13}; hvāraoxšni L_{18}.

143. (1) F_1 Pt_1 E_1; barāzaiti L_{18} P_{13} K_{15} H_4. — (2) F_1 E_1 P_{13} K_{15}; hanrəwnāiti Pt_1 L_{18}. — (3) F_1 Pt_1 E_1 P_{13} K_{15}; aδavaiš L_{18} H_4; adūiš J_{10}. — (4) F_1 Pt_1 E_1 L_{18} P_{13} H_4 K_{40} J_{10} Ml_2; paoiryš K_{15}; paōuirīš H_3. — (5) F_1 Pt_1 E_1 P_{13} $K_{15,40}$ H_3; hūbāmya L_{18} J_{10}. — (6) F_1 Pt_1 E_1 L_{18} P_{13}; K_{15} has xšaēštāiš, but the last š struck out. — (7) mainyū P_{13}.

144. (1) jiγaurvå̇ṇhəm F_1 Pt_1 E_1 K_{15}; jaγāur° L_{18}; jiγāur° P_{13}. — (2) cf. Ny. 2^{11}. — (3) dahyum F_1 (and daḣiyūm) Pt_1 E_1 L_{18} P_{13} K_{15} H_4 J_{10} both here and below.

145. (1) F_1 Pt_1; aiθyaējaṇha E_1; iθyae jaṇha K_{15}; L_{18} P_{13} H_3 Ml_2 abbreviate. — (2) J_{10} H_4 K_{40}; °āhu F_1 Pt_1 E_1 K_{15}; cf. Ny. 2^{12}.

ADDENDA

p. 8. A great advance in the understanding of the Rigvedic hymn to Mitra has meanwhile been achieved by Thieme, *Mitra and Aryaman*, pp. 39 *sqq*. Against Gdn.'s description of the hymn as 'ganz farblos' Thieme notes (p. 39) that it is 'as colourful and significant as any other hymn in the Rig Veda'. Gdn.'s notion that Mitra is 'the Friend' is rightly dismissed by Thieme as unsupported by the Rigvedic evidence (pp. 51, 58). The hymn is now clearly seen to portray the same god of contract as the Avestan Mithra Yašt, except in so far as a cosmic function is attributed to Mitra only (he 'keeps heaven and earth' in *v*. 1; see Thieme, p. 42). Of the similarities noted by Thieme between Mitra and Mithra the following are impressive: (*a*) Mitra is 'unblinking' (*v*. 1), Mithra 'sleepless and ever-waking' (st. 7); (*b*) Mitra is 'characterized by width' (*sapráthas*, *v*. 7), Mithra is 'as wide as the earth' (st. 95, cf. also st. 44); in both hymns the god (*c*) removes *aṃhas/azah-* (*v*. 2; st. 22; Thieme, p. 46), (*d*) induces joyfulness (*v*. 4; st. 34; Thieme, p. 51), and (*e*) must be invoked by name (*v*. 1, *bruvāṇás*; sts. 31, 54 *sqq*., 74, 137; cf. Thieme, p. 59). With regard to (*b*) one may also refer to *RV* 5.65.5 ('may we be under Mitra's protection, which is superlatively extended [*saprathastama*]'), and with regard to (*c*) to the preceding *v*. 4: 'Even from *aṃhu*, congestion, Mitra wrests spaciousness, free approach to one's dwelling', which recalls the 'spaciousness' of st. 5, and the preoccupation with dwellings in sts. 75 and 77. It is true that some of the above traits do not belong to Mit(h)ra alone,† but their convergence upon the same god in the Rig Veda and the Avesta supports Thieme's contention that Mitra is essentially the same god as his Avestan counterpart.

pp. 30 *sq*. Lüders' connection of Mitra with fire is supported by Thieme's ingenious suggestion that in *v*. 4 of the Mitra hymn the god is identified with fire (pp. 49 *sq*.; cf. also p. 84).

p. 31. Thieme thinks, as we do, that it is reasonable to suppose Mithra's worshippers would want 'something to look at' when praying to their god (p. 38). This something, Mithra's 'concrete manifesta-

---

† Thus, with regard to (*d*), Anāhitā, too, seems to induce *haomanaṇha-* in *Yt* 5.8, and the identical wording of *v*. 4 of the Mitra hymn is applied in the RV to Agni (3.1.21), Indra (6.47.13), and the Aṅgiras, etc. (10.14.6). As to (*e*), apart from Mithra the following divinities are said in the Avesta to (require to) be mentioned by name: Ahura Mazdāh (*Y* 5.3), the Aməša Spəntas (*Y* 15.1, 2 [= *Y* 51.22, see above, p. 165]), the Fravašis (*Yt* 13.79; cf. also *Y* 3.21, 22.27), Sraoša (*Y* 3.20), Vega (*Yt* 21.1), Sirius (*Yt* 8.11, 23–5), Fire (*Visp*. 9.5), and the waters and plants (*Yt* 13.79).

tion', was at a later stage the sun, the reason for this identification being according to Meillet and Thieme, p. 37, that both Mithra and the sun were thought of as seeing everything. However, in the Avesta Mithra is not yet a sun-god, but merely the light of daybreak, and the latter identification is all the less likely to have replaced an earlier one with the sun, as its beginnings can be traced to Rigvedic times. Thieme himself has pointed out (p. 69) the remarkable affinity of *Yt* 10.13 with RV 5.62.8: 'You, Mitra-Varuṇa, mount your gold-coloured, iron-propped seat at the flashing of dawn, at sunrise; from there you survey guilt and non-guilt.' I therefore see no reason to depart from the view that the later Iranian sun-god Mithra had developed from Mithra as representative of early daylight. Thieme scarcely does justice to sts. 13 and 142 *sq*. (in which the god behaves like the first morning light) when he dismisses the former passage as a 'suggestion', the latter as a 'comparison' (pp. 35 *sq*.).

In the present study we have not been concerned with the post-Rigvedic solar tendencies of the Indian Mitra, nor with the Indo-Scythian cult of Mihira as a sun-god which I. Scheftelowitz discussed at length in *Acta Orientalia*, XI, 293 *sqq*. We may note, however, that while the solar interpretation of Mithra on the part of the Indo-Scythian Sakas agrees with the Eastern Iranian development mentioned above, p. 41, the word for 'sun' in the language of the Khotanese Sakas is *urmaysde* (nom. sg.), and in Sanglechi *ormōzd*, both from *Ahura Mazdāh*. If it is rash to conclude from this terminology that contrary to the Avestan evidence Ahura Mazdāh was at an early stage a solar deity, it can scarcely be safer to by-pass the Avestan and Rigvedic evidence on Mit(h)ra, and antedate the identification of him with the sun.

p. 33. Thieme, p. 56, derives Mithra's dispensation of water (which survived into Western Mithraism, see Cumont, *TMMM*, I, 165 with notes 2 and 3) from his preoccupation with the covenant, rain being the result of the fulfilment by the gods of their part of the compact, the performance of the sacrifice being the part of men. Lüders, however, has shown that Varuṇa's bestowal of rain (which Thieme explains in the same way as Mithra's) is part of a far more intimate relation with water than is apparent in the case of Mit(h)ra. One may therefore prefer to see in Mithra's dispensation of water a relic of his former community of interests with Varuṇa, the god of oath and water.

p. 44. A trace of the dvandva *miθra ahura* can be seen in the name of the god Μεσορομασδης (Plutarch, *ad princip. inerudit.*, 3.780c), if S. Wikander is right in interpreting it as a compound of **Misa-* (as an OPers. form of Mithra's name) + *Ahuramazdāh*, see *Orientalia Suecana*, I (1952), 66 *sq*. Note, too, the legend ΜΙΘΡΑC ΩΡΟΜΑC ΔΗC on a tessera found at St. Albans, see H. Mattingley, *Numismatic Chronicle*, XII (1932), 54 *sqq*.

ADDENDA [pp. 45–155]

p. 45. After reading Thieme's account of Varuṇa's nature (pp. 54, 59 *sqq.*) I feel less inclined to stress the craftiness and 'grauenvolle Züge' (Lo., *Rel.* 273) of this god at the early Vedic stage, although Thieme, too, considers him to be 'dangerous' (p. 68). The guiding factor in the comparison of Mazdāh with Varuṇa must be the Aməša Spəntian structure of Mazdāh, which differentiates him so strongly from Varuṇa that nothing short of his bearing Varuṇa's name could make it credible that he had been evolved from the latter. There is in any case no compelling reason for assuming a genetic relationship between the two gods. What they have in common is sufficiently explained by the hypothesis that the Aməša Spəntian Mazdāh, sole god of a religious system which was growing on Iranian territory beside the inherited Varuṇa cult, had added to his own, unique, definition the essential traits of Varuṇa, thereby eventually ousting him.

p. 48, n. The latest attempt to grapple with Y 29.3 is Humbach's in *IF*, LXIII, 215. Although the author has made his the contention I put forward in *JRAS*, 1952, 174 and 177, that (*paitī.*)*mravaṭ* should not be a 3rd sg. pret., he achieves a very different result. The fact is that this stanza is utterly unfathomable, a real 'head-smasher', as one may well translate *sarəjā* in preference to Humbach's 'Schirmbrecher' ('Leidlosigkeit für die Kuh ist für [=verursacht] uns mit [=und] Aša kein Kopfzerbrechen'). There is nothing to show that this word, on which so much depends, is a compound, an -*an*- stem, or even a noun; one Iranian base *srg- possibly survives in Arm. *šaržem* 'to move', see Benv., *JAs.* 1936, 199, another, as G. Morrison once pointed out to me, in Pš. *šāyəl* 'to please', cf. Morg., *EVP*, 77. For the connection of Av. *sar*- with Ved. *śárman* Humbach should have quoted Andreas, *NGGW*, 1931, 323 *sq.* Similarly on p. 213, n. 8, of Humbach's article one misses a reference to *JRAS*, 1952, 176, where the connection of *būždyāi* in Y 44.17 with Ved. *bhūṣ*- had already been suggested.

p. 153, note 2[1]. As Thieme rightly points out (p. 80), the knave who in st. 2 wrecks 'the whole country' when he breaks a contract can only be the ruler of the country concerned.

p. 155. My suggestion that the Ossetic *idauæg*-s derive from ancient *artāvan*-s is vigorously opposed by G. Dumézil in *JAs.* 1956, 355 *sqq.*, who seems to have overlooked the saving clause regarding the initial ι · which I had entered in *BSOAS*, XVII, 484, n. 4, as well as the Greek references to ἀρταῖοι quoted above, p. 155. Without denying that the development I assumed is of an exceptional nature, I still prefer to trace the *idauæg*-s to a well attested Old Iranian notion, rather than rely with M. Dumézil on an invented one. [Note now Khwārazmian '*rδ'w* 'demon', see Henning, *Mitteliranisch* (*Handbuch der Orientalistik*), 117, n. 6.] It is also hard to share M. Dumézil's opinion, expressed in the same article, that the Ossetic *uæyug*-s, 'giants', descend from the ancient god *Vayu*: this Oss. common noun, as I indicated in *GMS*, §978, is likely to belong to the Sogd. (etc.) common noun *w'ywq* 'hunter', which Henning, *BSOAS*, XI, 471, n. 4, has connected with the Oss. verb *uain* 'to run'.

p. 157, n. §. For an attractive explanation of the notion of the 'path of Truth' Thieme, *Der Fremdling im Rigveda*, 118, n., should have been quoted.

p. 163, note 6². *haomayō gava* is interpreted by Thieme, *ZDMG*, 1957, 75 *sq.*, as a wrongly divided *haoma *yaogava*, with *-ao* changed to *-ō* in final position. The new word, representing an older **yava-gv-a-*† 'barley-milk', would be closely related to OInd. *yavāgū*. This is a very ingenious suggestion, but it does require the assumption that the original reading had been tampered with. Moreover, barley-milk is not otherwise mentioned in the Avesta, whereas the blending of milk and *haoma-* is a well-known Zoroastrian practice (cf. Bth., Wb., *s. vv. raēθwiškara-* and ⁵*gav-*, II, 3). Different again is K. Hoffmann's analysis, against which Thieme has raised valid objections, of *haoma yō gava* as 'with *haoma-* (instr.) which (nom.) (is [mixed]) with milk (instr.)', *MSS*, 8 (1956), 23.

p. 164, n. The YAv. form of the words *yeṅhē* and *vaṅhō* in the prayer (instead of expected Gāth. *yehyā* and *vahyō*) provides no valid reason for disbelieving the statement in *Y* 21.1 that Zarathuštra is the author of the *yeṅhē hātąm* prayer. Elsewhere, too, Gāthic words have been replaced by their YAv. equivalents in Gāthic quotations, cf. *drvantō Vend.* 5.62 (*drəgvantō Y* 31.20), *nmāne Vend.* 8.107, 14.18 (*dəmānē Y* 49.11), *arš mē Vend.* 19.10 (*ərəš mōi Y* 44.1), *yeṅhē Vend.* 20.8 (*yehyā Y* 31.4), *yeṅhē mē Y* 15.2 (*yehyā mōi Y* 51.22, see above, p. 165), and *mazda tava Y* 19.14 (*mazdā tavā Y* 53.9, see below, p. 329, n.). A similar case is *jiṯ. aša-* in *Vend.* 5.4, if taken, as Bth. thought, from *dōjīṯ. arəta-* in *Y* 53.9 (see on this word Burrow, *Sarūpa Bhāratī*, 5). It is noteworthy that the usual YAv. acc. pl. *tą* occurs in Gāthic only in *Y* 51.22, the stanza which is so closely allied to our prayer (see above, p. 165).

p. 169, n. Bth. actually took *hušitōiš* in *Y* 30.10 as an ablative, see *Wb.* 302, line 13.

p. 172, note 13³. A different etymology of Oss. *uændag* (Dig. *iuændæg*), as from **yavant-aka-*, has been suggested by G. Dumézil, *JAs.* 1956, 359.

p. 174, note 14¹. The association of *āfant-* with mountains in st. 14 and *Yt* 13.9 makes it tempting to derive from this OIr. word the name of the mountain in central Baškard which E. A. Floyer, *Unexplored Baluchistan*, 183, 194, spells *Aphen*, and which constitutes the second part of the name of the village *Garifin* (thus spelled by J. V. Harrison, *Geographical Journal*, 1941, 5). During the several months I spent in Baškard

† One should preferably start from Proto-Iranian **yava-gav-a-*, as **yava-gv-a-* would have resulted in YAv. **yao-va-*, cf. YAv. *hvōva-* from Gāth. *hvō.gva-*, or *aētava-* beside Ved. *étagva*.

ADDENDA [pp. 174-81

I never heard the two names pronounced otherwise than *Āhvęn* and *Garāhvęn*. This is in agreement with the phonology of the Pīzgī dialects spoken in the Kūh-i Āhven and Garāhven districts, in which OIr. intervocalic *f* becomes *hv*, as seen in *vahv-* 'to weave' < *vaf-*. The connection of the toponym *Āhven* with Av. *āfant-* occurred simultaneously to M. Benveniste and myself as we were looking through some of my Baškard material during the Munich Congress of Orientalists in 1957.

p. 178, note 20³. I am indebted for an improvement of the above interpretation to Dr M. A. Mehendale, who points out that the fact that the thrower is an *avi.miθri-* ought to be sufficient reason for the spear to turn back; accordingly *frāna ayanąm mąθranąm* may be expected to convey not the cause of the reversal of the spear's flight, but an additional discomfiture of the *avi.miθri-*: although he performs evil spells they are of no avail. Such a meaning can be obtained e.g. by taking *frāna*, as a preposition, to mean 'in spite of', or by replacing Bth.'s 'because of the abundance of evil spells' by 'with the abundance...', in the sense of 'with (=despite) all the evil spells'.

p. 179, note 24³. To Gr. ξύω also belongs Sogd. *pryš'w-* 'to scrape' in *P* 3.284: *šy...cnn pwγt'k'ry šyr'kw zp'rt pryš'wy* 'after boiling it (viz. the skull) he shall scrape it quite clean'.

p. 179, note 24⁵. *puθrāne* may alternatively belong to a stem *puθrān-*, as is argued by K. Hoffmann, *MSS*, 6 (1955), 35 *sqq.*

p. 180, line 2. MPers. *hwnsnd* (to which belongs NPers. *xursand*, see Henning, *ZII*, IX, 220, n. 2) is derived by Benv., *JAs*. 1933, 243, from **xᵛani-sanda-*. On Gāth. *sąndā* *Y* 51.14, and *sąs* *Y* 46.19 Humbach, though also assigning the forms to ¹*sand-*, has other views. *sąndā* is taken by him together with the preceding *ā* as a compound meaning 'ungefällig' (*IF*, LXII, 304, n. 1). This is manifestly impossible. On the other hand, Humbach's revised translation of *Y* 46.19 (*MSS*, 2 (1957), 30, n.), in which *mōi sąs* is rendered by 'erglänzest du mir' (better: 'you appear to me'), is an improvement on previous attempts.

p. 181, note 26¹. It must be admitted that Gdn.'s and Bth.'s rendering of *akatarəm sraošyaṇąm* has the advantage of allowing the same syntactical structure in each of the first four lines of st. 26. I now give it preference, on the assumption, however, that *akatara-* is not a comparative of *aka-*, but an adjective in the positive degree, coined by contaminating two synonyms, *aka-* and *ātara-* 'wicked' (see Bth., *Wb*. 1887). The latter has in *Y* 12.4 the very construction Bth. assumed for *akatarəm* in our stanza. Translate: 'who smashes the heads of evil gods, is wicked towards those who deserve punishment, punishes men false to the contract, engages the witches'.

The idiom *daiṅhaom...amāi daδāiti* for 'grants strength to the country' (*daiṅhave aməm*) has a parallel in *xšmākəm vohū manaṅhā vahmāi dāidī savaṅhō Y* 51.2 'through Good Mind grant the prayer of

(=for) profit to (me who am) yours'. This Gāthic line is interpreted differently by Humbach, *MSS*, 9, 76. *dā-* has perhaps the same construction in the *Ahuna Vairya* prayer, see below, p. 329, n., at the end.

p. 182, note 27[1]. The IE base may, alternatively, be reconstructed with *k*: **l/renk(h)-*. Sogd. *rxn-* is now known to have survived with the same meaning in Yaγn. *raxn-*, see M. S. Andreyev and E. M. Peshchereva, *Ягнобские тексты* (1957), 313. Av. *rąxšyant-* is translated by Thieme, p. 33, as 'about to act crookedly (?)', on the strength of the Slavonic base *ląk-*, which has a similar meaning. *apivaiti* in Thieme's opinion stands for *api-yuvati and means 'he compresses'. The replacement of *-yuv-* by *-iv-*, however, is otherwise only attested after consonants, cf. above, note 125[1]. *apivaiti* from *vā(y)-* (as Bth. thought) is not more surprising than *fraxštaite*, st. 98, from *stā-* (beside *fraxštāite*, st. 107), or *avazat̰*, *Yt* 19.81, from *zā(y)-*.

p. 182, n. †. For the *g* of *lagad*, a word which is often quoted as *lakad*, cf. the loanword *layata* in Pš., *lagat* in Bal.

p. 184, note 32[1]. Humbach, *MSS*, 2 (1957), 11, n., and 9 (1956), 70, translates *xšnuyå̄ nō...yasnahe* by 'hear our prayer'. The OPers. evidence, however, suggests that *xšnu-* means 'to hear' only when used with the preverb *ā-*, cf. Benv., *TPS*, 1945, 50. The semantic development assumed by Humbach ('spitzen'–'Ohren spitzen'–'hören' on the one hand, and 'schärfen'–'stärken'–'freundlich behandeln' on the other) inspires no confidence. *xšnu-* 'to sharpen', with infixed *n* (see above, note 24[3] with addendum), is to be kept distinct from *xšnu-* 'to satisfy' (with preverb *ā-* '(obey>) listen, hear'). The *n* of the latter is likely to belong to the base, whether or not one agrees with Benv. that this is ultimately **ĝen-* 'to know'.

garō nmāna- 'Paradise' was thought by Bth. to mean literally 'house of praise'; Herzfeld, supported by Zaehner, *Zurvan*, 214, suggested 'treasure-house', Burrow, *BSOAS*, xx (1957), 144, 'house of reward'.

p. 187, note 35[1]. A different explanation of °(*t̰*)*čaēša-*, as an intensive adj. from **čaičya-*, has been advanced by Thieme, p. 53, n. 39.

p. 187, note 35[2]. This is to assume two Iranian bases *yuz-*, one meaning 'to investigate', the other 'to be in commotion'. Thieme, p. 31, n. 17, tries to combine Bth.'s etymology of *yaoxšti-* (to √*yug-*) with the meaning urged by Gdn. He translates the word by 'alertness' and considers this meaning to have developed from 'state of being harnessed (*yuxta-*)'.

p. 193, note 39[4]. The sentence θwąm at̰ aēšąm (*scil. vačaṇhąm*) *paityāstārəmčā fradaxštārəmčā dadəmaidē* in *Y* 35.9 was translated by Bth.: 'dich (Zarathuštra) bestellen wir als Einschärfer und Lehrer dieser (Worte)'. Humbach, *IF*, LXIII, 216 *sqq.*, takes *fradaxštar-* as belonging to *fradaxšanā-* 'sling', and translates it as 'Schleuderer', that is 'Äußerer'. As a parallel to this unusual semantic development Hum-

ADDENDA [pp. 193–200

bach adduces the preceding *paityāstar-*, which in his opinion belongs to the base ²*ah-* 'to throw' and therefore means 'Entgegenschießer (=Aussender, Äußerer)'. That Avestan writers, who had at their disposal no fewer than four verbs for 'to speak', should have felt the need to supplement these with *two* verbs for 'throwing' is not very probable, especially as both *fradaxštar-* and *paityāstar-* are better explained otherwise. The former can safely be assigned to *daxš-* 'to teach', as Bth. had done, the latter either to *dā-*, with Bth., or to *ad-* 'to speak': if, as is likely, Bth.'s 'Wiederholer' (hence 'Einschärfer') is the correct translation, *pati* in the finite forms of *pati-ad-* (cf. also the perfect *paityāδa*, Bth., *Wb.* 840) 'to reply, respond' on the one hand, in *paityāstar-* 'he who says again' on the other, can be assumed to have the two connotations of German *wi(e)der* in *erwidern* and *wiederholen*. Whether YAv. *paiti.asti-, paityāsti-* is to be separated from *paityāstar-*, as Bth. wanted, or associated with it, as Humbach prefers, is difficult to say, as the contexts are far from clear. At any rate Humbach seriously underrates and misquotes Bth. when he writes of him (p. 216): 'um morphologische Probleme unbekümmert hält er...*asta-* für das P.P.P. von *stā-*'. Bth. printed °*a-sta-*, and justifiably interpreted *a-* as the shortened preverb, and *-sta-* as the outcome of **sthətó-*.

p. 194. Note that *ašivå* in Y 51.5, which is generally believed to be a strange gen. dual of *aši-*, is better explained as the nom. sg. masc. of *ašivant-* (cf. Gāth. *drəgvå*), possibly referring by implication to Sraoša.

p. 196, note 43[1]. Thieme, p. 28, translates: 'with slaying of a hundred for the slaying of fifty (whom the contract-deceiving mortals have killed); with slayings of a thousand for the slaying of a hundred; (etc.)', and comments: 'it is a retaliation in proportion to the crime committed, even in a precise proportion as would befit the contractual nature of the god'. It must be borne in mind, though, that this formula occurs frequently in the Avesta, in contexts where it is doubtful whether a compact is involved, cf. *Yt* 5.58, 8.61, 13.48; 5.54, 117, 9.31; *Vend.* 7.53 sq. Moreover, it is not safe to treat as irrelevant the difference between the sing. °*ynāi* and the plur. °*ynāiš*, as it may well reflect different connotations of °*yna-* in the two series of compounds. For the interpretation of *sataynāiš* as 'with a hundred killings' there is a parallel in *haptō.karšvōhva* (*Yt* 6.3) 'in the seven climes'. *hazaŋraynyāi* in Y 10.6 and *Yt* 13.45, which in any case has a different suffix, may mean either 'for the killing of thousands' or 'for a blow (aimed) at a thousand'.

p. 199, note 45[2]. For *yōi paurva...družinti* Thieme, p. 31, n. 17a, compares *Taittirīya Saṃhitā* 2.2.6.2, *yaḥ pūrvo 'bhidruhyati*.

p. 200, note 45[2]. Add Sogd. *rnβ-* 'to attack', Khot. *raph-* 'to fight', see Henning, *BSOAS*, XI, 719. Sogd. *r'f*, Khot. *rāha* (< **rāfa-*, see *GMS*, p. 249) 'illness' are assigned to this base by Bailey, *Dr S. K. Belvalkar Felicitation Volume*, 1 sq., who adds Rigvedic *raphitá*.

325

p. 202, note 46². Thieme's translation of Y 44.18 runs as follows: 'Wie werde ich durch die Wahrheit, die mir, o Weisheit, [durch dich] eingeblasen (inspiriert) wurde, (d.h. durch mein Gebet) diesen Lohn gewinnen: 10 Stuten mit einem Hengst und einem Kamel, [und (neben diesem Lohn)] Gesundheit und Leben, wie du sie geschaffen hast.' The mere necessity of referring *hyaṯ* to *ašā* instead of *taṯ miždəm*, its obvious antecedent, casts doubt on Thieme's new interpretation of *apivaitī* as 'eingeblasen', and the asyndetic co-ordination of the dvandva *haurvātā amərətātā* with either *taṯ miždəm* or *aspå...uštrəmčā* adds to the uncertainty. But the chief objection to Thieme's interpretation (on the assumption that the correct reading is *apivaitī*, and this form belongs to *vat-*) is that Bth.'s rendering of *vat-* by 'to understand' is supported by the MIr. evidence: Parth. *frwd-* 'to understand, know', *wywd-* 'to distinguish, separate', *hmwd-* 'to believe', *'mwst* 'faithful', Arm. *havat* 'faith', Khot. *haut-* 'to know, understand', cf. Henning, *Mir. Man.* III, 873, n. 3, *BSOAS*, XI, 74, n. 2, Bailey, *BSOS*, IX, 78. To these forms I would add the expression *p' mwd* in the Jewish Persian inscription of Tang-i Azao, which Henning hesitatingly interpreted as *pa uměd* 'in the hope (of God) = hoping in God', *BSOAS*, XX, 341 *sq.*: assuming a haplography of ' one could read *pa amwad*, and understand 'having faith in God'. An additional weakness of Thieme's explanation of *vat-* is that it forces him to account differently for the gen. in *fraēštəm apivatāite daēnayå*, *Vend.* 9.2, and *nōiṯ apivatāite daēnayå*, *Vend.* 9.47.

p. 207, note 55¹. St. 55 is translated by Thieme, pp. 28 *sq.*, as follows: 'If the mortals would worship me with a worship in which my name is spoken, as the others, [namely] the gods, worship [me] with a worship in which my name is spoken (that is: if they would call upon me in concluding a solemn contract), I should have come to the truthful men with the age of a definite (literally: "cut [off at both ends]") time (that is: "with the age of a particular contract, being born with it and ceasing with its expiration"), [with the age] of my own sun-like immortal life I should have come to [the age] of a definite [time].' This wording shows in Thieme's opinion that Mithra is not only the personification of contract, but 'manifests himself also in each single contract'. While it is quite likely that the poet thus conceived of Mithra, it is hard to see how the present stanza can be used to prove this. We have seen in note 55¹ that the words of the stanza do not reflect only Mithra's attitude, as they are elsewhere placed in Tištrya's mouth. There is, moreover, no reason to doubt that the 'cut time' refers to what the text says, viz. the 'coming' of a god in response to a call, and not the period of validity of a contract.

I also find it difficult to agree with Thieme's rendering of the words *yaθa anye yazatåṇhō...yazənti* by 'as the others, [namely] the gods, worship [me]'. For in *Yt* 8.24 the same protasis precedes the following apodosis: 'I (viz. Tištrya) would take on the strength of ten horses, camels, bulls, etc.', and in the following stanza Ahura Mazdāh states that he himself by worshipping Tištrya confers on him the strength of ten horses, camels, bulls, etc. From these two stanzas it appears that the worship offered by 'others, [namely] the gods' would have been sufficient to endow Tištrya with the required strength, without his having to solicit the worship of men. There follows that at least in Yt 8.24 the words yaθa anye yazatånhō...yazənti must mean 'as other gods are worshipped', and the same is likely to be the case wherever else this protasis occurs.

p. 210, note 61[3]. Benv., who against Bth. takes Av. *tāta-* as belonging to Oss. *tayun* 'to melt' (thus already W. Geiger, *Etymologie und Lautlehre des Afghānischen*, 189), assigns to the same base *tat̰.āpa-*, translating it 'aux eaux ruisselantes', see *BSL*, 1956, 44. Although this is possible, one has to bear in mind with Lo., *Deutsche Literaturzeitung*, 74 (1953), 401, n. 1, that Bth.'s derivation of *tāta-* and *tat-* from the base *pat-* 'to fall, fly' is supported by the Homeric notion of the διπετής ποταμός, on which see Lüders, *Varuṇa*, pp. 11, 141, n. 5.

p. 215, note 66[1]. In *Visp.* 7.2 Justice (*Arštāt*) is actually identified with the Religion, as Čistā is in *Yt* 16.1 (see above, p. 167).

p. 215, note 66[2]. A possible example of *g* disappearing in Ossetic in *secondary* contact with a following *u̯* is provided by *æruæd*, *iræd* 'bride-price', which can be derived from **arg(a)-u̯adū-*, cf. Skt. *argha*, Sogd. *'ry*. The order of the members of the compound would be as in *yæugæf*, etc., cf. *BSOAS*, XIV, 488, n. 2. A different explanation (< **adi-u̯adū-*) was offered by Bailey in *TPS*, 1946, 203. Cf. the Pš. word for 'bride-price', *walwar*, which represents **u̯adū-bara-*, see Morg., *NTS*, XII, 264.

With regard to Oss. *ræuæg*, as the word means not only 'light', but also 'lung', Benv., *BSL*, 1956, 40, has meanwhile reverted to Miller's derivation from **fravaka-*, with a view to connecting the word with Gr. πλεύμων. It seems to me inadvisable, for the sake of such a distant connection, to dissociate *ræuæg* from the usual word for 'light' in Eastern Iranian languages: on the one hand the base **pleu-* does not otherwise yield words meaning 'light' (cf. Pokorny, *IEW*, 836 *sq.*); on the other hand the lung is often described by words meaning 'light', as Benv. himself remarks (cf. also Engl. *lights*).

The identity of Oss. *ræuæd* 'calf' and Av. *fravaitī-*, to which Benv. refers on the same page, was first pointed out by Miller, *IF*, XXI, 332.

p. 221, note 71[6]. This is to take Pahl. *mrd'*, if this is how the word is to be read, as a historical spelling. For Khwārazmian -*'w* from -*u*- stems Henning, *Zeki Velidi Togan'a Armağan*, 436, quotes *nx'w* and *dmn'w*.

p. 223, note 78[2]. Thieme, p. 40, translates: 'thou observest those countries, [as to] which one has good keeping of its contract arranged', assuming that *yātayeiti* has the same meaning as Ved. *yātáyati* (on which see above, p. 7). What casts doubt on this assumption is that in the Rigvedic passages concerned the subject of *yat-* is a divinity, the object people, or a notion of guilt or debt, whereas Av. *yātaya-* has as subject people, and as object either *hubərəitīm* or Barsman-twigs (*N* 88, 97; in A. Waag's opinion, *Nirangistan*, 96, libations). My translation 'they strive' is an adaptation of Bth.'s 'sie befleißigen sich', lit. 'they set in motion (=encourage, institute, or sim.)'. In Iranian, *yat-* seems to refer to movement also in Sogdian (cf. *GMS*, § 539, n. 1):*yt-* 'to go', Chr. *yty*' 'happening, taking place, *event*', *ptyt-* 'to happen, take place', see Benv., *JAs.* 1955, 307; but Yaγn. *yat-* means 'to have, find, a place', see *Язн. тексты* (above, p. 324), 365.

p. 231, note 84[1]. To YAv. *jan-nar-a-* 'killing men', recognized by Duch., *Comp.* 199, as a compound of the *isə.xšaθra-* type, Humbach, *IF*, LXIII, 210, has plausibly added Gāth. *jə̄(n)-nar-* and *xrū-nar-* 'männerverletzend', both occurring in *Y* 53.8.

p. 237, note 92[2]. Thieme, who separates Ved. *bráhman* from all three: OPers. *brazman-*, MPers. *brahm*, and Av. *bərəg-*, proposes for the last, on the assumption of its being connected with Ved. *bŕhas(páti)*, the meaning 'Kraft', and for *bərəjaya-* 'begrüßen', as a development of 'jemanden mit Kraft versehen', see *ZDMG*, 1952, 129. In so doing he ignores the Pahl. translation *ārzū* and the obvious conclusions which its etymological relationship with Av. *bərəg-* invites.

p. 238. H. Humbach maintains, *MSS*, 11 (1957), 78 *sqq.*, that Gāth. *ahū...ratuš* in *Y* 29.6 and *Y* 27.13 (the *Ahuna Vairya* prayer) is not the forerunner of the YAv. title *aŋhuča ratušča*, as he considers the YAv. nom. sg. *aŋhu(ča)* to be due to an early Zoroastrian misinterpretation of the two Gāthic passages, in which *ahū* is, in his opinion, the instr. sg. of *ahu-* 'Lebenskraft'. Two arguments are adduced by Humbach: (i) an alleged parallel in the YAv. sentence *vanāt aša akə̨m drujim Yt* 19.95, where *aša* is generally believed to stand for *ašə̨m* under the influence of Gāth. *ašā drujə̄m və̄ŋhaitī Y* 48.1; and (ii) a new interpretation of *aŋhə̄uš* in *Y* 33.1, according to which this word is not the gen. of *aŋhu-* 'life', but of our word for 'judge'.

Neither argument has force. *aša* in *Yt* 19.95 may well be an instr., with *Astvat̰.ərətō* as implied subject of *vanāt̰*, the wording of *Yt* 19.96, contrary to Humbach's belief, being incapable of invalidating this interpretation; but even if we allow that the ending of *aša* is wrong, having been prompted, in *one* isolated passage, by the wording of *Y* 48.1, this would be a very different matter from Humbach's unbelievable assumption: is it likely that as a result of somebody's mistaking an instr. for a nom. in one or two Gāthic passages, the correct nom. of what must have been a very common word in the YAv. language was forgotten and replaced by the instr.? As to Humbach's second argument, one might dismiss it simply because there is nothing to show that his trans-

ADDENDA [pp. 238–41

lation of *aŋhāuš* by 'Lebensherr' deserves preference over Bth.'s 'Leben'; but even if Humbach were right as regards the meaning, in what way could the existence of a gen. *aŋhāuš* (beside the acc. *ahūm* and the dat. dual *ahubya*) disprove that the corresponding nom. was *ahū*?

With the restoration of Gāth. *ahū* to its previously unchallenged nom. sg. function, Humbach's elaborate reinterpretation of *Y* 29.6 and the *Ahuna Vairya* prayer, which, as far as the prayer is concerned, has been adopted to a large extent by Duch. (*IIJ*, II (1958), 66 *sqq.*), becomes unnecessary.† Duch., who resorted for guidance to the Zoroastrian exegesis of the prayer as handed down in *Y* 19, has failed to point out that the co-ordination of *ahūm(ča)* with *ratūm(ča)* in *Y* 19.12 means that the commentator took *ahū* and *ratuš* in the prayer to stand in the same case, namely, as shown by *ratuš*, the nom. sg.

p. 241, line 4. The explanation which Benv. has advanced of Sogd. B. *'yt'w*, Man. *xtw*, *BSL*, 1956, 51 *sq.* (from **haxtāva-*, to Av. *haxta-* 'qualified'), breaks down over the final -(')*w*, which in Benv.'s interpretation would be part of the Sogdian stem. Such a stem would be heavy, and would appear intact before the abstract suffix *-yā(k)*, in the same way as Man. *xwt'w*, Chr. *xwdw* 'king' (*xut/ddw*) preserves its *-dw* in B. *ywt'wy'kh* 'kingship' (which in Man. would appear as **xwt'wy*', in Chr. as **xwdwy*'). But the abstract of *xtw* 'judge' is in Man. *xty'k*, which proves that the stem preceding *-y'k* is light (see *GMS*, § 1110), and the *-w* of *xtw* an ending (from the Sogdian point of view). This ending has been surmised in *GMS*, § 1190, to represent the formative of an old *-u*-stem.

† I am inclined, however, to agree with these two scholars that there is no compelling reason to refer *ahū*, *ratuš*, and *yim* to Zarathuštra, who is not even mentioned in the prayer. We know from the last line of *Y* 53.9 that Ahura Mazdāh looks after the poor by means of his *xšaθra-*. The same will be stated in slightly different terms in *Y* 27.13 if *yim* is referred to Ahura Mazdāh. Moreover, in view of the common Av. formula *xšaθrəm vairīm*, the predicate *vairyō* may well refer to *xšaθrəm*, too. If then we bear in mind that *dazdar-* 'doer' (to IIr. *dhā- 'facere'*, see Bth., *Wb.* 716, IV) may be a convenient differentiation from *dātar-* 'giver; creator', and that the construction of *xšā(y)-* with the gen. may have been extended to *xšaθrəm* (thereby inviting the use of the dative *mazdāi*... *ahurāi.ā* in the possessive function which the corresponding *tavā*, referred to by the commentator of *Y* 19.14, has in *Y* 53.9), the following translation can be suggested:

yaθā ahū vairyō aθā ratuš ašāṯčīṯ hačā
vaŋhāuš dazdā manaŋhō šyaoθananąm aŋhāuš mazdāi
xšaθrəmčā ahurāi.ā yim drigubyō dadaṯ vāstārəm.

'As the (*or* an) *ahu* is to be chosen (in accordance with Truth, to be the doer of the actions of Good Mind), so the (*or* a) *ratu* (is to be chosen) in accordance with Truth, (to be) the doer of the actions of Good Mind, and Ahura Mazdāh's power over the world (is to be chosen, [because he, Mazdāh, is the one]) whom they have given as shepherd to the poor (*or* to whom they have given [= entrusted] the poor as to a shepherd; cf. above, pp. 323 *sq.*, on the construction of *dā-*).' The three lines thus appear to answer the question: 'what is it that is to be chosen?'.

p. 242, note 92[7]. Add Sogd. *šyry pty-cyh* (with final *-h*) 'before a good thing', Henning, *BSOAS*, XI, 475[20]. Another Iranian adjective which goes back to a comparative is NPers. *sirih* 'beautiful', from *srayah-*, see Bth., *Sitzb. d. Heidelberger Ak. d. Wiss.*, 1919, No. 10, p. 28. H. W. Bailey has meanwhile returned in *BSOAS*, XXI, 45 *sq*. to his combination of Sogd. '*myδry* with Agnean *mṣa-*. He no longer holds that the *miši which '*myδry* in his opinion represents is the word for 'seed, field' attested in Khot. *miṣṣa-*; instead he now traces Sogd. '*myδry* and Agn. *mṣa-* to a different OIr. **miša-*, to which he assigns the meaning 'group', regarding it as an *-s-* derivative of the base **mik̂-* 'to mix'. It does not seem to me that the case for the 'chieftain of the local genii' is thereby strengthened, especially since after the removal of Khot. *miṣṣa-* the claim that Agn. *mṣa-* is a loanword from Iranian rests exclusively on the assumption that Sogd. '*myδry* is a pseudo-historical spelling.

p. 254, note 104[7]. H. W. Bailey, *Rocznik Orientalistyczny*, XXI (1957), 68, has also in the meantime suggested that Av. *sanaka-* belongs to *san-* 'to rise', but he interprets the word as meaning 'high ground..., as the place where the Raṇhā river rises'. The Yaγn. past stem of *san-* is given as *sata-* in Ягн. тексты (above, p. 324), 318.

p. 255. Add note **106**[1]: *nōiṯ mašyō...stē* was taken by Bth. to mean literally 'kein sein könnender Mensch'; in rejecting this interpretation of *stē* as an infinitive Benv., *Inf.* 37, declared the word to have been inserted by a commentator against grammar. I follow Lo. in translating *stē* by 'in (the) existence'. The form can be explained as a loc. of *sti-* (*stē* < *stiya, with the long *ē* of monosyllables), treated as an -*ī*- stem (cf. OPers. *hara^buvatiyā*).

p. 261, note 112[3]. Thieme, p. 45, translates the crucial lines as follows: 'Wonderful are Contract's means of refreshing (*frayanå*: by which he brings what the RV calls *práyas*), who comes to that country; as he brings when he is well kept [means of refreshing that are] broad, deep for cattle pasture (the "means of refreshing" obviously are rivers which are a condition for rich pasture).' It is, however, risky to attribute the meaning of *práyas* to a supposed Av. cognate which is *not* derived from the corresponding *-as-* stem **frayah-*, the more so as 'broad and deep refreshments' for 'b. and d. rivers' would be a somewhat daring metonymy.

p. 263, n. §. The final sentence of *Ny* 1.7 indicates no closer association of the sun with Mithra than the almost identical final sentence of *Yt* 13.81 does with Ahura Mazdāh or his Fravaši. The same applies to the invocation of the sun in *Yt* 6.6, which, however, may have been prompted by the last word of *Yt* 6.5 (quoted above, p. 227).

p. 282, note 136[1]. For this stanza Thieme, p. 37, proposes the following translation: 'For whom (Contract) white horses yoked to the chariot

ADDENDA [pp. 282–99

will draw [the chariot] towards the place of a man (*šē*) if he offers him (Contract) pourings [of sacrificial food]—by the one golden wheel [of that chariot] even the skies are all-luminous.' The construction Thieme implies, with the second °*šē* supplying the clue to the identity of the subject of the preceding clause, is unnatural, and would have to be contemplated only if it were impossible to understand *avi...maēθanəm* as a complement of *baraiti*, and both °*šē*-s as referring to the same person. We need not therefore concern ourselves with Thieme's further tentative suggestion that here and in st. 90 (where Thieme interprets the last lines as Da. had done, see above, p. 234) the sun looks after Mithra's interests. On the closing sentences of *Ny* 1.7 and *Yt* 6.5, which Thieme adduces in this connection, see the preceding addendum to p. 263.

p. 290, note 143². Alternatively the whole *karde* 34 may be a quotation of Ahura Mazdāh's words, as we have assumed for *karde* 33. If this were so, since sts. 137 and 138 clearly report an utterance of Mazdāh's, the same should perhaps be assumed for st. 139. The god's speech would then run from st. 137 to st. 143. My reason for excluding st. 139 from Mazdāh's speech was the same as in the case of st. 3, viz. the appearance in it of the god's name.

p. 294, note 143¹⁰. The translation 'majestic' of *xšaēta*- (against Bth.'s 'strahlend'), which conforms to Gdn.'s 'mächtig', Lo.'s 'König', Henning's 'Herr' (*Mir. Man.* I, 187, n. 6), is justified by the meaning of Sogd. *xšyδ* 'lord' (*GMS*, §269), and the later Iranian words which correspond to the Av. fem. *xšōiθnī*-, viz. Arm. *ašxēn* and Oss. *æxsīn* 'lady', see Bth., *Sitzb. d. Heidelberger Ak. d. Wiss.* 1920, ii, 29, n. 4, and cf. *BSOAS*, XVII, 481, n. 4. The situation is somewhat complicated by the meaning 'reddish, of a fiery colour' of Arm. *ašxēt*, and '(evening, morning) glow' of Oss. *æxsid* (Digor *æxsed*), both of which presuppose OIr. **xšaita*-. One may assume with von Wesendonk (above, p. 293), p. 31, that 'aus der Vereinigung mit *hvar*- hat sich für *xšaēta*-, das zu dem Verbum *xšay*- "vermögen" gehört, der Sinn "leuchtend, strahlend" ergeben'.

p. 299, note 145². Note I. Scheftelowitz's suggestion, *Acta Orientalia*, XI, 325, n. 4, that the MPers. Rām who was identified with the wind-god Vayu has nothing to do with the Avestan Rāman, but derives from the Babylonian storm-god Ramman.

INDEXES

INDEX LOCORUM

Avesta	*page* 334
Old Persian inscriptions	337
Rig Veda	338

WORD INDEX

Indo-European	339
Reconstructed Old Iranian	339
Avestan	339
Old Persian	345
Middle Persian	345
Parthian	346
Sogdian	346
Khotanese	347
Khwārazmian	347
New Persian	347
Pašto	348
Balūčī	348
Ossetic	348
Yaγnōbi	349
Indo-Iranian frontier languages	349
Other Iranian dialects	349
Vedic and Sanskrit	349
Buddhist Sanskrit	350
Armenian	350
Greek	350
Latin	351
Other Indo-European languages	351
Semitic	351
Elamite	351

SUBJECT INDEX 352

INDEX LOCORUM

AVESTA

A 3.4 237
   8 184
   10 *sqq.* 246
F 12 183
H 2.9 292 *sq.*
   10 221
N 1 237
   11 209
   22 284
Ny 1.7 52 n., 263, 330
   8 263 n.
   16 163
Vend. 1 175 *sq.*
   1.19 253 with n.
   4.2 *sqq.* 26 n.
   5.4 178, 260 *sq.*
   7.27 183
   8.34 178
     37 200, 247
     107 247
   10.18 *sq.* 239, 242 *sq.*
   13.45 200, 287
     46 287
     47 255
     48 200
     49 196
   18.11 177, 221
     14 *sq.* 62
     16 257, 293
     65 183
   19.13 168 n.
     13–16 211 *sq.*
     15 179, 212, 269
     27 247 n.
     28 171, 291 with n.
     34 216
     35 211, 212
   21.5 172, 285

Vič. 2 229
Visp. 2.4 238 *sq.*
   7.2 215, 327
   11.7 154, 155
   16.1 223
Vyt. 1; 3 181
Y 4.25 *sq.* 164
   9.1 207, 269
   15 214 n., 248 *sq.*
   11.1 190, 191 n., 272
   16.7 153 *sq.*
   17.18 266
   19.12 329
     14 329 n.
     18 265 *sq.*
     19 188 n.
   21.1 *sq.* 163 *sq.*
   27.13 328, 329 n.
   29.2 238
     3 48 n., 269, 321
     6 55, 56, 238, 328
     7 169 n., 215, 238
     8 *sq.* 238
   30.3 46
   31.9 168 n.
   32.3 213 n.
   33.1 328
     14 68 n.
   34.10 168
   35.9 324
   36.6 293
   38.5 180
   39.3 164
   41.1 269
   42.2 54
   43.12 194
   44.7 168 n.
     9 243
     10 242

16	214	6	330
17	321	7.3	263
18	10, 201, 326	8.0	210
19	49	7	217
45.4	168	8	187
46.5	49	11	207
19	180, 323	24	326 *sq.*
47.3	55, 56	25	53, 327
48.1	328	35	172
49.5	215	38	217
51.2	323 *sq.*	9.1	170
5	325	10.0	299
7	56 n.	1	51, 195, 211, 212, 238
10	169	2	26 with n., 321
12	284 *sqq.*	3	26 n., 59, 228, 331
14	180, 323	5	54, 229, 287
22	164 n., 165	6	25, 240
53.8	328	7	26, 31, 319
9	329 n.	9	25 n., 59, 161
57.2	54, 57	13	31, 35, 39, 205, 234, 320
10	223	14	322
21	196, 205, 283	16	52, 202, 229, 271
23	211	17	26 n., 298
24	211, 235	18	27, 197, 297 *sqq.*
28	270	20	196, 271, 323
29	174	21	25 n.
58.1	224	22	319
60.7	217, 228 *sq.*	24	26, 30, 203
62.5	210	25	49 n.
68.4	163	26	25 n., 26 n., 160 n., 323
11	162	27	27, 52
71.13	174	28	223, 225
Yt 4.9	209	29	30, 235
5.7	220 n.	30	32, 161 n., 223, 225, 229, 283, 294
17	53	31	319
78	245	32	199, 204, 324
86	209	33	43, 229
87	208	34	25 n., 319
94 *sq.*	63, 200	35	26, 33, 66
102	189	36	206
126	220 n., 273	37	178, 249
129	178		
130	208		
6.5	227, 294, 330		

38 25 n., 26 n., 34 n., 54, 225, 230, 233, 262
39 25 n., 271
40 192
41 59
42 *sq.* 206
43 325
44 204, 225, 226, 245, 319
45 26 n., 166, 190, 191, 325
46 55
48 25 n., 26 n., 181, 182, 194 n., 206, 262, 271
50 52, 205 n., 225, 226
51 35, 59, 204
52 59, 194
53 19 n., 51
54 19 n., 55, 319
55 326
60 26, 30, 179
61 27, 30, 32, 51, 195, 249, 327
62 26 n.
64 58, 196, 292
65 32, 216, 228, 229
66 58, 59, 205, 249
67 37, 39, 244, 271, 294
68 30, 58, 59, 62 n., 167, 168, 207, 215, 219, 222, 276, 290
69 49 n., 216, 218
70 59, 62 n., 168, 194, 217, 271
71 25 n., 181, 197, 231
72 26 n., 197
73 51, 206
74 207, 319
75 319
76 58, 217
77 226, 235, 319
78 25 n., 27, 261, 266, 327 *sq.*
79 59, 70
80 26 n., 55, 223, 235

81 59, 224, 225
82 26 n., 187
83 *sq.* 206
84 54, 232
85 34 n., 54, 230
86 25 n., 34 n., 54, 230, 262, 271
87 177, 297 *sqq.*
88 59, 235
89 231, 234, 272, 273, 292
90 35, 294
91 26, 234 *sq.*
92 53, 58, 59, 205, 215, 223
93 235, 236
95 31, 38, 39, 319
96 279, 281
97 219, 281
98 53, 216, 218
99 219, 234, 243, 244, 261, 266, 271
100 25 n., 30, 39, 59, 70, 271, 276
101 27, 178, 244, 271
103 31, 55, 216
104 26 n., 291 n.
105 252 *sq.*
107 25 n., 26 with n., 43, 187, 231, 254 n., 271
108 32, 52, 63, 178, 229, 260
109 26 n., 27, 60, 160 n., 185, 202
110 29 n., 178, 181, 194
111 26 n., 27, 60, 160 n., 185, 202, 258, 259
112 32, 203, 204, 214, 330
113 44, 66
115 296
116 26 with n.
117 26, 28
118 35, 204, 238
119 25 n., 51, 67, 238

INDEX LOCORUM   AVESTA

120	59, 161 n., 234, 269	30	227 *sq.*
121	238, 269	38	158, 188 *sq.*
122	51	45	161
123	52, 53, 204, 287	46–48	159 *sqq.*, 166
124	39, 168 n., 204, 271	49	154
125	30, 62 n., 277 n.	54	172
126	25 n., 39, 58, 59, 166, 167, 215, 234, 249, 271, 278	56	172
		76	249
		81	10 n., 330
127	59, 60 *sq.*, 62 n., 168, 202, 271	89	213 n.
		92	237, 242
128	25, 160 n.	94 *sq.*	27, 215
129	25, 271, 279	107	228 n.
131	280	136	196
132	245, 279, 280	14.15	168
133	244, 271	46	187, 209
134	219	47	195
135	53, 159, 218, 248	63	194 n., 202, 203
136	25 n., 35 *sq.*, 204, 330	15.2	53
137	17, 51, 197, 224, 225, 234, 238, 287, 319, 331	40	261
		54	259
		16.1	167, 327
138	331	17.2	10 n., 168 n., 217 *sq.*
139	12 n., 51, 59, 331	6	52, 226
140	25 n., 52, 53, 204, 238, 289	7	208
		16	69, 195
141	26, 31, 37, 43, 159, 179, 217, 256, 261, 291 n.	17	217 n.
		61	232
142	31, 39, 52, 61, 234, 320	19.11	273
		34 *sqq.*	59
143	9, 26 n., 35, 36, 52, 53, 58, 59, 161 n., 216, 217, 222, 227 n., 238, 283, 331	36	159
		42	166, 219 (n.) *sq.*
		43–44	36, 254
		51	59
145	35, 42, 44, 261, 273	54	158
146	299	81	214 n.
11.3	159	89	273
14	170, 195	93	186
12.18–19	253	95 *sq.*	328
23; 25	205 with n.		
13.3	37, 195 n., 217		

OLD PERSIAN INSCRIPTIONS

Beh. I, 30   260
      91   171
    IV, 37   177

9	173, 322
17	154
18	259 *sq.*, 261

44   184
64   286
70 sqq.   197 sq.
77 sq.   198
V, 19; 35   156, 250 sq.
NR, a, 1-3   293
Susa, k, 4   15
Daiva, 51 sqq.   153, 156

RIG VEDA

3.59   8, 319
5.47.3   283
62.8   320
65.4-5   319

WORD INDEX

INDO-EUROPEAN

*(e)ni, 278
*dher-n-gh-, 267
*ĝen-, 324
*ĝhustu-, 220 n.
*ḱen-, 254
*ḱes-, 233
*ḱHem-, 274
*kreu-, 189
*lekr̥/n̥-to-, 182 n.
*l/renk(ʷh)-, 182, 324
*lubh-, 237
*okʷ-, 278
*os-, 281
*pleu-, 327
*pr̥ḱs(ó)-, 224 n., 233
*rebh-, 215
*sḱ(h)en-, 254
*(s)pen-, 183
*sthǝtó-, 325
*u̯er/l-, 162

RECONSTRUCTED OLD IRANIAN

*abi-dasta(na)-, 220 n.
*ǎvya-, 200 n.
*āpavant-, 201
*čam-, 184
*dug-gva-, 230
*havya-, 200 n.
*jī(va)θra-, 260
*lakata-, 182 n.
*nǎxar/n-, °aka-, 277
*niba-, 282
*pati-dastāna-, 220 n.
*pati-mauga-, 203
*piti-, 231
*ptar-, 270 sq.
*ra(n)x-, 182

*rāfa-, 325
*rāθi-, 222
*span-, 183
*sr̥g-, 321
*star-, 254 n.
*tāra-, 277 n.
*θam-, 174
*θrāya-, 209
*u̯arsa-, 265
*vid-, 250
*xrāh-, 188
*xšu-, 179, 323, 324
*xᵛani-sanda-, 323
*zu-, 272

AVESTAN

aēm, 200 n.
aēva.čaxra-, 282
¹aēš-, see vi-isa-
aoδa-, 253 sq.
aka-, 277 n.
akatara-, 181, 323
aya-, 283
axšnušta-, 258
ad-, 325
aδara.dāta-, 269
aδka-, 220 n.
apa.xᵛanva-, 187
apana-, 36 n.
ap(aya)-, 232
apavant-, 201
apaskaraka-, 201
apāvaya-, 200 n.
aipi, 199
apišman-, 255
aibijarǝtar-, 226 n.
aiwi.druxti-, 255
aiwištar-, 254 n.
aiwyāxš-, 278

AVESTAN

afrakatak-, 219 sq.
aṇhaēna-, 281
anaipi.pārəmna-, 247
ainika-, 278
ainižbərət°, 200 n.
anu.mainyāi, 283 sq.
anu.sastrāi, 283 sq.
anya-, 286
anyāθā, 169
amaēniγna-, 158
amərəxti-, 273
amərətāt-, 10 n., 197
ayaṇhaēna-, 281
av-, 201, 250
ava-, 269, 293
¹*avant-*, 152
²*avant-*, 201
arədra-, 215
arənaṯ.čaēša-, 186, 265, 324
aurvaθa-, 170
aurvant-, 172
arəza-, 166
arəzah-, 166
arəzahī-, 217
arəzō.šūta-, 187
aršan-, 232
arštāt-, 197, 286 sq.
as, 248 sq.
asan-, 283
asaya-, 218
ast(aēna)-, 281
astvaṯ.ərəta-, 181, 328
aspa vīra, 250
ắsna-, 171
aspya.payah-, 163
asrū.azan-, 191
azgati-, 228 n., 285
aša-, 6, 153 n., 156, 194
ašavan-, 153 sqq.
ašavasta-, 163, 298
aša- vahišta-, 237
ašəmnō.vid-, 192
aši-, 194, 325
ašivant-, 194, 325

WORD INDEX

°*ašta-*, 197
ašti-, 199, 265, 285
ašya-, 194
ahū-, 240 sq., 328 sq.
ahuraδāta-, 50, 169 with n.
ahurānī-, 45
ahurō.mazdāh-(?), 212
āgrəmati-, 226 n.
ātara-, 323
ātərəxᵛarənah-, 278 n.
āiθi-, 188
āfant-, 172 sqq., 201, 322 sq.
āfyeiδyāi, 174, 201
āyu-, 268
āsna-, 158
āzūti-, 169 n., 215
ərəδwō.zənga-, 210
ạiθyā-, 182
ạs-, 171, 197
iθắ, 254
ima-, 254
iškata-, 174
uγra.bāzu-, 196
uxšyaṯ.urvara-, 32
upa.θwaršti-, 207
upaṇha-, 278
upamana-, 167
upara.dāta-, 269
upairi.spāt-, 183, 275
upa.haxta-, 231 n.
upāiri.saēna-, 174
urvāpa-, 174
uš(ah)-, 288, 291
uši, 291
uštāna-, 167, 221
kaēta-, 156
kamərəδō.jan-, 259
kayaδa-, 156 sq.
kavi-, 185 sq.
karapan-, 186
karša-, 210
kəša-, 282 n.
kạsaoya-, 200 n.
gaēθya-, 171

340

WORD INDEX

AVESTAN

gatō.arəza-, 166
gayō.dā-, 32, 43
³gav-, 232
gava-, 176
gavasna-, 280
gavašayana-, 176
gar-, 193, 226 n.
garō nmāna-, 324
gātu-, 189 n., 208 n.
gāθrō.rayant-, 203
γnā-, 267
xayəuš, 181
xratu-, 256
xrą̇ṇhaya-, 188
xrūnar-, 328
xrūm(y)a-, 189, 191
xšaēta-, 294, 331
xšaθra-, 208, 276, 297
xšayamna-, 178
xšąnmə̄nē, 178
xšnu-, 324
čixra-, 217
čiθra-, 213 sq., 260, 261
čimāne, 179, 184
jafra-, 262
jə̄nar-, 328
jiṯ.aša-, 322
tā, 222
taṯ (taδa), 263
taṯ.āpa-, 210, 327
tanu.kəhrp-, 181
tanu.drug-, 180 sq.
tanu.pərəθa-, 247
tanumąθra-, 180 sq.
tāta-, 327
tiži.asūra-, 219
tiži.dąstra-, 219
daēnā-, 242 sq.
dag-, 181
daxš-, 325
daiṅhupati-, 245, 261, 266, 296–9
daiṅhusasti-, 233, 297 sq.
daiṅhuš.aiwištar-, 253 (n.) sq.
³dar-, 275

darəyō.haxəδrayana-, 170 n.
darəna-, 191
darši-, 275
dasvar-, 198
dašina-, 244
dā̆hyuma-, 265 sq., 298
dā-, 224, 255, 269, 323 sq., 325, 329 n.
dānu-, 189 n.
²dāman-, 188 n.
dāmi-, 9 n., 166–9, 211
dāmi.δāta-, 210 sqq.
dārā-, 245
dāsmanī-, 198 sq.
də̄jīṯ.arəta-, 322
dərəta-, 275
disu-, 255 with n.
dūγδō.vā-, 231
dučiθra-, 260 sq.
duš.kərət-, 262
dužbərənt-, 203
dužniδāta-, 255
dvā(čina)-, 230 sq.
drang-, 266 sq.
drəgvant-, 156
driγu-, 231
θanvarətan-, 279 sq.
θamnahvant-, 174
θātar-, 174
θwāša-, 215
θwōrəštar-, 54 sqq.
iθyajah-, 178
θrāyavan-, 209
θrāyō, 209
paoiri̯-, 290
paxruma-, 189
pat-, 179, 327
patarəta-, 270 sq.
paiti.paršti-, 184, 242
paiti.pā(y)-, 177
paiti.biši-, 203
paiti.fraxštar-, 184, 242
paiti.niš.harətar-, 229
paitī.vyādā-, 180

AVESTAN WORD INDEX

paiti.sənda-, 180
paiti.zan-, 160, 192
paityāstar-, 325
paθana-, 262
pay(ah)-, 231, 283
pavant-, 201
par-, 245 sqq.
para, 196
para.paθwant-, 179
parō.kəvid-, 250
parō.nmāna-, 287
paurva-, 199 sq.
paršu-, 219
paršuya-, 219
pasča.nmāna-, 287
pā(y)-, 174, 222
pāyu-, 54–7, 67
pāra-, 245
pāirivāza-, 276 sq.
pāšna-, 201
pərəti-, 285
pərəθa-, 247
pəšō.tanū-, 245 sqq.
pouruta-, 174 sq.
pouru.spaxšti-, 170
piθana-, 170
piθe, 230 sq.
puθran-, 179, 323
puθrō.dā-, 32, 228
baēvarə.pati-, 160 n., 189 n.
baoδa-, 159
baoδah-, 159
bar-, 177
barō.baoδa-, 159 sq.
barō.zuš-, 220 n.
barəziš(ta)-, 183
barəša-, 170
bāδa, 206 sq., 218
bāiδištəm, 207
bāmaniva-, 282
bāmya-, 290 sqq.
bāzuš.aojah-, 181
bərəg-, 236 sq., 328
bərəjyąstəma-, 237

būšyąstā-, 248
būždyāi, 321
fraorəti-, 163
fraorat̰, 166
frayrāti-, 285, 293
fraxšti.dā-, 215
fraxšnin-, 166
fratəmat/δāt-, 197, 297 sqq.
fradaxš-, 181, 192 sq., 324 sq.
frat̰.āpa-, 210
framanah-, 185
framainya-, 224
frayana-, 261, 330
fravaitī-, 327
fravaz-, 271
fravaši-, 163
fravi-, 250
frasāna-, 36 n.
frastanv-, 177
²*fraspāt-*, 183, 275
frazainti-, 158, 210, 237
frazuš-, 220 n., 273
fraša-, 224 n., 233
frašna-, 261
frādərəsra-, 218
frāpa-, 210
frāy-, 261
frā.vərəsa-, 265
frāzušta-, 219 (n.) sq.
frāšma-, 233
frāšmi-, 233
frəna, 178, 323
fryāna-, 261
fšuyant- vāstrya-, 207
naēδa, 177, 221
naviθya- (naoθ-), 264
nar-, 68 n., 152
³*nas-*, 180
nixš(a)ta, 277 sq.
niδāta-, 213
niδātō.barəzišta-, 183
nipašnaka-, 201
ništarətō.spaya-, 183
nižbar-, 159

WORD INDEX

AVESTAN

nižbərət°, 200 n.
nmānya-, 265 sq., 298
mati-, 224
³*man-*, 177, 224
manaoθrī-, 221
manavantya-, 223 sq.
manō.marəta-, 284
mainivasah-, 170, 274
¹*manya-*, 224, 283
²*manya-*, 284
mainyava-, 170
mainyu.tāšta-, 294
mayah-, 205
mairya-, 152
**mazăt/δ(āt)-*, 197
mazant-, 152, 196, 213
mazdaδāta-, 50
măyā-, 205
māyu-, 205
māh-, 289
mərəzu-, 221
mą̇ṇha-, 289
miθwa-, 287
miθra-, 26, 30, 40 n., 41 n.
miθrahe vāxš, 283, 284
miθrō.aojah-, 253
miθrō.drug-, 153
miθrō.rašnu-, 37
yā, 222
yaoxtivant-, 187
yaoxšti-, 187, 324
yaod-, 167 n.
¹*yaona-*, 172, 207 sq.
²*yaona-*, 208, 237
yaoz-, 167 n., 187, 324
yaoždā-, 242 sq.
yat- (cf. *yātaya-*), 275
yaṯ, 232
yasna-, 16 n.
yaz-, 271
yātaya-, 327 sq.
yu-, 208
yuγō.səmi, 274
vaēkərəta-, 176

vaēd-, 263 n.
vaēδa-, 250
vaēδayana-, 166
vaēiδi-, 288
vaēsma-, 233
vačō.marəta-, 284
vat-, 326
vaδū-, 327
vaδrya-, 208
vaf-, 323
⁴*van-*, 222
vavazāna-, 273
var-, 162 sq.
varənya-, 218
varəsa-, 265
vasō.yaona-, 207
vaz-, 271
vahmō.səndah-, 179
vā(y)-, 324
vərəθra-, 158-62
vərəθrayna-, 158, 162, 217
vərəθra.baoδa-, 159
vouru.ašta-, 197
vouru.gaoyaoiti-, 43, 65 n., 68, 151, 299
vi-isa-, 200 sq.
¹*vīδāta-*, 196
²*vīδāta-*, 214
vīδāti-, 196
vīδāraya-, 182
vīθiš-, 230
vīra-, 182
vīsūpti-, 261, 267
vīspati-, 261
vīspəmāi, 163
vīsya-, 265 sq., 298
vyāxana-, 166
vyāna-, 213
raočah-, 293
rayu-, 215
ratu-, 240 sq., 329 with n.
raθa-, 191, 222
raθaēštar-, 170
raiθya-, 191

AVESTAN

rapti-, 200, 285, 325
rašnu-, 53, 223
rənjišta-, 215
rąxšyant-, 181 sq., 324
urvaiti-, 184
urvizō.maiδya-, 191
sak-, 257 sq.
saxta-, 256 sq.
satayna-, 325
san-, 36 n., 254, 330
sanaka-, 253 sq., 330
sand-, 179 sq., 323
sar-, 321
sarəjā, 321
sasti-, 297
sastra-, 284
sazdyāi, 180
sātar-, 185, 259
sāna-, 36 n., 254
sāstar-, 259
sa̰h-, 179
simā-, 274, 275
simōiθrā-, 274 sq.
supti-, (261,) 267
suptiδurənga-, 266 sq.
sūra-, 288
stā(y)-, 183
stəhrpaēsah-, 26 n., 283, 294
sti- (stē), 330
stig-, 220 sq.
²span-, 183
sparəya-, 280
spā-, 275
spənta-, 157, 172, 173 n.
sraēštāiš (dāmąn), 291, 292 sq.
srao°, 183
sraoša-, 193
sraošāvarəz-, 62
sraošyă-, 181, 193
srayah-, 330
sravaṇhu-, 183
sravašəmna-, 178, 183
srifa-, 264
srū-, 281

WORD INDEX

srvant-, 183
srvō.zana-, 281
srvī.stāy-, 280 sq.
zaēna-, 245
zaēnahvant-, 210, 291
zaotar-, 272, 284
zaoθrā̆-, 272
zafar-, 280
zantuma-, 265 sq., 298
zaraθuštrō.təma-, 265 sq.
zarənya-, 245
zuš-, 219 (n.) sq.
zyā-, 188 sq.
šanman-, 178 sq., 183, 192
haēnā-, 187
haomayō.gav-, 163, 322
haxa-, 281 n.
haxti-, 281 n.
haxəδra-, 170 n., 226 sqq.
haxman-, 229
hačimna-, 37, 217, 231
²had-, 171
haδō.gaēθā-, 267
haiθīm.ašavagan-, 190 sq.
haθrāka-, 215 sq.
hama-, 259
hamaēstar-, 192
hamō.xšaθra-, 259 sq.
haya-, 200 n.
²hav-, 187
³hav-, 267
hava-, 200 n.
haurvatāt-, 10, 197
hazaŋraynya-, 325
hāmō.taxma-, 274
hāvišta-, 157 n., 268
hąm.ivā-, 274, 275
hąm.vaintī-, 223
hąm.varəti-, 59, 162
hindu-, 253
huxšnuta-, 179, 324
huδāman-, 188 n.
hupatarəta-, 270
hubərət-, 203, 261 sq.

WORD INDEX

hubərəta-, 222, 261 sq.
hubərəti-, 222, 262
hunu-, 267
huyāγnā-, vii n., 267 sq.
huzaēna-, 210, 291 with n.
hvapah-, 207
hvarəxšaēta-, 292, 294
hvāpah-, 208, 237
hvāyaona-, 157
hvāyaozda-, 167 n.
hvāraoxšna-, 31, 61, 288 sq.
hvō.γžaθa-, 157
xᵛaēui°, 189 n.
xᵛan-, 187
xᵛaniraθa-, 176
xᵛainisaxta-, 256 sq.
xᵛaini.starəta-, 189 n.
xᵛar-, 273
xᵛādaēna-, 157
xᵛāθrō.disya-, 32 n., 223, 229
xᵛāpaθana-, 157
xᵛāpara-, 237
xᵛāstra-, 299
xᵛīti-, 218

OLD PERSIAN

aʰuramazdāh-, 212
aurmzdahdᵘgm, 184
aθiy, 171
anušiya-, 298 n.
avaθāštā, 197
avaparā, 196
arta-, 153 n., 156
artāvan-, 153 sqq.
arštā-, 286 sq.
āgariya-, 226 n.
āšnai, 171
xšnu-, 324
taumā, 198 n.
daθ-, 198
daršam, 275
dinǰ-, 188
duškrta-, 262
θakata-, 258

AV., OLD PERS., MIDDLE PERS.

θatagu-, 175
θāti, 185
patiprsa-, 184
baga-, 60 n.
brazman-, 236, 328
fratama-, 298 with n.
Fravarti-, 162
fraša-, 224 n., 233
marīka-, 153
*yazna-, 16 n.
yāvai, 250 sq.
vratiya-, 184
hamapitar-, 260
hamiθriya-, 192
hubrta-, 262

MIDDLE PERSIAN

'dyšm, 233
ahramōy, 203
ahrav, 153
'mhr'spnd, 10
'pwrdg, 245
ārzū(k), 236 sq., 328
'whrmyzdbg, 61
'wšt'pykly, 183
'wšyb'm, 291
'st'n-/'std, 177
b', 207
b'(y)st'n, 207
brahm, 236, 328
cyhr, 214
dātaβar, 53, 223
duš-hēm, 255 n.
frašēmurv, 219
frāšm, 233
hs'z-, 257
hwnsnd, 179, 323
kēδ, 156
mēnōk, 170
mērak, 153
mhr'spnd, 10
miθrāndružān, 153
mrd', 221, 327
nyr'myšn, 176

ōštāp, 183
phypwrs-, 184
plš, 261
prmyn-, 185
ps(')z-, 257
puhl, 247
rp-, 200, 325
sač-, 258
srwšhr'y, 194 n.
sz-, 257
tanāpuhl, 247
tarāz, 282
ušbām, 291
Varan, 218
wyn'rt, 196
wysndkyh, 179
xrat, 256
x^{v}āpar, 237
xwybš, 181

PARTHIAN

'mwst, 326
'wsxt, 258
b'dyst'n, 207
cyhr, 214
drwxtmyhr, 153
frh'h, 178
frmnywg, 185
frwd-, 326
frwyn-, 166
h'mcyhrg, 214
h'mgyh, 267
hmwd-, 326
hw'bs'gyft, 257
myhr, 41
myhryzd, 40, 58, 61
n'šmy, 233
nys'ž-, 257
pdbwrs-, 184
ps'c-, 257
psynd-, 179
pw'g, 277 n.
rf-, 200, 325
s(')c-, 257

sn-, 254
srwšyft, 193
sxt, 258
wywd-, 326
ywdy-, 167 n.
zxrwb, 207

SOGDIAN

'γt'w, 240, 329
'myδry, 35, 40 n., 240 sqq., 330
'ns'c-, 257
'ns'wr, 219
'rγ, 327
'rtwspy', 163
'yw'rδkw, 215
β'δ, 206
βnš, 170
βrwz-, 271
βsys-, 257
cytk, 157
δ'm'yγt, 35, 53, 238, 240 sqq.
d'r, 245, 280
δβr-, 171 n.
δrγwšk-, 231
δšny, 244
δxš-, 181 n.
frwyδ-, 263 n.
frwz-, 271
fs'c-, 257
γw'w-, 187
qš'wrzyy, 157 n.
kwt-, 173
qwy'q, 186
mrδ'spnd, 10
myšyy βγyy, 34, 40 n., 41
n'γn, 277 n.
nšmy, 233
nwk, 264
p'δδ, 179
p'r(h), 245
pδynd, 182
prγš'w-, 323
prwyδ-, 263 n.
pšn', 201

WORD INDEX

pšyn-, 192
pt'ych, 35, 241, 330
ptβs-, 184
pts'c-, 257
ptsynd-, 179
ptšng, 192
ptycyh, 330
ptyt-, 328
ptz'n-, 160
r'f, 325
r'y-, 203 n.
rnβ-, 325
rxn-, 181 sq., 324
ryncwk, 215
yt-, 328
s'c-, 257
syty', 258
sn-, 254
srwšrt, 195 n.
st-, 254
t'r, 277 n.
tm'yγ, 241 sq.
w'ywq, 322
wδyr-, 182
wrs, 265
x'w-, 187
xrwm, xwrm, 189
xšyδ, 331
xtw, 240, 329
xwsndy', 179
yδw, 288
zyn-, 188

KHOTANESE

ābei'sa, 265
bid-/bist-, 192
bilsahai, 265
biṣṭa-, 268
bulj-, 236
dairä, 245
daśtąnya, 220 n.
ggei's-, 265
grra(ta), 256
ggūysna-, 280

SOGD., KHOT., KHWĀR., NEW PERS.

gyūna-, 207
haut-, 326
hāva-, 250
miṣṣa-, 330
nasa, 180 n.
nauhä, nūha, 264
ṇiysuna-, 272
paysān-, 160
pūś-, 184
phan-, 173 n.
raph-, 325
rraysga-, 215
rāha-, 325
sata-, 254
sāj-/sīya-, 257
skyätä, 258
ttāra, 277 n.
urmaysde, 320
vahī, 231 n.
ysin-, 188
ysuma-, 272

KHWĀRAZMIAN

angēθ, 267
'rδ'w, 321
dmn'w, 327
karb-, 186
nx'w, 327
ufān-, 173 n.

NEW PERSIAN

āstīn, 220 n.
čam(īdan), 184
čihr, 214
falāxan, 192
farasp, 275
gavazn, 280
gurs, 265
jašn, 16 n.
kašāvarz, 157 n.
lagad, 182 n., 324
mihr, 41 n.
nāxun, 277 n.
nōk, 264

palindīn, 182
parvāz, 277
pasand-, 179
pāšīdan, 219
rōz, 293
sarfāk, 264
saxš, 258
saxt, 256 sq.
saz-, 257
sān, 185
sāxtan, 257
sēčīðan, 257
sirih, 330
sitadan, 177
sīm, 274
šahr, 208 sq., 297
zōš, 220 n.

PAŠTO

əzmār, 152
cēr, 214
daryal, 156
yəšai, 245
layata, 324
mayan, 205
mīna, 205
nūk, 277 n.
pežandəl, 160
pōr, 245
saž, 258
šāyəl, 321
tṣ̌əl, 219
walwar, 327
weṣ̌tə, 265
wraẓ̌, 170
yūn, 207
zwamna, 272
zwaẓ̌, 200 n.

BALUČI

bušk, 170
lagat, 324

OSSETIC

afædz, 172
afon(æ), 172 sq.
ahæ, 282 sq.
ami, 285
ard, 153 n.
ænsur, 219
æruæd, 327
æstæiron, 254 n.
æxsid, 331
æxsin, 331
barc, 170
cædžindz, 182
cægindzæ, 182
dauæg, 155, 321
fat, 179
fætæn, 262
fes, 283
fīston, 220 n.
idauæg, 155, 321
iræd, 327
kehog, 176
kŭdz, 173 n.
nĭv, 282
nivahæ, 282 sq.
nivesinæ, 283
nix, 277 sq.
nixagæ, nixæg, 277 with n.
ræstæg, 173 n.
ræsug, 218
ræuæd, 327
ræuæg, 215, 327
ssir̂, 219
stær(on), 254 n.
sūdzin, 181 n.
tayun, 327
tæn, 279 n.
tærnix, 277 with n.
tehun, 221
uælindzæ, 182
uændag, 172, 322
uæyug, 321 sq.
udin̂, 167 n.

urdugistæg, 210
usm, 173 n.
xeŭ, xiŭ, 189 n.
zust, 220 n.

YAΓNŌBI

azōn, 191
daxš-, 181 n.
raxn-, 324
san-, 254 with n., 330
yat-, 328
živar-, 159

INDO-IRANIAN FRONTIER LANGUAGES†

anaxno (Y.), 277 n.
avlǎsto (Y.), 220 n.
åwusp (Y.), 275
dogər (W.), 277 n.
fräspïy (Y.), 275
ju-saxa (K.), 258
mī (Z.), 36 n.
mīra (Y.), 41 n.
mrīg (O.), 153
narxax (S.), 277 n.
naxk (O.), 277 n.
ormōzd (S.), 320
prusnaw- (O.), 219
puši-žeri (K.), 214
rānjk (W.), 215
rəwəz- (W.), 271
sām (M.), 274
san- (W.), 254
šəxs- (S.), 258
wəšp (W.), 275
zöl (S.), 220 n.

OTHER IRANIAN DIALECTS

amwad (Jew. Pers. inscr.), 326
Āhven (Baškardi), 322 sq.
dašn- (Galinqaya), 181 n.

simāk (Sīstāni), 274
šahr (Baškardi), 208 sq.
vahv- (Baškardi), 323

VEDIC AND SANSKRIT

ánīka, 278
apaskŕ-, 201
apāṣṭhá, 265
argha, 327
avaskŕ-, 201
avratá, 170
áśman, 283
aṣṭi, 285
ásthi, 281
ápas, 237
ápi, 237
āpṛ-, 246
udán, 254
upamāna, 166, 167
upala, 283
urúgavyūti, 65 n.
ūrdhvajānu, 210
r̥tá, 6, 153 n., 156
r̥távan, 153 sqq.
kavi, 185
kāyādhavo, 157
kṣan-, 179, 192
kṣā-, 233
godúh, 230
cam-, 184
citrá, 213 sq.
tigmá, 220
traya(s), 209
tsar-, 183
ditsú, 255
dhar-, 275
nakh(ar)á, 277 with n.
nibha, 282
niṣṭha, 278
páyas, 231
pāvaká, 277 n.
pī-, 231, 283
pr̥kṣá, 224 n.

† K(howar), M(unji), O(rmuri), S(anglechi), W(axi), Y(idγa), Z(ɛbaki).

pṛṣ-, 219
práyas, 330
prāv-, 250
bṛ́haspáti, 328
bráhman, 236, 328
bhūṣ-, 321
márya, 152
māyā́, 187 n.
mās(a), 289
mitrá, 30, 41 n., 167
mith-, 192
mithyā́, 287
mī-, 287
yat- (cf. yātaya-), 275
yavāgū, 322
yātaya-, 7, 8 n., 30, 327 sq.
yugaśamyá, 274
yukti, 187 n.
raphitá, 325
lakuṭa, 182 n.
vádhri, 208
vāvas/śāná, 273 sq.
vāvahi, 274
vi-iṣ-, 201
vidh-, 192
vṛtrá, 162, 167
vyàṃsa, 261
vyāpṛ-, 246
vratá, 184
śak-, 256, 257
śátru, 185
śámyā, 274
śárman, 321
śastrá, 284
śā-, 185
śíprā̀, 264
sákthi, 281 n.
sagh-, 228 n.
sahantya, 224
sā́rathi, 222
súbhṛta, 262
sū-, 267
sváyukta, 157
hṛ-, 226 n.

hetí, 245
hótṛ, 272, 284

BUDDHIST SANSKRIT

avadranga, 267

ARMENIAN

ah, 188
amehi, 192
ašxēn, 331
ašxēt, 331
ba(r)š, 170
čitak, 214
despak pʻaṙac, 283
havat, 326
henum, 183
hraša(kert), 224 n.
manavand, 223
meł, 152
oskr, 281
part, 245
patšač, 257
petmog, 203
sametik, 275
samik, 274, 275
šaržem, 321
štap, 183
taraz, 282
yuzem, 187, 324

GREEK

ἀρταῖοι, 155, 321
γαυνάκης, 151
δέω, 275
διπετής, 327
ἔκηλος, 157
ἠϋκανός, 291
ἤπιος, 237
καθήκει, 257
κάμαξ, 274
καυνάκη(ς), 151
κρατύς, 257
λάξ, λάγδην, 182 n.
Μεσορομασδης, 320

WORD INDEX GREEK, LAT., OTHER I.-E., SEM., ELAM.

ξύω, 179, 323, 324
ὄνυξ, 277
ὀστέον, 281
πέριξ, 277
πλεύμων, 327
Πρηξάσπης, 224 n.
προσήκει, 257
πτερόν, πτέρυξ, 270
στερέω, 254 n.
στίγμα, στίζω, 220
Τισσαφέρνης, 214
υἱύς, 267
Φραόρτης, 162
Φρασαόρτης, 224 n.
ὤψ, 278

LATIN

antae, 182
firmus, 275
iacere, 275
imperium, 297
lacertus, 182 n.
nec, 177
os, 281
prō(nus), 178
scando, 254
species, 213
valeo, 162

OTHER INDO-EUROPEAN
LANGUAGES

двойник (Russ.), 167
guss (MIrish), 220 n.

hardus (Goth.), 257
hrōrian (OE), 188
ḫašša (Luvian), 281 n.
kaveś (Lydian), 185 n.
ląk- (Slav.), 324
lights (Engl.), 327
Lob (Germ.), 237
love (Engl.), 237
милый (Russ.), 41 n.
mṣapantin- (Tokh.), 241, 330
ость (Russ.), 265
passen (Germ.), 257
se, soy(ä) (Tokh.), 267
spa/innan (OHG), 183
wi(e)der (Germ.), 325

SEMITIC

'drng (Aram.), 266 sq.
'zt (Aram.), 158
di-na-a-tú (Akk.), 286
dwškrt' (Aram.), 262
hngyt (Aram.), 267
kašadu (Akk.), 171 n.
mzdyzn (Aram.), 16 n.
ph, pypywt (Hebr.), 280
pitqudu (Akk.), 226 n.
pwm' (Syr.), 245, 280

ELAMITE

kam-ba-ti-ia, 176
NU.MAN, 198 n.
pir-ra-tam-ma, 298

SUBJECT INDEX

Aēšma, 247 *sq.*
Ahriman, *see* Angra Mainyu
Ahuna Vairya prayer, 324, 329 with n.
ahura, 49 n., 50, 51, 59
Ahura Mazdāh, 9, 12, 54 *sq.*, 69, 195, 217, 236; in Persis, 15 *sq.*, 20; his relation to Varuṇa, 44 *sqq.*, 321; the waters his daughters, 45; as *ahū-ratu*, 53, 238 *sq.*; not an Aməša Spənta, 12 n., 286; Av. *ahurō.mazdāh-* (?), 212; called *dāmi-*, 169, 211; creates Mithra, 210 *sq.*; worships Mithra, 23, 52, 53, 66, 287, 290; builds Mithra's abode, 204; speaks in the Mithra hymn, 238, 269, 273, 287, 289 *sq.*, 331; identified with the sun, 320; referred to in the *Ahuna Vairya* prayer, 329 n. *See also* ΩPOMACΔHC.
ahū-ratu-ship (cf. *ratu*), 47, 53, 205, 223, 236, 237–42, 328 *sq.*
Aməša Spəntas: 'incremental Immortals', 157; as organs, emanations, children, shapes, of Mazdāh, 9 n.; as elements, 10 *sqq.*, 47; as Entities, 9 n., 11, 164 *sq.*; their collective name, 10, 165; their number, 11, 165; their gender, 164 *sqq.*; headed by Spənta Mainyu, *q.v.*; position of Truth among them, 46; do not replace IE gods, 48 n.; have Mithra as brother, 205; build his abode, 204; consider him (and Zarathuštra) an *ahū-ratu* and 'purifier' of the Religion, 236 (242); assist Ǝrəxša, 217

Anāhitā, 3, 20, 53, 63, 66, 200 *sq.*, 269
Angra Mainyu (Ahriman), 9, 12, 15, 46 *sq.*, 56, 269; kills the bull, 62–7
Apąm Napāt, 27 *sq.*, 59 *sq.*, 62
'Απαρύται, 175
Arimanius, 63
Ārmaiti, 10, 37, 59, 69, 168 *sq.*, 195 with n.
Arštāt, *see* Justice
Artaxerxes I, 18, 21, 22
Artaxerxes II, 20, 21
Aryana Vaējah (cf. Chorasmia), 14–20, 49, 50, 176, 284, 299
ásura, 50 n., 51
Aša, *see* Truth
Aši, 10 n., 58, 62, 168 n., 194, 195, 205, 215, 217 *sq.*, 226, 228 *sq.*, 269
Attis, 68 n.
Āturfarnbāg, 60 *sq.*, 278 *sq.*
Avestan: archaic form of the language, 20, 23; faulty grammar, 24
Bactria, 176
Baškard, 208, 322 *sq.*
Boar, 62, 168, 169, 219, 220 *sq.*
Bull, 62–6
Buner, 218
Būšyąstā, 248, 257
Cautes, Cautopates, 68, 69, 71 *sq.*, 151 *sq.*
Chorasmia (cf. Aryana Vaējah), 14 *sq.*, 16 n., 176, 185; its political and religious administration, 296–9

SUBJECT INDEX

Činvat Bridge, 204, 205, 247 n., 286
Čistā, 58, 166 sq., 215, 276, 327
Cock, 62
Cow: looked after by Mithra, 34 n., 42 sq., 54, 230, 232 sq.; a 'virtual Truth-owner', 190 sq.; its 'Fashioner' (gə̄uš tašan-), 48 n., 55 sq.
Cyrus, 14, 15, 24 n.
Daēvas, 51, 63 sq., 201, 218 sq., 244, 248
Dāmōiš Upamana, 62 n., 166 sqq., 218, 276
Dānus, 189 n.
Darius, 14 sqq., 64, 156, 251
Daylight, 293
δημιουργός, 40 sq., 57, 211 sq.
Drug, see Falsehood
Dualism, 9, 12, 46 sq.; physical and ethical, 48
Entities, see Aməša Spəntas
Falsehood, 9, 12, 46; with Darius, 15; possession of, 156
Farnbāg, 61, 278 sq.
Fire, 7, 30, 31, 59, 60 sq.
Fortuna, 62, 218
Fortune, see xᵛarənah
Fravašis, 59, 154 sq., 161, 163, 223, 227 sq., 249, 260, 266
Γάβαι, 176
Gāndhāra, 174 n., 175 sq.
Gayōmart, 69 sq.
gə̄uš tašan, see Cow
Grammar: lapses in the Mithra Yašt, 25 n.; Gāthic forms replaced by YAv. equivalents, 322
  phonology: shortening of $ā$ before antevocalic n, 167, 178, 179, 261; before antevocalic y, 205, 232; before antevocalic $u̯$, 250, 273; changes of vowel quantity in compounds, see compounds (below); quantitative metathesis, 277 n.; replacement of -i- by -ai-, 189; of -(i)y- by -ay-, 261; initial hu- lost in Oss., 282; -$avyam$ > -$aēm$, 200 n.; -mm- > -nm-, 178; $gu̯$ > $u̯$ in Av., 231, 322 n.; in Oss., 215, 327; -$n/myuv$- > -n/miv, 274; initial (s)t-, 221; s instead of z in derivatives of $dhā$-, 255 n.; interchange of postvocalic t:δ:$ṯ$, 197; of -$ạm$:-$ạ$, 221, 261 sq.; of $č$:j, 261; haplology, 151, 279
  morphology: nom. sg., 290, 291, and loc. sg., 330, of -$ī̆$- stems; Gāthic abl. in -$ōiš$, 169 n., and loc. in -$ō$, 285, of -i- stems; nom. sg. of OPers. -ant- partc., 199; neuter of $mazant$-, 196; $yahi$, old loc. sg. fem. of rel. pron., 213; 2nd pl. pret. act. of -$nā$- present stems, 188
  stem formation: -ti- stems in Oss. and Sogd., 173 n.; -$ā(u)$- with -u- stems, 181, 221, 327; -$ān$- with -an- stems, 179, 323; -$ī$- fem. of them. adjectives, 261; -ta- suffix added to nouns, 270; suffix -$tara$-, 253; -s- extension, 186, 265; fut. pass. part. in -ata-, 200 n.; inchoative in -$āsa$-, 283; them. treatment of pres. stems in -$ā$-, -$ā(y)$-, 324; infinitive $s(a)ite$, 189
  compounds: lengthening of a in cp. junction before $u̯$, 157 with n.; changes of vowel quantity in Oss. cpp., 277 n.; fem. -$ā$-, 231, and dual -$ā$-, 268, in cp. junction; cp.

SUBJECT INDEX

Grammar: compounds (*cont.*)
thematization, 183, 230, 237, 274; gen. with cpp., *see* syntax, below; *pati* governing the second member in cpp., 203; cpp. with governing base as first member, 230 *sq.*, 328; determinative cpp. with pass. past part. as second member, 269; participial cpp., 203; inverted determinative cpp. in Oss., 327; bahuvrīhis involving metaphoric identification, 180 *sq.*; trans. and intrans. meaning of negative cpp., 255
syntax: *ta-* as article, 160 n.; rel. pron. as article, 154 n., 279; rel. clause with dem. antecedent omitted, 161 with n., 224, 294; instr. subject in YAv., 188 with n.; adverbial instr., 197, 268 *sq.*, 278; gen. absolute, 194 n., 207; gen. depending on one member of a compound, 160 n.; *hăma-*compounds with gen., 216, 260; dat. in apposition to gen., 292; plur. in apposition to *nomen plurale tantum*, 293; dem. pron. 'eius' suffers case attraction, 285; rel. pron. suffers gender attraction, 289; nom. of adj. in adverbial function, 199; *-tar-* nouns governing the acc., 272; construction of *dā-*, 323 *sq.*, 329 n.; *pati + ni* as joint preverbs, 229; impersonal sentences, 203; group inflection, 236
semantics: middle of *dā-*, 224; *naēδa* comparable to Lat. *nec*, 177; indefinite meaning of *čim*, *čiš* after negative, 221; *yaṯ* as comparative conjunction, 232; old comparative used as positive, 241, 330; collective function of suffix *-tāt-*, 298; 'press':'oppress', 183; 'pass':'suit', 257; 'desire':'praise', 237; 'satisfy':'obey':'listen', 324; 'strong':'hard', 256 *sq.*; 'light':'lung', 327; 'appearance':'race', 213
style: hendiadys, 202; concatenation, 234, bottom; double negation, 200 n., 247; word order in *aspa vīra*, 250

Hamster, 255
Haoma, 59, 233 *sq.*, 269, 273
Harā, 39, 204 *sq.*
Heracles, 62, 219
Hermes, 62, 206
Herodotus, 3, 15 with n., 17, 18, 175
Hindukuš, 174 n., 175
Hystaspes, *see* ¹Vištāspa
Indo-Scythians, 320
Indra, 33 *sq.*
Justice (Arštāt), 53, 59, 215, 260, 286 *sq.*, 327
Kavis, 185 *sq.*
Κυρέσχατα, 175 n.
Libations, 184, 201
Magi, 16 *sq.*, 20–2, 64
μαγοφονία, 16
Manicheism, 40 *sq.*, 58, 206, 210
Mazdāhism, 16, 18 *sq.*; monotheistic, 47, 56; pantheistic, 50 *sqq.*, 63
Mazdayasnian(s), 16, 18, 22, 63
Media(n), 14, 17, 20, 21, 296
Metre, 24, 160
Mihira, 320
Mithra
summary of functions, 42 *sq.*; god of contract, 26 *sqq.*, 34 *sq.*, 43 *sq.*, 184, 319; giver of life, 32 *sq.*, 64; judge of the world, 35,

SUBJECT INDEX

47, 53, 223, 239, 240, 247 *sq.*; warlord, 33; light of daybreak, 30 *sq.*, 288, 320; as sun-god, 35 *sq.*, 39 *sq.*, 70, 320, 330; as 'purifier' of the Religion, 236, 239 *sq.*, 242 *sq.*; as *ahū-ratu*, 236; as δημιουργός, 41, 57, 211 *sq.*; as μεσίτης, 195, 228; not a sky-god, 36 *sqq.*; not a daēva, 49, 51; not replaced by Vohu Manah, 48 n.
created by Ahura Mazdāh, 211 *sq.*; son of Ahura Mazdāh, 195; his attributes compared with Mazdāh's, 52 *sq.*; the dvandva *miθra ahura*, 44, 49–52, 55, 239, 263 *sq.*, 320; his association with Spənta Mainyu, 49, 57; with Justice, *q.v.*; with Rāman, 299; his relation to Apąm Napāt, 27 *sq.*, 59 *sq.*; his pledge to the Religion, see Religion; has Aši (*q.v.*) as charioteer, 217, 222; his retinue, 58 *sq.*
takes care of the house, 223, 229; concerned with material welfare, 32; dispenses water, 32, 42 *sq.*, 65 n., 210, 320; kingmaker, 60; inspects the country, 296; endowed with own light, 31, 61; averse to blood sacrifices, 67 *sq.*, 271; his concern for the cow, 42 *sq.*, 54; his abode, 204 *sq.*, 226; his itinerary, 39, 244; his ubiquity, 216, 244 *sq.*
his one-wheeled ('star-decked', 26 n., 36, 294) chariot, 35 *sq.*, 279 *sq.*, 281 *sq.*, 290; his mace, 33, 227, 245, 279 *sq.*; his bow, 279 *sq.*; his 1000 perceptions, 26, 187; his 1000 ears and 10,000 eyes, 26, 256; his insight (*xratu-*), 43, 52, 256; his spies, 26 n.
in Herodotus, 3; in Manichean sources, 40 *sq.*; the 'condensed hymn' to Mithra, 44, 52 n., 263; prayers in which his name is mentioned (*miθrahe vāxš*), 183, 283, 319 with n.; the triads in which he appears, 193–5; meets and addresses Zarathuštra, 269; worshipped by daēva-worshippers, 63; etymology of his name, 28
Mithras: giver of light, 29, 61; of life, 29, 62; derives from Mithra, 61 *sq.*, 68; stands behind the sun, 32, 61; dispenses water, 320; his connection with Ahriman, 63 *sq.*; kills the bull, 62–7; τριπλάσιος, 61, 69–72; as *Saošyant*, 273; mentioned together with ΩΡΟΜΑCΔΗC, 320
Mitra: Mitanni reference to, 4, 28; in the RV, 4 *sq.*, 8, 29 *sq.*, 319; his union with Varuṇa, 7, 29 *sq.*, 52, 57; his cosmic function, 319; resides in fire, 30 *sq.*, 319; joins in killing Soma, 64 *sq.*; as the 'Friend', 8 n., 30, 319
Mitrāvaruṇā, 4, 29, 44, 57
Monotheism, 12, 46 *sq.*
Moon, 64 *sq.*, 228, 288 *sq.*
Nabarzes, 68 n.
Nairyō.saṃha, 40 *sq.*, 59, 62, 205 *sq.*, 216
Nama, 68
Oceanus, 60, 62
ΩΡΟΜΑCΔΗC, 320
Πακτύες, 175 n.
Paradise, 154, 184, 204, 324
Pārəndi, 58, 215, 286

355

SUBJECT INDEX

Παροπάνισος, 174 n.
Παροῦται, 175
Parthians, 40 sqq.
Persa, 68
Persians, 40 sq.
Persis, 16, 17, 18, 19, 50
Raγa, 265
Rāman, 59, 299, 331
Ramman, 331
Rašnu, 37, 39, 59, 69 sq., 168 n., 193 sqq., 205, 215, 223, 225 sq., 276, 286
ratu (cf. *ahū-ratu*-ship), 236, 265 sq., 298
Religion (Daēnā), 27 sq., 58, 195 with n., 268, 286; her *alter ego*, Čistā, *q.v.*; her relationship with Mithra, 215, 217, 236; invested with *ratu*-ship, 236; identified with Justice, 327; purification of the R., 43, 242 sq.; individual R., 242 sq., 292
Sakas, 320
Saošyant, 273
Satavaēsa, 210
Σατταγύδαι, 175
Sebesio, 68, 273 n.
Snāviδka, 35 sq.
Sogdian(a), 34, 40 sq., 176
Soma, 64 sq., 233
Spənta Mainyu, 9, 15, 41, 46 sq.; chief Aməša Spənta, 12, 166; as θwōrəštar- and gə̄uš tašan-, 55 sqq.; has Ahura Mazdāh as father, 9; builds Mithra's chariot, 58 n., 59, 294
Spirit: Incremental, *see* Spənta Mainyu; Fiendish, *see* Angra Mainyu, Arimanius
Spitamas, 22
Sraoša, 49, 59, 62, 69 sq., 193 sqq., 205, 223, 248, 270, 283, 325
Staota Yesnya, 273

Sun, 35, 38–40, 228, 234, 292 sq., 320, 330
Tištrya, 53
Truth (Aša), 9, 10, 12, 153 n.; with Darius, 15, 251; its two-fold position, 45 sq.; as 'bridge', 48; possession of T., 153–6; virtual Truth-owners, 190; Paradise its seat, 154; straightest path of T., 157, 322
Tvaṣṭar, 56
θεογονία, 17, 20
Θwāša, 59, 215
Varan, 218
Varəna, 218
Varuṇa, 5 sq., 7, 29 sq., 32, 45; displaced by Ahura Mazdāh, 46, 49 sq., 321; *Vouruna, 47 sqq., 57, 239
Vayu, 53, 286, 299, 321 sq., 331
Vendidad, 22 n.
Vərəθraγna, 50, 59, 62, 193 sq., 219, 286; his *alter ego*, Dāmōiš Upamana, *q.v.*
Vīspe Ratavō, 273
¹Vištāspa (father of Darius), 15, 16 n.
²Vištāspa (Kavi), 16 n., 49 sq., 185, 299
Vohu Manah, 10, 48 n.
Waters, 45, 59, 217 n., 249
Xerxes, 17, 51, 63, 64, 153 sqq., 298
x^varənah (fortune), 52, 59 sq., 186 n., 217, 283; identified with fire, 61, 278
Yajur Veda, 64
Yeṅhē hātąm prayer, 9, 163 sqq., 322
Yima, 69 sq., 225
Zarathuštra
his dates, 9; his doctrine, 9 sqq., 47; not mentioned by Herodotus, or in Achae-

menian records, 15, 16;
familiar to fourth-century
Greeks, 22; as Magus, 21; in
Zoroastrianism, 23 *sq.*; not
referred to in the *Ahuna
Vairya* prayer, 329 n.; author
of the *Yeṅhē hātąm* prayer,
163 *sqq.*; as *dah́yuma* and
Ratu General, 265 *sq.*, 298; as
ahū-ratu, 53, 237 *sqq.*, 242
his language, 20, 24; his attitude to Mithra, 48 *sq.*, 65,
67; to *kavi*-s, 185; 'purifies' the Religion, 239 *sq.*, 242
sq.; condemns blood sacrifices, 67; puts to flight the
daēvas, 214 n.; addressed by
Mithra, 238, 269; by Ahura
Mazdāh, 238, 269, 287, 289
sq.; by other gods, 269

Zarathuštrian: priests, 13 *sq.*,
17, 20, 21, 50; non-Z. priests,
20, 25; non-Z. circles, 49 *sq.*.

Zoroaster, 21 *sq.*, 69 *sq.*

Zoroastrian: religion, 13, 18, 20,
21; priests, 17 *sq.*, 284; scripture,
20, 50; calendar, 18; parts of the
Mithra Yašt, 24 *sq.*; organization of the Z. Church, 298 *sq.*

Zoroastrianization, 23 *sq.*, 25, 58